Chemometrics & Data Handling

Chemometrics & Data Handling

Editor:

Asad Rahman

Delve Publishing

www.delvepublishing.com

Chemometrics & Data Handling

Editor: Asad Rahman

708 3rd Avenue, 6th Floor

New York, NY 10017

United States of America

www.delvepublishing.com

ISBN: 978-1-68095-892-8

Library of Congress Control Number: 2016951753

About the Editor

Asad Rahman

Asad Rahman obtained MSc Chemical Engineering from University of Nottingham in 2012. Currently He is working towards PhD in Energetic Material Synthesis and its Performance Analysis in Micro Thrusters. His research interests are in Ultrasound based particle treatment, Energetic material synthesis and Micro thruster applications.

Contents

	List of Contributors	*xi*
	Preface	*xv*
	Introduction	*xvii*
Chapter 1	**A Guide to Generalized Additive Models in Crop Science Using Sas and R**	**1**
1.	Introduction	1
2.	Smoothing Functions	2
3.	Steps in Generalized Additive Modelling	3
4.	Analyses in R	7
	Conclusions	10
	Acknowledgements	11
	References	11
Chapter 2	**The Chemeqlb.LNG and the Poolingr. LNG Model (2 Basic Learning Models on Lingo)**	**13**
1.	Model:	13
2.	Sets:	13
3.	Enddata	14
Chapter 3	**An Introduction to the "Special Volume Spectroscopy and Chemometrics in R"**	**17**
1.	Introduction	17
2.	An Outline of the Contributions	18
	Conclusions and Outlook	18
	References	18
Chapter 4	**Principal Component and Partial Least Squares Regression in R**	**21**
1.	Introduction	21
2.	Example Session	23
3.	Formulas and Data Frames	26
4.	Fitting Models	27
5.	Choosing the Number of Components with Cross-Validation	27
6.	Inspecting Fitted Models	28
7.	Predicting New Observations	29
8.	Further Topics	30
	References	32
Chapter 5	**Timp: An R Package for Modeling Multi-Way Spectroscopic Measurements**	**33**
1.	Introduction	33
2.	The Implementation of Timp	36
3.	User-Accessible Functions	39
4.	General Model Options	40
5.	Kinetic Models	41
6.	Spectral Models	48
7.	Extension of Supported Model Types	52
8.	Conclusions	53
9.	Acknowledgments	54
10.	A. Partitioned Vs. Unpartitioned Variable Projection Algorithms	54
11.	B. New Nls Options	55

12. C. Data Formats For Input Into Timp ... 55
13. C.1. Time Explicit Format ... 55
14. C.2. Wavelength Explicit Format ... 56
References ... 56

Chapter 6 A Mathematical Model for the Industrial Hazardous Waste Location-Routing Problem ... 59
1. Introduction ... 59
2. Literature Review ... 60
3. Mathematical Model ... 61
4. Proposed Solution to Multiobjective Problem ... 63
5. Analysis and Application for Province of Iran ... 64
Conclusion ... 66
References ... 66

Chapter 7 Optical Nondestructive Visual and Near-Infrared Methods for Food Quality and Safety ... 69
1. Introduction ... 69
2. Review of Hyperspectral/Multispectral Methods for Different Food Groups ... 77
Conclusions ... 82
Appendices ... 83
Acknowledgments ... 95
References ... 95

Chapter 8 Effects of Temperature and Other Atmospheric Conditions on Long-Term Gaseous Mercury Observations in the Arctic ... 107
1. Introduction ... 107
2. Methods ... 108
3. Results and Discussion ... 109
Conclusions ... 111
Acknowledgements ... 112
References ... 115

Chapter 9 Chemometrics in Fingerprinting by Means of Thin Layer Chromatography ... 119
1. Introduction ... 119
2. Fingerprinting: Classical And By Means Of Chemometrics ... 119
3. Creating Fingerprints ... 120
4. Chemometric Pretreatment ... 120
5. Chemometrics Itself: Unsupervised And Supervised Techniques ... 121
Conclusion ... 121
References ... 122

Chapter 10 Ethanol Production from Waste Potato Mash by Using Saccharomyces Cerevisiae ... 125
1. Introduction ... 125
2. Experimental Section ... 126
3. Results and Discussion ... 128
Conclusions ... 132
Acknowledgments ... 132
References ... 132

Chapter 11 Integrated Electricity Planning Comprise Renewable Energy and Feed-in Tariff ... 135
1. Introduction ... 135
2. Materials and Methods ... 136

3. Results 139
4. Discussion 139
Conclusion 139
Acknowledgment 140
References 140

Chapter 12 Statistical Optimization of Fermentation Conditions for Cellulase Production from Palm Oil Mill Effluent 141
1. Introduction 141
2. Materials and Methods 142
3. Results 142
4. Discussion 144
Conclusion 144
Acknowledgment 144
References 144

Chapter 13 Process Simulation of Oxy-Combustion for Maximization of Energy Output Using Aspen Plus 147
1. Introduction 147
2. Validation of the CLC Process Simulation with Experiment 148
3. Investigation of the Effect of Various Parameters on the Energy Output of the CLC Process Simulation 149
Conclusions 153
Appendix 154
Reference 166

Chapter 14 Volatile Constituents in Radix Flemingiae Macrophyllae and a Substitute by Gas Chromatography-Mass Spectrometry and Chemometric Methods 167
1. Introduction 167
2. Results and Discussion 168
3. Experimental 174
Conclusions 174
Acknowledgments 174
References 174

Chapter 15 Rapid Quantitative Analysis of Ethanol and Prediction of Methanol Content in Traditional Fruit Brandies Using Ftir Spectroscopy and Chemometrics 177
1. Introduction 177
2. Materials and Methods 178
Conclusions 183
Acknowledgements 183
References 183

Chapter 16 Simultaneous Spectrophotometric Determination of Rabeprazole Sodium and Domperidone Maleate in Capsules by Chemometric Methods 185
1. Introduction 185
2. Experimental 186
3. Results and Discussion 187
Conclusions 192
Conflict of Interests 192
Acknowledgments 192
References 192

Chapter 17 **Near-Infrared Spectroscopy as an Analytical Process Technology for the on-Line Quantification of Water Precipitation Processes During Danhong Injection** **195**

1. Introduction 195
2. Materials And Methods 196
3. Results And Discussion 198

Conclusions 205

Conflict Of Interests 205

Acknowledgments 205

References 205

Chapter 18 **Quantitative Analysis of Adulterations in Oat Flour by Ft-Nir Spectroscopy, Incomplete Unbalanced Randomized Block Design, and Partial Least Squares** **209**

1. Introduction 209
2. Materials and Methods 210
3. Results and Discussion 210

Conclusions 212

Conflict of Interests 212

Acknowledgment 212

References 212

Index **215**

List of Contributors

Josefine Liew
Swedish University of Agricultural Sciences, Department of Crop Production Ecology, Box 7043, SE-750 07 Uppsala, Sweden.
Valiguard AB, SAI Global Scandinavia, Box 5609, SE-114 86 Stockholm, Sweden.

Johannes Forkman
Swedish University of Agricultural Sciences, Department of Crop Production Ecology, Box 7043, SE-750 07 Uppsala, Sweden.

Katharine M. Mullen
Vrije Universiteit Amsterdam

Ivo H. M. van Stokkum
Vrije Universiteit Amsterdam

Bjørn-Helge Mevik
Norwegian University of Life Sciences

Ron Wehrens
Radboud University Nijmegen

Katharinc M. Mullen
Vrije Universiteit Amsterdam

Ivo H. M. van Stokkum
Vrije Universiteit Amsterdam

Omid Boyer
Department of Industrial Engineering, Islamic Azad University of Najafabad, Najafabad, Isfahan, Iran

Tang Sai Hong
Department of Mechanical and Manufacturing Engineering, Faculty of Engineering, University Putra Malaysia, 43400 Serdang, Selangor, Malaysia

Ali Pedram
Department of Mechanical and Manufacturing Engineering, Faculty of Engineering, University Putra Malaysia, 43400 Serdang, Selangor, Malaysia

Rosnah Bt Mohd Yusuff
Department of Mechanical and Manufacturing Engineering, Faculty of Engineering, University Putra Malaysia, 43400 Serdang, Selangor, Malaysia

Norzima Zulkifli
Department of Mechanical and Manufacturing Engineering, Faculty of Engineering, University Putra Malaysia, 43400 Serdang, Selangor, Malaysia

Jarmo T. Alander
Department of Electrical and Energy Engineering, University of Vaasa, 65101 Vaasa, Finland

Vladimir Bochko
Department of Electrical and Energy Engineering, University of Vaasa, 65101 Vaasa, Finland

Birgitta Martinkauppi
Department of Electrical and Energy Engineering, University of Vaasa, 65101 Vaasa, Finland

Sirinnapa Saranwong
Bruker Optics KK, 1-4-1 Shinkawa, Chuo-ku, Tokyo 104-0033, Japan

Timo Mantere
Department of Electrical and Energy Engineering, University of Vaasa, 65101 Vaasa, Finland
Bruker Optics KK, 1-4-1 Shinkawa, Chuo-ku, Tokyo 104-0033, Japan

A. S. Cole
Air Quality Research Division, Environment Canada, 4905 Dufferin St., Toronto, Ontario M3H 5T4, Canada

A. Steffen
Air Quality Research Division, Environment Canada, 4905 Dufferin St., Toronto, Ontario M3H 5T4, Canada

Łukasz **Komsta**
Department of Medicinal Chemistry, Medical University of Lublin, Jaczewskiego 4, 20-090 Lublin, Poland

Gulten Izmirlioglu

Ali Demirci
Department of Agricultural and Biological Engineering, Pennsylvania State University, University Park, PA 16802, USA
The Huck Institutes of Life Sciences, Pennsylvania State University, University Park, PA 16802, USA

Ho Wai Shin
Process System Engineering Centre (PROSPECT), Faculty of Chemical Engineering, University Technology Malaysia, 81310, Skudai, Johor, Malaysia

Haslenda Hashim
Process System Engineering Centre (PROSPECT), Faculty of Chemical Engineering, University Technology Malaysia, 81310, Skudai, Johor, Malaysia

Jamal I. Daoud
Bioenvironmental Engineering Research Unit, Faculty of Engineering, International Islamic University Malaysia, Gombak, 50728 Kuala Lumpur, Malaysia

Md. Zahangir Alam
Bioenvironmental Engineering Research Unit, Faculty of Engineering, International Islamic University Malaysia, Gombak, 50728 Kuala Lumpur, Malaysia

Xiao Zhang
Department of Mechanical Engineering & Materials Science, Washington University in St. Louis, 1 Brookings Drive, St. Louis, MO 63130, USA.

Subhodeep Banerjee
Department of Mechanical Engineering & Materials Science, Washington University in St. Louis, 1 Brookings Drive, St. Louis, MO 63130, USA.

Ling Zhou
Department of Mechanical Engineering & Materials Science, Washington University in St. Louis, 1 Brookings Drive, St. Louis, MO 63130, USA.

Ramesh Agarwal
Department of Mechanical Engineering & Materials Science, Washington University in St. Louis, 1 Brookings Drive, St. Louis, MO 63130, USA.

Shen-Yu Cheng,
School of Chemical and Material Engineering, Quzhou College, Quzhou 324000, China

Yan Xie,
School of Chemical and Material Engineering, Quzhou College, Quzhou 324000, China

Xiao-Liang Feng
School of Chemical and Material Engineering, Quzhou College, Quzhou 324000, China

Lan-Fang Huang
School of Chemical and Material Engineering, Quzhou College, Quzhou 324000, China

Teodora Emilia COLDEA
University of Agricultural Sciences and Veterinary Medicine, Faculty of Food Science and Technology, 400372 Cluj-Napoca, Romania

Carmen SOCACIU
University of Agricultural Sciences and Veterinary Medicine, Faculty of Food Science and Technology, 400372 Cluj-Napoca, Romania

Florinela FETEA
University of Agricultural Sciences and Veterinary Medicine, Faculty of Food Science and Technology, 400372 Cluj-Napoca, Romania

Floricuța RANGA
University of Agricultural Sciences and Veterinary Medicine, Faculty of Food Science and Technology, 400372 Cluj-Napoca, Romania

Raluca Maria POP
University of Agricultural Sciences and Veterinary Medicine, Faculty of Food Science and Technology, 400372 Cluj-Napoca, Romania

Mira FLOREA
University of Medicine and Pharmacy "I. Hatieganu", Dept. of Community Medicine, Cluj

Marc Ohmenhaeuser
[1]Chemisches und Veterinäruntersuchungsamt (CVUA) Freiburg, Bissierstraße 5, 79114 Freiburg, Germany

Yulia B. Monakhova
Chemisches und Veterinäruntersuchungsamt (CVUA) Karlsruhe, Weissenburger Straße 3, 76187 Karlsruhe, Germany
Bruker BioSpin GmbH, Silberstreifen, 76287 Rheinstetten, Germany
Department of Chemistry, Saratov State University, Astrakhanskaya Street 83, Saratov 410012, Russia

Thomas Kuballa
Chemisches und Veterinäruntersuchungsamt (CVUA) Karlsruhe, Weissenburger Straße 3, 76187 Karlsruhe, Germany

Dirk W. Lachenmeier
Chemisches und Veterinäruntersuchungsamt (CVUA) Karlsruhe, Weissenburger Straße 3, 76187 Karlsruhe, Germany
Ministry of Rural Affairs and Consumer Protection, Kernerplatz 10, 70182 Stuttgart, Germany

Xuesong Liu
College of Pharmaceutical Sciences, Zhejiang University, Hangzhou 310058, China

Chunyan Wu
College of Pharmaceutical Sciences, Zhejiang University, Hangzhou 310058, China

Shu Geng
College of Pharmaceutical Sciences, Zhejiang University, Hangzhou 310058, China

Ye Jin
College of Pharmaceutical Sciences, Zhejiang University, Hangzhou 310058, China

Lianjun Luan
College of Pharmaceutical Sciences, Zhejiang University, Hangzhou 310058, China

Yong Chen
College of Pharmaceutical Sciences, Zhejiang University, Hangzhou 310058, China

Yongjiang Wu
College of Pharmaceutical Sciences, Zhejiang University, Hangzhou 310058, China

Ning Wang
School of Material Science and Engineering, Tianjin Municipal Key Lab of Fiber Modification and Functional Fiber, Tianjin Polytechnic University, Tianjin 300389, China

Xingxiang Zhang
School of Material Science and Engineering, Tianjin Municipal Key Lab of Fiber Modification and Functional Fiber, Tianjin Polytechnic University, Tianjin 300389, China

Zhuo Yu
School of Material Science and Engineering, Tianjin Municipal Key Lab of Fiber Modification and Functional Fiber, Tianjin Polytechnic University, Tianjin 300389, China

Guodong Li
School of Material Science and Engineering, Tianjin Municipal Key Lab of Fiber Modification and Functional Fiber, Tianjin Polytechnic University, Tianjin 300389, China

Bin Zhou
School of Material Science and Engineering, Tianjin Municipal Key Lab of Fiber Modification and Functional Fiber, Tianjin Polytechnic University, Tianjin 300389, China

Preface

Chemometrics is a term for describing analytical methods used for processing large number of data for better understanding. Svante Wold and Bruce Kowalski coined the term Chemometrics in 1970s.

The most troubling issue as a researcher of any field of science is to identify the correct set of data. In addition, one must satisfy the most important questions of, what, why, when and how. It is important, because, a reviewer will raise questions over selection of parameters and what procedure was used to identify parameters of significance. To resolve these issues, Chemometrics are used. It is a Mathematical and Statistical method used for better understanding and utilization of data in all science fields. It helps in optimization, statistical analysis and presentation of refined data. In very simple terms, Chemometrics analysis tells us about the significant data set out of thousands of data sets. It involves statistical analysis, economic analysis, and material and property analysis to identify the correct set of significant data set. Therefore, many methods and analytical tools are available for this purpose. Furthermore, it is based on mathematics and statistics; hence, if any given software can perform such operations, then it is valid for Chemometrics. In the end, it all depends on the end user and its requirements.In this book, several methods and their applications are discussed. In the beginning, few papers are presented which gives an overview of different Chemometric tools and then examples from different areas of chemistry, chemical engineering and pharmacy are presented.

Editor
Asad Rahman

INTRODUCTION

Chemometrics refers to the process of analyzing information from data often based on many variables. It can be used to see patterns in how data are organized, measure chemical properties of materials, and use mathematical models to classify information. In addition to analyzing the light spectrum of a substance or measuring acidity, chemometrics is often used for managing product quality and in manufacturing pharmaceuticals. Statistics and mathematics are typically used to accomplish such tasks. Methods such as exploratory data analysis, regression, and classification often help to group large quantities of data. Chemometrics has developed over the past decade from a fairly theoretical subject to one that is applied in a wide range of sciences. The word chemometrics was coined in the 1970s, and the early development went hand in hand with the development of scientific computing, and primarily involved using multivariate statistical methods for the analysis of analytical chemistry data. Early chemometricians were likely to be FORTRAN programmers using mainframes and statistical libraries of subroutines. The earliest applications were primarily to analytical chemical datasets, often fairly simple in nature, for example a high performance liquid chromatography (HPLC) cluster of two or three peaks or a set of ultraviolet (UV)/visible mixture spectra.

Traditionally chemometrics was regarded as very much based around laboratory and process instrumental analytical chemistry. Using partial least squares to calibrate the ultraviolet (UV)/Visible spectrum of a mixture of pharmaceuticals to determine the concentrations of each constituent is a typical application. There are numerous similar types of application in a wide variety of spectroscopies such as infrared (of all types), atomic spectroscopy, mass spectrometry, and in chromatography such as HPLC and gas chromatography-mass spectrometry (GC-MS). This area, whilst perhaps not very glamorous, is still very widespread and still the majority of reports on the use of chemometrics, in the academic literature, are to instrumental analytical chemistry where it is very successful and often regarded as essential especially in both experimental design and the interpretation of multivariate spectroscopic data.

However, the applications have diversified substantially over the last few years. There were certain economic driving forces that related partially to the organizations that employed the first pioneers of the subject. In the 1970s and 1980s there were not many dedicated academic chemometrics positions, so the 'parents' of the subject would have described themselves as organic chemists or analytical chemists or pharmaceutical chemists, etc., rather than chemometricians. Only in the mid 1980s and 1990s came a small trickle of dedicated chemometrics posts in academia, with industry perhaps moving faster [industry tends to wait until the academic seed has been sown but then can move much faster as there is no need to maintain departments for long periods so new staff can be rapidly hired (and fired of course) as the need arises]. In the 1970s and 1980s, food chemistry was one of the main early growth areas, probably because of the tradition of sensory statistics and a good interface between data analysis and analytical chemistry, hence a large development especially in the use of NIR and food analysis, and an important economic driving force especially in Scandinavia and Southern Europe. Process chemistry, also has a vintage in the 1980s with the establishment, over two decades, of several centers, pioneered first in the University of Washington (Seattle). Chemical engineers and analytical chemists work closely with manufacturing industry, spectroscopy including NIR (again) and the monitoring of processes has led eventually to developments, such as the Process Analytical Technology (PAT) initiative. A third and important influence was pharmaceutical chemistry, with a good link to hospitals, pharmacy and clinical applications, especially the use of HPLC: analytical chemists in Benelux who were trained as chromatographers first and foremost were responsible for the development of chemometrics within the area of pharmaceutical chromatography. These major growth points seeded major research groups, in Scandinavia, Benelux and the USA in the late 1970s and early 1980s allowing substantial growth in applications especially to HPLC and NIR as applied to pharmaceuticals, food and process analysis, and also the development of the intellectual driving force, first commercially marketed software and main principles.

During this period other people in industry quietly followed the academic lead: in a subject such as chemometrics, it is hard to disentangle the 'name' which is often trumpeted by academic groups who try to obtain students, grants, papers, theses and conference presentations, and people in industry that have less need to invent a name or write papers but get on with the job. Several industries notably the petrochemical industry also applied chemometric principles in these early days. There also was a separate development in environmental chemistry, some impinging on chemometrics, some being published as environmetrics, so diluting the focus in the analytical chemistry literature.

With the 1980s wrapping up and the early 1990s commencing, chemometrics became much more widespread. Individual researchers in academic positions, some with prior training in some of the more established groups, and others who read papers and books that had been written over this early period, dabbled with chemometrics, and so the applications spread not just to dedicated research groups but to investigators working in different environments. The most notable expansion even to this day is the biological and medical area, with the enormous increase in development of chemical analyses such as GC-MS, mass spectrometry (MS), nuclear magnetic resonance (NMR), optical spectroscopy and pyrolysis, and the tradition of computationally oriented biologists. The interface between chemometrics and bioinformatics is also a more modern growth point over the past 5 or 6 years, especially as biological data (e.g. gene and protein structure) is increasingly correlated with metabolomic data (metabolomics) obtained from chemical analyses. Other niche but important growth points include forensics (the use of chemical and spectroscopic information to determine origins of samples), multivariate image analysis (especially of pharmaceutical tablets), chemical engineering (far beyond the original process analytics), and a small but expanding interest in materials (including thermal analysis). All these applications have in common that potentially a large amount of multivariate data can be obtained, usually very rapidly, from instruments and there are major data analytical problems in their interpretation which is aided by chemometrics. The spectral properties of different chemicals, which include how they can be identified using light and color, are often analyzed through chemometrics. Other types of measurements can be made as well. The scientific principles of the field are in many cases applied to spectrometers and other laboratory instruments. Scientists who use such equipment usually need a strong background in algebra and other applied mathematics.

What chemometrics generally adds to scientific processes is a way to speed up the classification of large amounts of data. Disparate data can be associated and organized out of hidden patterns. Outlying details are often used to find patterns as well as ongoing trends in a specific process. Complex sets of data can be mathematically grouped by using concepts such as principal component analysis as well as hierarchical cluster analysis.

Data can be grouped using these methods, and the variables that lead to different results are often determined as well. Scientists can also focus on specific properties that may be common to multiple sets of data. Results of known measurements can be compared to a variety of information by using regression analysis. In chemical engineering, methods called partial least squares and principal component regression are sometimes applied to make clearer measurements of data. They are often used in process control and to monitor production in a variety of industries.

These three classification methods in chemometrics are typically used to predict what group a sample belongs to. Data on similar samples can be used to make grouping decisions and predictions. Chemometrics is often applied to the calibration of spectroscopes, modeling of processes for monitoring systems and detecting faults, as well as designing scientific instruments. Data collection is typically simplified to save time, by avoiding the need to go through large quantities of information manually. Variations are often applied to other areas of science as well as engineering and business.

What is Chemometrics?

The development of the discipline chemometrics is strongly related to the use of computers in chemistry. Some analytical groups in the 1970s were already working with statistical and mathematical methods that are ascribed nowadays to chemometric methods. Those early investigations were connected to the use of mainframe computers. The notation chemometrics was introduced in 1972 by the Swede, Svante Wold, and the American, Bruce R. Kowalski. The foundation of the International Chemometrics Society in 1974 led to the first description of this discipline. In the following years, several conference series were organized, e.g., Computer Application in Analytics (COMPANA), Computer-Based Analytical Chemistry (COBAC) and Chemometrics in Analytical Chemistry (CAC). Some journals devoted special sections to papers on chemometrics. Later, novel chemometric journals were started, such as the Journal of Chemometrics (Wiley) and Chemometrics and Intelligent Laboratory Systems (Elsevier). An actual definition of chemometrics is:

– the chemical discipline that uses mathematical and statistical methods, (a) to design or select optimal measurement procedures and experiments, and (b) to provide maximum chemical information by analyzing chemical data.

The discipline of chemometrics originates in chemistry. Typical applications of chemometric methods are the development of quantitative structure activity relationships or the evaluation of analytical–chemical data. The data flood generated by modern analytical instrumentation is one reason, that analytical chemists in particular develop applications of chemometric methods. Chemometric methods in analytics is the discipline that uses mathematical and statistical methods to obtain relevant information on material systems. With the availability of personal computers at the beginning of the 1980s, a new age commenced for the acquisition, processing and interpretation of chemical data. In fact, today every scientist uses software, in one form or another that is related to mathematical methods or to processing of knowledge. As a consequence, the necessity emerges for a deeper understanding of those methods. The education of chemists in mathematics and statistics is usually unsatisfactory. Therefore, one of the initial aims of chemometrics was to make complicated mathematical methods practicable. Meanwhile, the commercialized statistical and numerical software simplifies this process, so that all important chemometric methods can be taught in appropriate computer demonstrations. Apart from the statistical–mathematical methods, the topics of chemometrics are also related to problems of the computerbased laboratory, to methods for handling chemical or spectroscopic databases and to methods of artificial intelligence. In addition, chemometricians contribute to the development of all these methods. As a rule, these developments are dedicated to particular practical requirements, such as the automatic optimization of chromatographic separations or in prediction of the biological activity of a chemical compound.

The Computer-based Laboratory

Nowadays the computer is an indispensable tool in research and development. The computer is linked to analytical instrumentation; it serves as a tool for acquiring data, for word processing and for handling databases and quality assurance systems. In addition, the computer is the basis for modern communication techniques such as electronic mail or video conferences. In order to understand important principles of computer usage some fundamentals are considered here, i.e., coding and processing of digital information, the main components of a computer, programming languages, computer networking and automation processes.

Analog and digital data

The use of digital data provides several advantages over the use of analog data. Digital data are less noise sensitive. The only noise arises from round-off errors due to finite representation of the digits of a number. They are less prone to, for instance, electrical interferences and they are compatible with digital computers.

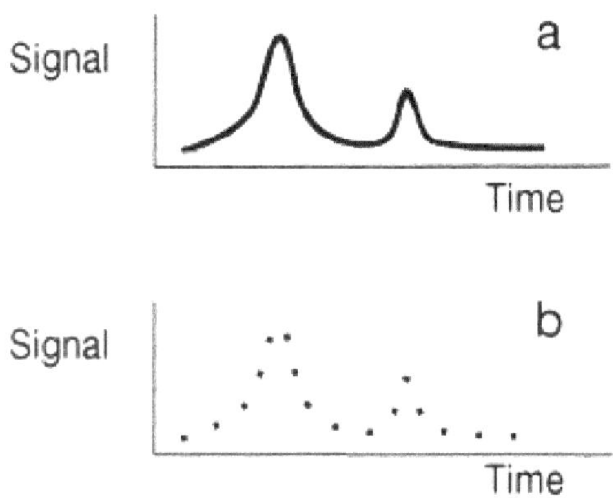

Figure. 1 Signal dependence on time of an analog (a) and a digital detector (b)

As a rule, primary data are generated as analog signals either in a discrete or a continuous mode (Fig. 1). For example, monitoring the intensity of optical radiation by means of a photocell provides a continuous signal. Weak radiation, however, could be monitored by detecting individual photons by a photomultiplier. Usually, the analog signals generated are converted into digital data. This is carried out by an analog-to-digital converter as explained below.

Binary versus decimal number system

In a digital measurement, the number of pulses occurring within a specified set of boundary conditions is counted. The easiest way to count is to have the pulses represented as binary numbers. In this way only two electronic states are required. To represent the decimal numbers from 0 to 9 one would need 10 different states. Typically, the binary numbers 0 and 1 are represented electronically by voltage signals of 0.5 V and 5 V, respectively. Binary numbers characterize coefficients of the power of 2, so that any number of the decimal system can be described.

The R package chemometrics

Multivariate data analysis is the simultaneous observation of more than one characteristic. In contrast to the analysis of univariate data, in this approach not only a single variable or the relation between two variables can be investigated, but the relations between many attributes can be considered. For the statistical analysis of chemical data one has to take into account the special structure of this type of data. Very often, classic model assumptions are not fulfilled by chemical data, for instance there will be less observations than variables, or correlations between the variables occur. Those problems motivated the adaption and development of several methods for chemometrics. The statistical environment R (R Development Core Team 2006) is a powerful tool for data analysis and graphical representation. The programming language behind it is object oriented. As part of the GNU project (Galassi et al. 2009), it is an open source software and freely available on http://cran.r-project.org. The fact that anyone can assist in improving the code and adding functions makes R a very flexible tool with a constantly growing number of add-on packages. An introduction to R can be found in Venables and Smith (2002). Varmuza and Filzmoser (2009) wrote a book for multivariate data analysis in chemometrics, and contributed to the R framework with a function package for corresponding applications. The package contains about 30 functions, mostly for regression, classification and model evaluation and includes some data sets used in the R help examples. The work at hand is a vignette for this R package chemometrics and can be understood as a manual for its functionalities. It is written with the help of Sweave (Leisch 2002), a reporting tool which allows for LATEX as well as R code and output to be presented within one document. For details on R vignettes and how to extract the R code from the document see

Leisch (2003). The aim of this vignette is to explain the relevant methods and to demonstrate them based on practical examples. To get started, the package chemometrics and other internally required packages can be loaded simultaneously by the command

Principal Component Analysis

Functions discussed in this section:

- pcaCV
- nipals
- pcaVarexpl
- pcaDiagplot

The Method

Principal component analysis (PCA) is a method for the visualization of complex data by dimension reduction. Besides exploratory data analysis also prediction models can be created using PCA. The method was introduced by Pearson (1901); Hotelling (1933) made further developments. The high number of variables in multivariate data leads to three main problems:

1. Graphical representation of the data is not possible for more than three variables.
2. High correlation between the variables makes it impossible to apply many statistical methods.
3. Many variables contain only very few information.

PCA is able to avoid all of these problems by transforming the original variables into a smaller set of latent variables which are uncorrelated. Data transformed in such a way can then be used by other methods. The latent variables with the highest concentration of information form lower dimensional data which can be visualized graphically. Noise is separated from important information.

Instrumental Response and Data Processing

Instrumental response in analytical study usually necessitates some form of mathematical manipulation to make the manipulations more meaningful to chemists. For instance, the absorbance of a peak deduced from its ultraviolet--visible spectrum can be transformed into a chemical variable of concentration of the related compound. Similarly, the HPLC peak height or area of a component can be correlated to its concentration. Considering the latter case, if the HPLC peak of a one-component system is free from overlapping and other interference, then the concentration of the component can be determined when the relevant calibration curve is available. In this case, only one concentration or chemical variable is involved and the system is regarded as a univariate one by chemometricians. Figure 1.3 shows the chromatogram of the ginsenoside extract of an American Ginseng sample as acquired by the HP HPLC model 1100 instrument. It consists mainly of several peaks corresponding to the ginsenosides of Re, Rg1, Rf, Rb1, Rc, Rb2, and Rd. The peaks with retention times greater than 15 min are well separated from one another and are free from impurities. Hence, each one can be considered as a onecomponent system that can be treated using univariate methods under an ideal situation. Yet, the profile around 6 min shows two overlapping peaks that should be classified as a multivariate system in data analysis. Usually, analytical chemists try their best to adjust the experimental conditions to separate these overlapping peaks. Yet, it is time-consuming and not easy to achieve this, especially for complicated systems such as herbal medicine, which has many constituents. Thus, different multivariate analysis techniques have been developed to accomplish the target.

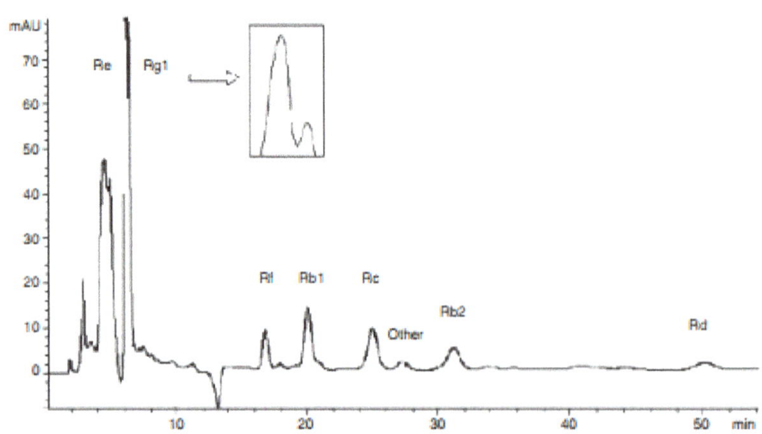

Figure. 2 The HPLC chromatogram of the ginsenoside extract of an American Ginseng sample.

The powerful computers currently available considerably help in this respect. The multivariate approach is currently considered as one of the developing areas of chemometrics for tackling problems for complicated chemical systems such as in the example cited above.

CHAPTER 1

A GUIDE TO GENERALIZED ADDITIVE MODELS IN CROP SCIENCE USING SAS AND R

Josefine Liew, Johannes Forkman

Keywords: Generalized additive mode, polynomial regression, Sonchus arvensis, GAM.

INTRODUCTION

Generalized additive models (GAMs), introduced by Hastie and Tibshirani (1986), relax and extend generalized linear models. GAMs are useful for applications where polynomials fail to describe the observed curvature in the data because they include splines or local regression smoothing (LOESS) functions (Cleveland 1979). The output of a GAM analysis is mainly graphical; the explicit resulting smoothing function is complicated since it is not a function of any easily interpreted parameters (Venables and Dichmont 2004). Although primarily used for exploratory purposes, GAMs can also be used for approximate statistical inference, e.g., hypothesis testing. GAMs are said to be non-parametric (Yee and Mitchell 1991) or semi-parametric (Guisan et al. 2002), which refers here to the lack of a particular functional form of the relationship between the dependent variable Y and the explanatory variable X. GAMs have been available in some software since the early 1990s (Hastie 1992, Hilbe 1993), are available in, for example, the SAS System (SAS Institute 2008) and the open source environment R (Wood 2006a).

GAMs are rarely used in agronomy, although they could be useful in various applications. This article offers a starting point for further study of GAMs for readers who have basic knowledge of statistics as a complement to other fields of research. We specifically describe how GAMs can be used for comparing effects of experimental treatments, since this information is often lacking in other reviews. An example illustrates how GAMs can be fitted using the SAS System and R. Fitting GAMs is compared with fitting polynomials

GAMs can be explained with reference to linear models (LMs), generalized linear models (GLMs), and additive models (AMs). The LM, with a single response variable, y, and p explanatory variables takes the form:

$$y = \beta_0 + \beta_1 x_1 + \beta_2 x_2 + \ldots + \beta_p x_p + e, \quad (1)$$

where β_0 is the intercept and $\beta_1, \beta_2, \ldots, \beta_p$ are the regression coefficients, i.e., the slopes of the explanatory variables. In many applications, there is only a single explanatory variable, for example time. In this case, Eq. 1 reduces to a simple linear regression model $y = \beta_0 + \beta_1 x_1 +$ e. For statistical inference, it is assumed that the error terms e, are (i) normally distributed with expected value 0; (ii) have the same variance; and (iii) are independent. It is also assumed that (iv) the values $x_1, x_2, \ldots, x_p$

Citation: Josefine Liew , Johannes Forkman, A guide to generalized additive models in crop science using SAS and R, Communications in Biometry and Crop Science Vol. 10, No. 1, 2015, Pp. 41–57 are fixed, i.e., have no variance, and are known.

In LMs, the mean, μ, can be written as: $\mu = \beta_0 + \beta_1 x_1 + \beta_2 x_2 + \ldots + \beta_p x_p$, i.e., it is a direct linear function of the parameters. In GLMs, a transformation, $g(\mu)$, of the mean is a linear function of the parameters. Thus, a transformation of the mean is modeled instead of the observations. GLMs extend LMs to link functions other than the identity. Assumptions (i) and (ii) for LMs are relaxed. GLMs consist of three components: the random component, which is the response variable and its probability distribution; the systematic component, which is a linear function of the explanatory variables; and the link function, which links the random and the systematic components. With p explanatory variables, a GLM can be written as:

$$y = \beta_0 + s_1(x_1) + s_2(x_2) + \ldots + s_p(x_p) + e \quad (3)$$

Combination of GLMs and AMs gives a GAM:

$$g(\mu) = \beta_0 + s_1(x_1) + s_2(x_2) + \ldots + s_p(x_p) \quad (4)$$

In contrast to AMs, GAMs are not restricted to the normal distribution, but can be used for any probability distribution from the exponential family. Just as with the other models, GAMs require explanatory variables to be measured without errors. Additionally, high correlation of explanatory variables causes inferential problems because individual effects of the explanatory variables cannot be separated.

SMOOTHING FUNCTIONS

LMs partition data into "model + error", while smoothing functions separate data into "smooth + rough" and strive to minimize the rough part as much as possible (Hastie and Tibshirani 1990). Many types of smoothing functions are available, but all rely on the same principles:

- Each observation in the dataset is predicted from a regression model fitted to the surrounding observations.
- The resulting graph, i.e., the curve of the smoothing function, is smooth.
- The smoothness of the curve is controlled by a smoothing parameter λ.

LOESS functions work as follows: For each value x^i of the explanatory variable, a neighborhood (also called window, band or span) is defined. Within these neighborhoods, low-order polynomials are fitted using weighted regression. First- or second-order polynomials are generally used. To emphasize that weighted regression is used rather than ordinary regression, LOESS is sometimes called LOWESS. Values close to are weighed higher than values far from x^i. This fitting process may be repeated several times, during which observations with large residuals are down-weighted. Given the polynomial degree, the smoothness is controlled by the band width, which functions as the smoothing parameter. The band width is usually defined as the proportion of the data used in the local regression fit. For example, this proportion may be set to λ = 0.25, indicating that 25% of the data are used in each local regression. Values larger than λ = 0.25 are often preferred, but the choice depends on the data. A higher value of the smoothing parameter λ gives a smoother fit than a lower value. In practice, the smoothing parameter is frequently determined by trial and error based on visual inspection of the fitted curve compared with the observations. Undesirable patterns, as results of inadequate fits, can be detected in plots of residuals against fitted values. The obstacle to straightforward application of GAMs is that the same set of data can give rise to different interpretations because the result will depend on the degree of smoothing, which is chosen by the analyst.

Cubic splines are third-degree polynomials that are fitted in adjacent neighborhoods connected at so-called knots. At these knots, the polynomials are constrained to have the same derivatives, so that the polynomials join smoothly. The degree of smoothness can be controlled by the number of knots, with a large number of knots giving a smoother curve than a small number. The knots need not be equally spaced on the -axis. In intervals with many knots, the curve becomes more flexible than in intervals with few knots. Cubic splines can be chosen on the basis of visual inspection of fitted curves and residuals.

Nowadays, cubic splines are usually fitted by optimizing a penalized least squares criterion. In the case of a single explanatory variable, this criterion, which should be minimized, can be written as:

$$\sum_{i=1}^{n}(y_i - s(x_i))^2 + \lambda \int (s''(x_i))^2 dx \, . \quad (5)$$

Here $s''(x_i)$ is the second order derivative of the smoothing function s. The penalized least squares criterion (Eq. 5) consists of two terms. The first one measures closeness to observations while the second one penalizes high curvature, giving the curve a smooth appearance. If λ = 0, then Eq. 5 reduces to the standard least squares criterion, according to which the smoothing function,s(x), should be chosen such that the error sum of squares is minimized. It would result in a highly curvaceous graph that joins all observations. Since a smooth curve is preferred, a positive λ value is always used. The second-order derivative in Eq. 5, which measures curvature, is squared and integrated over the -axis. Thus, a high value of the smoothing parameter λ penalizes curvature more than a lower value and results in a smoother curve. At one extreme, i.e., when λ → ∞, the penalty term dominates, so that the graph of the optimal function is a straight line. At the other extreme, i.e., when λ → 0, the penalty term becomes unimportant and the optimal curve wriggles up and down, tracing the observations.

The smoothing parameter λ is usually determined indirectly through the choice of effective degrees of freedom, which can be thought of as the "tune" that determines curviness (Jones and Almond 1992). The number of effective degrees of freedom is analogous to the number of degrees of freedom of a LM, which is the number of linear constraints or, for the error, the difference between the number of observations and the number of linear constraints. Most computer software has options for specifying the preferred number of effective degrees of freedom. The more effective degrees of freedom the smoothing function uses, the rougher and more complex the curve becomes. The choice of the appropriate level of smoothing, by specifying the effective degrees of freedom, is among the most crucial steps in fitting GAMs. Smoothness and fit must be balanced. When the model includes several smoothing functions, they can have varying numbers of effective degrees of freedom. The number of effective degrees of freedom can also be chosen automatically, using cross validation. Through cross validation, points (s_i, y_i) are left out, one at a time, and the smoothing function at s_i is estimated based on the remaining n- 1 data points. Cross validation sum of squares for the smoothing parameter λ is computed for a range of λ. The preferred estimate of λ is that which minimizes this sum of squares. However, cross validation may be time consuming, even with fast computers. Instead, a weighted version of the full cross validation procedure may be used (SAS Institute 2008). The performance of automatic procedures for choosing the smoothing parameter is sometimes poor, which in many cases is evident, taking into account the biological interpretation when inspecting the resulting plots. This will be further described in the sections "How to assess model fit" and "Example". In practice, a smoothing parameter is commonly chosen so that the number of effective degrees of freedom is around 3–5.

Smoothing functions are easily fitted for models with one or two explanatory variables. However, models with several explanatory variables, sometimes in interaction (e.g. the effect of the dose of a herbicide may depend on the seed rate of the crop), are common. Smoothing is possible in such cases too. Consider a model with p explanatory variables and one response variable. If p= 1, smoothing is done in two dimensions, plotting the response variable against the explanatory variable. If p= 2, there are three dimensions, and the graph of the smoothing function is a two-dimensional curved surface. If p>, 2, it is difficult to visualize the graph. Variants of surface smoothers have been developed for p<1 explanatory variables. The thin-plate regression spline (Wood 2003) is a p-dimensional version of the cubic spline fitted through penalized least squares. Thin-plate regression splines are isotropic, i.e., they are not affected if the co-ordinate system of the explanatory variables is rotated. The assumption of isotropy (uniformity in all directions) is reasonable when the variables are on the same scale, for example when two explanatory variables measure geographical location in two dimensions. However, isotropic smoothing functions are less suitable when the explanatory variables measure different things, for example when one variable measures distance and the other time. Using an isotropic smoothing function, the result will depend on the units chosen, for example whether the explanatory variables are measured in hours and meters, or weeks and inches (Wood 2006a). In such situations, anisotropic smoothing functions may be preferable. In particular, tensor-product smoothing functions (Wood 2006b), constructed from univariate smoothing functions, are frequently used.

STEPS IN GENERALIZED ADDITIVE MODELLING

How to Specify Models

For controlled experiments, pure GAMs (such as $g(\mu) = \beta_0 + s(x)$ for a single explanatory variable) are rarely adequate. Since experiments usually aim at comparing treatments, treatment factors must be included in the model. A GAM will then comprise more parameters and several non-parametric smoothing functions.

If $g(\mu)$ as a function of is generally increasing or decreasing in x not as a perfect straight line, but as a bending curve around a straight line, this can be modeled as a combination of a straight line and a smoothing function. With a single explanatory variable, this model is written as:

$$g(\mu) = \beta_0 + \beta_1 x + s(x)\,. \tag{6}$$

Here, the parametric linear effect $\beta_1 x$ is separated from the effect of the smoothing function s(x). This makes it possible to test the null hypothesis that the model is linear against the alternative that the model is non-linear. By definition, the smoothing function is centered around 0, so the resulting flexible curve becomes centered around the estimated line. If the term $\beta_1 x$ were omitted from Eq. 6, the smoothing function s(x) would describe the sum of all linear and non-linear effects of x. Higher-order polynomials, for example a quadratic curve or a cubic curve, can be fitted combined with a smoothing function only when the dataset is large.

Categorical explanatory variables can be included as factors in GAMs. It is convenient to use dummy variables for coding the factor levels. Thus, let &- denote a dummy variable that takes the value 1 when the observation belongs to the jth level of the factor, and 0 otherwise. For an experiment with two treatments (i.e., a treatment factor with two levels), Eq. 6 can be augmented into

$$g(\mu) = \beta_1 d_1 + \beta_2 d_2 + \beta_3 x + s(x)\,, \tag{7}$$

which is a model with treatment-specific intercepts and a smooth curve around a straight line. According to Eq. 6, the curve of the first treatment is parallel to the curve of the second treatment. This model is analogous to the linear model for analysis of covariance assuming parallel slopes. Eq. 7 can easily be extended to a model for experiments with J> 2 treatments, through the use of J dummy variables and J+ 1 parameters: $\beta_1, \beta_2, \ldots, \beta_{J+1}$. It is usually not necessary to construct a data set with dummy variables, because statistical software packages do this work automatically if informed that the explanatory variable is a factor. However, for interpretation of parameter estimates it is essential to know exactly how the specific statistical software package codes the factor levels.

A model with two treatment-specific curves, i.e., a model with two curves that are not parallel, can be specified as:

$$g(\mu) = \beta_1 d_1 + \beta_2 d_2 + \beta_3 d_1 x + \beta_4 d_2 x + s_1(x) d_1 + s_2(x) d_2\,. \tag{8}$$

Eq. 8 describes a varying-coefficient model (Hastie and Tibshirani 1993). This model is in effect similar to an analysis of covariance model with dissimilar slopes. Note that the smoothing functions s_1 and s_2 should be functions of x and not of the products xd_1 and xd_2. This is because when d_1 is zero, the observation should not be included in the fit of the smoothing function s_1 (and similar, when d_2 is zero, the observation should not be included in the fit of S_2).

We can use two types of models with two continuous explanatory variables. The first uses two separate smoothing functions for each variable, as here:

$$g(\mu) = \beta_0 + s_1(x_1) + s_2(x_2)\,, \tag{9}$$

where x_1 and x_2 are two continuous explanatory variables, for example longitude and latitude. The second one uses a single smoothing function for both variables:

$$g(\mu) = \beta_0 + s(x_1, x_2)\ . \quad (10)$$

While Eq. 9 does not allow for interaction between x_1 and x_2 Eq. 10 does. Such interaction means that the effect of one variable depends on the value of the other. For example, grain yield of a certain wheat variety depends on nitrogen fertilization, but also on the available soil water content: since water is needed for uptake of nutrients, the effect of fertilization depends on the soil water content. So, nitrogen fertilization interacts with soil water content in terms of affecting grain yield. Eqs 9 and 10 can easily be extended to any number of continuous explanatory variables. For models including many variables, however, fitting may be difficult.

In agricultural science, it is common to perform comparative experiments, in which treatments (e.g. soil preparation methods) are compared with each other. The treatments can be represented in the model through the use of dummy variables. The experiment might also include some continuous explanatory variables (e.g. clay and soil water content) that could possibly interact with each other. One of the scientific questions might be whether a smoothing function of these two continuous variables is different for the two soil preparation methods. Using dummy variables d_1 and d_2 for the two soil preparation methods, the appropriate statistical model is:

$$g(\mu) = \beta_1 d_1 + \beta_2 d_2 + s_1(x_1, x_2) d_1 + s_2(x_1, x_2) d_2 . \quad (11)$$

Under the null hypothesis that there is no difference between the two treatments in the smoothing functions x_1 and x_2 Eq. 11 reduces to Eq. 10. If the experiment compares 3 treatments, then the model from Eq. 11 will be expanded to include 2 3 terms.

How to Fit Models

"The aim in a good smooth is to capture fully the underlying trend, while not paying too much attention to underlying noise." (Jones and Almond 1992, p. 437)

Shipley and Hunt (1996) compared LOESS and cubic spline smoothing functions. They concluded that both methods "represent versatile and accurate ways of form-free curve fitting" that "capture main trends". In their study, LOESS gave slightly better results. However, nowadays cubic spline smoothing functions are more popular, since their accessibility and flexibility make them "the best option" (Zuur et al. 2009).

The gam procedure of the SAS System and the gam function of the gam package in R fit LOESS and cubic spline smoothing functions, using the backfitting algorithm (Hastie and Tibshirani 1986, 1990). The gam function of the mgcv package in R fits cubic spline smoothing functions using direct penalized likelihood maximization (Wood 2006). With the gam procedure of SAS and the gam function in the mgcv package in R, the number of effective degrees of freedom can be determined through cross validation. The gam function of the mgcv package can use cross validation when the scale parameter is unknown, and a procedure that minimizes an estimate of the expected mean square error when the scale parameter is known (Wood 2006a). This function can also fit the smoothing function through maximum likelihood or restricted maximum likelihood.

Treatment-specific curves of varying-coefficient models can be fitted one at a time. Thus, to fit Eq. 8, $g(\mu)$ is divided into two parts: $g(\mu) = g_1(\mu) + g_2(\mu)$, where:

$$g_1(\mu) = \beta_1 d_1 + \beta_3 d_1 x + s_1(x) d_1\ ; \quad (12)$$

$$g_2(\mu) = \beta_2 d_2 + \beta_4 d_2 x + s_2(x) d_2\ . \quad (13)$$

Eqs. 12 and 13 can be fitted to the observations of the first and second treatment, respectively. The varying-coefficient model can be divided into as many parts as the treatment has levels. Thus, with 3 levels, also 3 smoothing functions $(s_1, s_2, \ldots, s_t)$ must be fitted. These smoothing functions might need different degrees of smoothness. Users of the mgcv package in R can conveniently use the by option of the gam function for fitting a varying-coefficient model in a single step. In the gam procedure of SAS, isotropic thin-plate splines ("spline2") can be fitted for p= 2 explanatory variables. In the gam function of the mgcv package in R, isotropic thin-plate regression splines and anisotropic tensor product smoothing functions can be fitted for p>1 explanatory variables (using functions s and te, respectively). Note that it might not be necessary to fit smoothing functions for all of the model's explanatory variables. Some of the explanatory variables can be included in parametric terms while some other in the smoothing functions. Since we prefer simple models than complex ones when the fits are approximately the same, the number of smoothing functions should as a rule be kept as small as possible.

How To Assess Model Fit

In LMs and GLMs, parameter estimates constitute a concise description of the fitted linear function. But in GAMs, the fitted curve cannot be described by a small set of parameter estimates. Instead, it is described graphically by plotting the fitted values against the values of the explanatory variable. To show how well the curve fits the observed values, they can be included in such a plot. It is easy for models with one explanatory variable, but is more complicated for models with several explanatory variables. Partial residuals are useful for investigation of explanatory variables (Guisan et al. 2002, Wood 2006). The partial residual is the sum of the residual for the whole model and the fitted smoothing function for the explanatory variable of interest. In partial residual plots, partial residuals are plotted against the values of the explanatory variable. For example, the partial residual plot for the first explanatory variable x_1 is a plot of partial residuals, i.e., $s(x_1) + e$ against x_1. Such a diagram illustrates the relationship between the dependent variable and the explanatory variable, similarly to a scatter plot. When the fit is good, partial residuals are randomly distributed around the curve that is described by $s(x_1)$.

Deviance (5) is twice the difference between the log likelihood (log 6) of the saturated model, which corresponds to a perfect fit, and the log likelihood of the fitted model. Deviance can be viewed as a generalization of sum of squares, since for a normal distribution, deviance equals the error sum of squares. For varying-coefficient models such as Eq. 8, deviance can be computed as the sum of the deviances of the treatment-specific parts. For example, the deviance of the fit of Eq. 8 is the sum of the deviance of the fits of Eqs. 12 and 13. For GAMs, as for GLMs, it is possible to determine approximately whether a particular explanatory variable is significant or not, by comparing deviances (Hastie and Tibshirani 1990). The difference in deviance between a model omitting the explanatory variable to be tested (i.e., the "reduced model") and a model including the explanatory variable to be tested (i.e., the "full model") is compared with a chi-square distribution with degrees of freedom equal to the difference in degrees of freedom between the two models. Thus:

$$\chi^2 = D(Reduced) - D(Full) \qquad (14)$$

is approximately chi-square distributed with degrees of freedom equal to the difference between the degrees of freedom (i.e., the sum of effective degrees of freedom and the number of all regression coefficients) of the full and the reduced model i.e. $df(Full) - df(Reduced)$. Wood (2006) justified the use of this method for penalized regression spline smoothing functions. When performing these tests, the two models should be nested (i.e., the terms of one model should be a subset of the terms of the other model). Consequently, to test for interaction between two continuous explanatory variables, the model described by Eq. 10 cannot be directly compared with the model described by Eq. 9. For hypothesis testing, the model including the interaction term must also include the main effects terms. Thus, the model

$$g(\mu) = \beta_0 + s_1(x_1) + s_2(x_2) + s(x_1, x_2) \qquad (15)$$

can be compared with the model specified by Eq. 9. When doing this, one must ensure that the two models are indeed nested. This requirement is satisfied if a tensor product smoothing function is used for the bivariate term in Eq. 15, provided that the marginal smoothing functions of the tensor product smoothing function are of the same type as the univariate smoothing functions in Eq. 9 (Wood 2006).

Overdispersion

Overdispersion occurs when the variance in the observations is greater than expected according to the assumed distribution (Olsson 2002). For Poisson distribution, variance theoretically equals mean μ, while for Binomial distribution, variance theoretically equals $n\mu(1-\mu)$, where + is the number of independent Bernoulli (i.e., yes/no) trials. In practice, deviance or sum of squared Pearson residuals should be compared with a chi-square distribution with degrees of freedom equal to the number of error degrees of freedom. Usually, these quantities agree as to whether overdispersion is present or not.

The most used methods for taking overdispersion into account are the same as for GLMs (Thurston et al. 2000). In the quasi-likelihood method for Poisson and binomial data, the dispersion parameter ϕ, which for these distributions theoretically equals 1, is estimated from the data. The dispersion parameter can be estimated as either the deviance divided by the number of degrees of freedom for error or the sum of squared Pearson residuals divided by the number of degrees of freedom for error (McCullagh and Nelder 1989, p. 328). Statistical tests are adjusted using the estimated dispersion parameter. If ϕ^* is the estimator of the dispersion parameter ϕ, then the statistic

$$F = \frac{D(Reduced) - D(Full)}{[df(Full) - df(Reduced)]\phi^*} \qquad (16)$$

can be compared with an F-distribution. In the numerator, the number of degrees of freedom equals $df(Full) - df(Reduced)$, as previously defined. In the denominator, the number of degrees of freedom equals the number of degrees of freedom for error of the full model. In a GAM, the number of degrees of freedom for error equals the number of observations minus the sum of the number of regression coefficients and the number of effective degrees of freedom for the whole model. Zuur et al. (2007) suggest F-tests for comparing models when overdispersion is present and provide an example of the use of a quasi-Poisson GAM.

The quasi-likelihood method cannot correct a pattern of residuals (Thurston et al. 2000). When a plot of deviance residuals against fitted values indicates that variation increases with the mean, a negative binomial distribution can be tried instead of the Poisson distribution.

Bayesian Methods

Bayesian methods for smoothing spline functions were introduced by Wahba (1983) and Silverman (1985), and are commonly used for assessing uncertainty in fitted GAM curves. The penalty term of the penalized least squares criterion (Eq. 5) is used as the Bayesian prior distribution, but with a minus sign in front of it so that that smooth curves are believed to be more likely than wiggly curves. Given this prior distribution for the model and the distribution for the data, the Bayesian posterior distribution for the model is the penalized least squares criterion (Eq. 5). The fitted smoothing spline function is the posterior mean in this Bayesian formulation. The confidence in the fitted curve can be expressed by Bayesian credible bands, which are Bayesian analogs to confidence bands, plotted around the fitted curve (Wood 2006a). This option is provided by the gam procedure of the SAS System and the gam function of the mgcv package in R. Marra and Wood (2012) studied the performance of this Bayesian method and showed, through simulation, that Bayesian credible bands have coverage close to nominal levels. By default, the gam function of the mgcv package in R uses the Bayesian covariance matrix for hypothesis testing of smoothing functions and parametric terms, but the frequentist covariance matrix is also suitable for testing model terms for equality to zero (Wood 2006).

Example

The dataset used in the following example is a sub-sample from a dataset previously used by Andersson et al. (2013) to study the seasonal emergence pattern of defoliated plants of perennial sow-thistle (Sonchus arvensis L.) with intact root systems. Similar experiments have previously been analyzed using categorical data analysis (Brandsæter et al. 2010), which models time as a classification factor. The Andersson et al. (2013) dataset consists of 128 observations from two northern populations (N1, N3) and two southern populations (S1, S3) of perennial sow-thistle, made on 12 test dates during one year (2008), as follows:

25 observations from N1 (from only nine test dates)

32 observations from N3

36 observations from S1

35 observations from S3

At each test date, three plants (if available) per population were exhumed, defoliated and placed under forcing conditions in a climate room for four weeks. Then, the numbers of emerged shoots were counted. The objective was to study (i) whether the emergence of shoots varies over time, (ii) whether the two regions differ in this variation over time, and (iii) whether the four populations differ in this variation. The plants were grown in a splitplot design. The experimental field was divided into three blocks with four rows per block, each row constituting a main plot in a split-plot design. Populations were randomized to main plots (i.e. rows) within blocks, and test dates were randomized to subplots within main plots. The observational unit was plant, and there was no repeated measurement on the same sampling unit. In the climate room, the plants were placed in a completely randomized manner and then moved around during the four weeks. The complete experiment in Andersson et al. (2008) was analysed in SAS version 9.2 (SAS Insitutute, 2008), using the gam procedure. In the example provided here, new analyses have been done on the subset of 128 observations (i.e., the dataset) and complemented with analyses in R.

Analyses in Sas

The analyses in SAS were performed in version 9.4 of the SAS System (SAS Institute 2008). To assess the importance of the main plot error variance, a split-plot model was fitted:

$$\log(y_{ijk}+1)=\alpha+b_i+\beta_j+\gamma_k+\delta_{jk}+a_{ij}+e_{ijk}\,, \quad (17)$$

where α is an intercept, b_i a random block effect (Q = 1, 2, 3), β_j - a fixed population effect (R = 1, 2, 3, 4), γ_k a fixed date effect (S = 1, 2, ..., 12), δ_{jk} a fixed effect of the population-by-date interaction, P- a random main plot effect, and e_{ijk} a random residual error. The random effects and the residual errors were assumed to be independent and normally distributed with expected value zero. This model was fitted in SAS using the mixed procedure.

To fit polynomials, we used the genmod procedure for generalized linear models. The Poisson distribution was used with log link. To adjust for overdispersion, we used the quasilikelihood method. The dispersion parameter was estimated as the sum of squared Pearson residuals divided by the number of degrees of freedom for error. We investigated three objective procedures for determining the best number of terms for the polynomials: (i) forward selection, (ii) backward selection, and (iii) the Akaike information criterion. To find the best polynomial model using forward selection, a polynomial of degree zero (i.e., a horizontal straight line) was fitted first. Next, a polynomial of degree one, i.e., a straight line, was fitted. The slope of this line being non-significant, the polynomial of degree zero was selected; otherwise a polynomial of degree two was fitted and the significance of the secondorder term of that polynomial was tested. This second-order term being non-significant, the polynomial of degree one was chosen; otherwise a polynomial of degree three was tested. In this way, a new term was added to the polynomial until accepting the model because of a non-significant last term. To find the best polynomial model using backward selection, a polynomial of degree seven was initially fitted. When the genmod procedure did not converge or the term of the highest order was not significant according to the adjusted F-test, this term was removed from the model. This step-wise reduction of the model was repeated until the model converged and the term of the highest order was significant. The genmod procedure in SAS offers the Akaike criterion, a popular tool for model selection. The Akaike criterion value, which should be as small as possible, is twice the difference between the number of parameters and the log likelihood (log 6). To find the best polynomial according to the Akaike criterion, all polynomials of degree smaller than eight were investigated. The selected polynomial had the smallest value of the criterion

GAMs were also fitted, using the gam procedure, with generalized cross-validation and with four effective degrees of freedom. In SAS, it is the default option and the program automatically includes a parameter expressing the linear effect of the explanatory variable, like in Eq. 6. Since this parameter uses one degree of freedom, the smoothing function is fitted with three degrees of freedom. To investigate effects of time, region and population, four models were compared: (i) a model with an intercept (i.e., with a mean only), (ii) a GAM with a single spline term (Eq. 6); (iii) a GAM with two spline terms, one per region (Eq. 8); and (iv) a GAM with four spline terms, one per population. For these analyses, we used SAS with the default four effective degrees of freedom method. Box 1 presents SAS code for population-specific curves. Due to overdispersion, we needed to employ adjusted approximate F-tests (Eq. 16) for inference. Because the gam function in SAS does not provide them, we calculated them in Microsoft Excel 2010, based on the output of the GAM analysis in SAS (Tables 1 and 2). The p-values were computed using the fdist function in Excel.

Table 1: Deviance output in SAS for n=128 observations analyzed with GAM using smoothing functions with four effective degrees of freedom. The complexity of the models, i.e., the number of degrees of freedom (df), increases downwards in the table.

Model	Deviance	Df
Only intercept	1963.44	1
Time	575.928	5
Time x Region	544.574	10
Time x Population	443.193	20

Table 2: Results of approximate F-tests (Eq. 16) based on the deviance output in Table 1.

	Deviance	df	MS	F-ratio	*p*-value
Only intercept (R)	1387.511	4	346.878	74.08	<0.001
Time (F)	575.928	123	4.682		
Time (R)	31.355	5	6.271	1.36	0.244
Time x Region (F)	544.574	118	4.615		
Time x Region (R)	101.381	10	10.138	2.47	0.011
Time x Population (F)	443.193	108	4.104		
Time (R)	132.737	15	8.849	2.16	0.012
Time x Population (F)	443.193	108	4.104		

The table includes four comparisons of reduced models (R) with full models (F). For the reduced models, the Deviance column displays the difference in deviance (Table 1) between the two fitted models. For full models, the Deviance column displays the deviance of the fitted model. For the reduced models, the degrees of freedom (df) column displays df(full model) – df(reduced model). For the full models, df = n – df(full model), where n = 128. Mean Squares (MS) – Deviance/df, and F-ratio = MS (reduced model)/MS (full model). The F-ratio is compared with an F-distribution with degrees of freedom according to the df column, providing an approximate p-value

ANALYSES IN R

The analyses in SAS were complemented and compared with some analyses in version 3.1.1 of R (R Core Team 2012). In R, GAMs were fitted with the gam function of the mgcv package, both with generalized cross-validation and with four effective degrees of freedom. The results were compared to the same analyses in SAS. For populations N3 and S3, the model according to Eq. 6, with three effective degrees of freedom for the smoothing function, did not fit the data well in R. For this reason, models without any parameters expressing linear functions (available in R, but not in SAS) were fitted in R (i.e., using Eq. 5 with $\beta_1 = 0$). In these analyses, the smoothing function was fitted with four effective degrees of freedom. It was obtained by giving the options k = 5 and fx = TRUE in the s function, which is used within the gam function. Box 1 provides R code for fitting of population-specific curves with four effective degrees of freedom.

In the split-plot analysis, the block variance, the main plot error variance, and the subplot error variance were estimated as 0.0020, 0.0072 and 0.3016, respectively. The main plot error variance was small compared with the subplot error variance.

Figure 1 shows the results of the population-wise analyses to find the best polynomial model in SAS. Using forward selection, polynomials of degree one were chosen for populations N3, S1, and S3. However, using backward selection, polynomials of degree six were needed for populations N3 and S3, and a polynomial of degree four was needed for population S1. Use of forward and backward selection resulted in the same conclusion only for population N1, for which a third-order polynomial fitted the data best. For populations N3 and S3, backward selection gave predictions far from real observations. Obviously, these curves did not fit the data well. The AIC criterion resulted in polynomials of small orders: degree one for population N1, degree two for populations N3 and S3, and degree zero for population S1. Because of the logarithmic link function, the polynomials of degree one did not look like straight lines when presented on the original scale (Figure 1). Figure 2 shows the GAM fits using the generalized cross-validation methods of SAS and R. The two software packages produced the same curve only for population N1. For the other populations, SAS gave curves with much larger fluctuations than R. Curves fitted with generalized crossvalidation in SAS failed to represent the biological phenomena in the studied example. The generalized cross-validation in SAS resulted in curves that showed pronounced peaks in areas without any observations.

Box 1: R and SAS code for fitting population-specific curves with four effective degrees of freedom.

R code

```
Model.Pop <- gam(shoot ~ factor(Pop) + s(skord, k = 5, fx = T,
by = factor(Pop)), family = poisson, data = Four.pop)
summary(Model.Pop)
Model.Pop$deviance
```

SAS code

```
proc gam data = PopS1 ;
      model Shoot = spline(Skord) / dist = Poisson   ;
      output out = Pred ;
run ; * Deviance =   170.05406604, 5 df, 36 obs ;

proc gam data = PopS3 ;
      model Shoot =  spline(Skord) / dist = Poisson   ;
      output out = Pred ;
run ; * Deviance =   119.79814787, 5 df, 35 obs ;

proc gam data = PopN1 ;
      model Shoot =  spline(Skord) / dist = Poisson   ;
      output out = Pred ;
run ; * Deviance =   57.236398516, 5 df, 25 obs ;

proc gam data = PopN3 ;
      model Shoot =  spline(Skord) / dist = Poisson   ;
      output out = Pred ;
run ; * Deviance =   96.104344253, 5 df, 32 obs ;

      * Total deviance = 170.05406604 + 119.79814787 + 57.236398516 +
                    96.104344253 = 443.193, 20 df. ;
```

Figure 3 shows polynomials of degree four and GAMs using four effective degrees of freedom. Besides population N1, SAS and R gave different results. For populations N3, S1, and S3, the SAS curves obtained using four effective degrees of freedom (Figure 3) fitted the observations better than the corresponding SAS curves fitted using generalized crossvalidation (Figure 2). Differences between fits based on four effective degrees of freedom and generalized cross-validation were less pronounced in R. All models illustrated in Figure 3 captured the biologically interesting decrease in emergence in September, followed by a period of low emergence until the rate recovered again in November–December. For population N3, the polynomial seemed to predict too sharp a bend towards the end of the period, at about 45 emerged shoots. The curves produced by R were similar to the polynomials, but with less pronounced fluctuations. In our opinion, the curves obtained using SAS were more realistic from the biological point of view

Deviance output and the approximate F-tests resulting from the GAM analysis in SAS are shown in Tables 1 and 2. The effect of time was clear because a model with a single curve fitted the data significantly better ($P < 0.001$) than a model with only an intercept. A model with region-specific curves did not fit the data significantly better ($P = 0.244$) than the model with a single curve. However, a model with population-specific curves fitted the observations significantly better ($P = 0.011$) than a model with region-specific curves. The model with population-specific curves was significantly better ($P = 0.012$) than the model with only a single curve. Thus, the best model, according to this analysis in SAS, can be described by the four solid curves shown in Figure 3. It can be noted that the corresponding analysis using R led to the same conclusion, i.e., population-specific curves fitted the observations significantly better than a single or two region-specific curves.

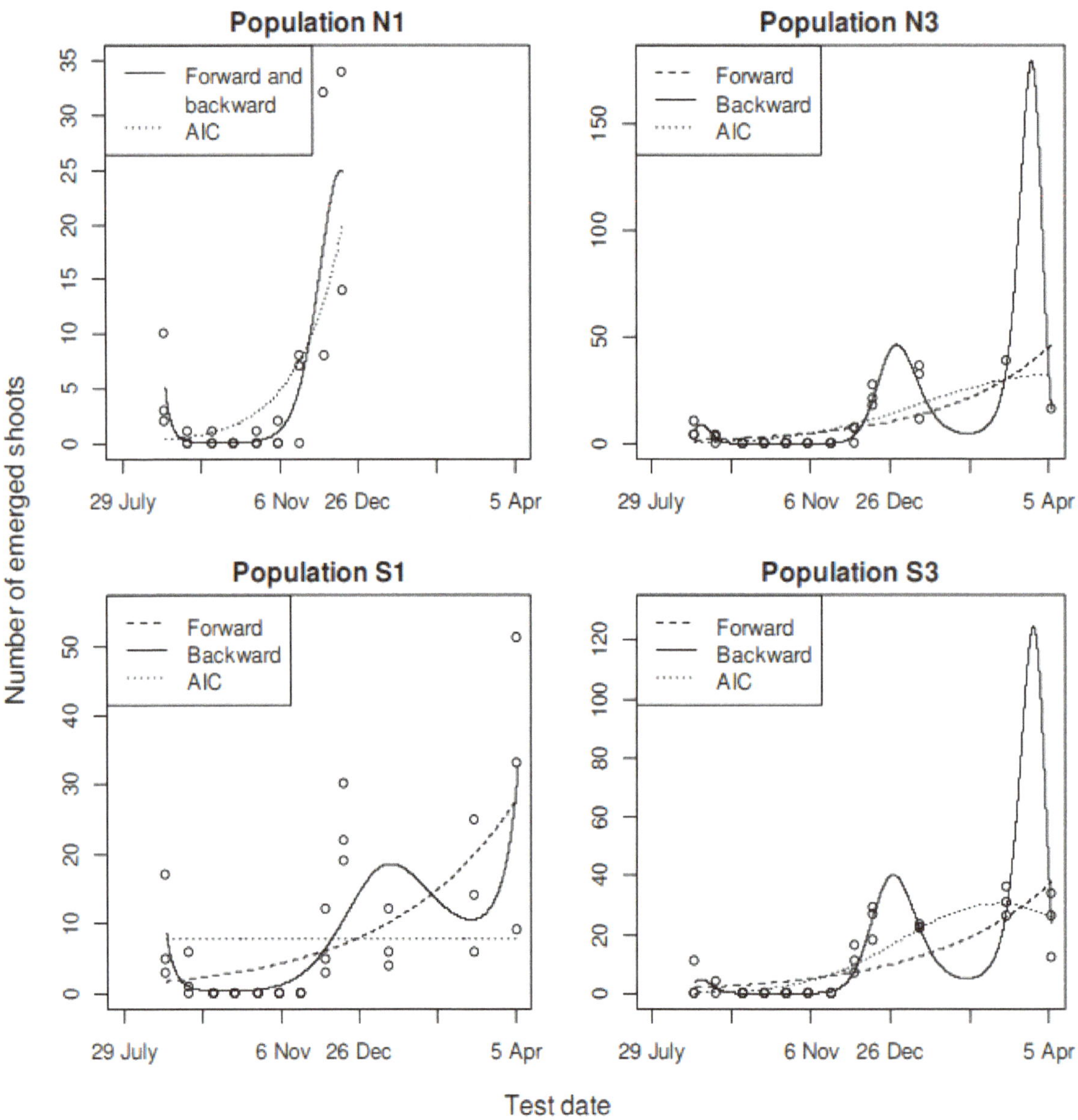

Figure 1: Best polynomials determined by forward and backward selection and using the Akaike information criterion for perennial sow-thistle populations N1, N3, S1, and S3.

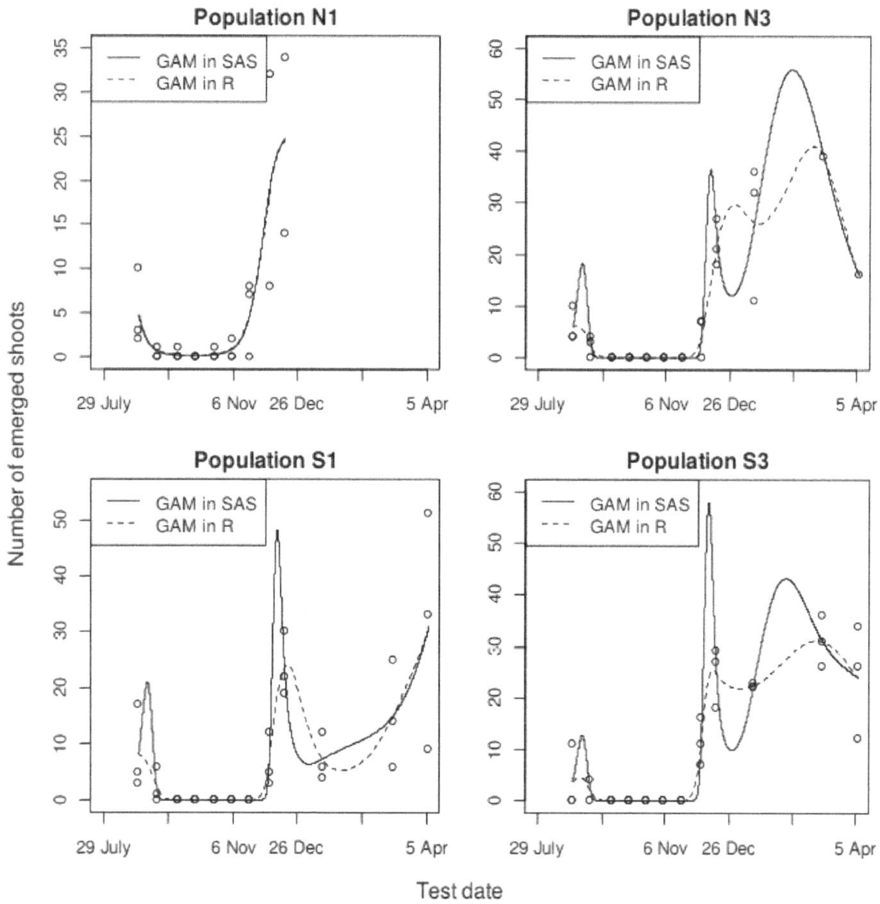

Figure 2: Population-specific GAMs using the generalized cross-validation method of SAS procedure gam for perennial sow-thistle populations N1, N3, S1, and S3.

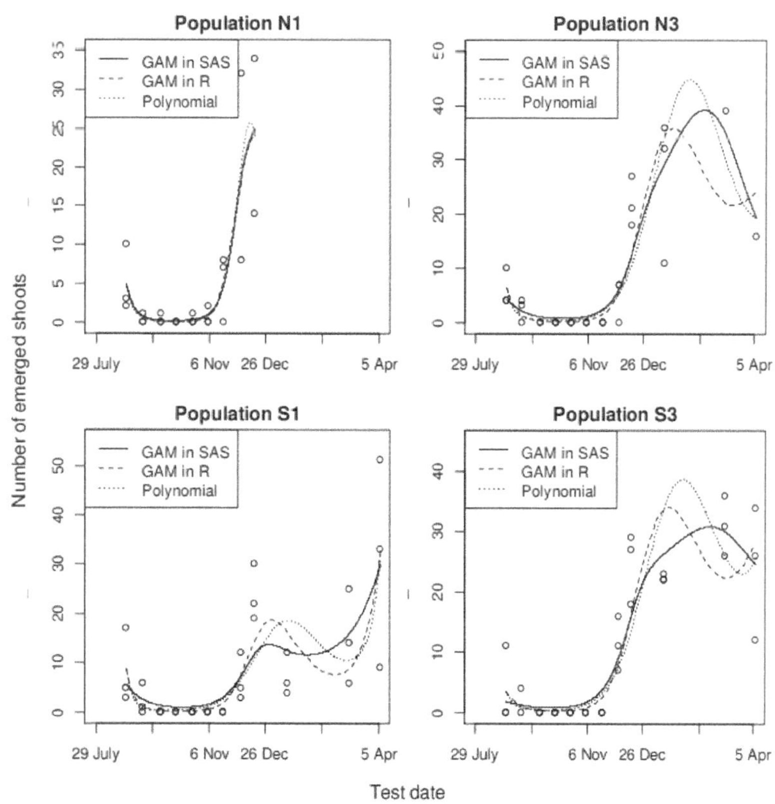

Figure 3: Fourth-degree polynomials compared with GAMs fitted in SAS and R with four effective degrees of freedom for perennial sow-thistle populations N1, N3, S1, and S3.

CONCLUSIONS

This paper deals with GAMs used for comparing experimental treatments when the response variable is a function of one or more explanatory variables. Although there are good introductions to GAMs and other statistical models using the open source software R (such as Zuur et al. 2009 and Everitt and Hothorn 2010), there is, to our knowledge, nothing similar for SAS (apart from SAS Institute 2008). Publications using GAMs for statistical analysis of field experiments are rare. The present paper added a comparison of GAM analyses performed in SAS and R.

The GAM curves fitted in SAS and R can differ. In the example provided here, generalized cross validation performed poorly in SAS, since it produced curves with sharp bends in areas without observations. This is worrying. With a fixed number of effective degrees of freedom, however, SAS produced curves that were slightly more realistic than those produced by R. Based on a small example we will not generalize our conclusions on which of the software program is better. However, it should be pointed out that the functions in R have many more options than the gam procedure of the SAS System.

The example provided here illustrated that the procedure of selecting the degree of smoothness is not well established yet. A similar problem sometimes arises when fitting linear models. For example, a third-degree polynomial may fit data significantly better than a second-degree polynomial although the second-degree polynomial does not fit the data significantly better than a straight line. Furthermore, information criteria such as the Akaike information criterion may lead to other conclusions on the model choice than hypothesis testing using forward or backward selection. In the example provided here, the analysis supported the recommendation of Hastie and Tibshirani (1990) to use 3–5 effective degrees of freedom to estimate the smoothing parameter. We believe that in general the number of effective degrees of freedom must be determined from data. The fit of the curve should always be investigated graphically. Cross-validation methods should not be relied upon blindly.

Statistical tests for comparing GAMs are the same as for GLMs. In GLMs, the chi-square test (Eq. 14) and the F-test (Eq. 16) are approximate. However, the performance of the tests for GAMs is more questionable, since the number of degrees of freedom is not exact. There is a risk that the tests are too generous in providing significant results, leading to an over-fitted final model. The curves fitted with penalized maximum likelihood are not maximum likelihood solutions. To overcome these problems, Wood (2006) proposed testing hypotheses using un-penalized fits of functions with fixed smoothing parameters. At the same time, however, fitted curves based on cross-validation and penalized likelihood method are reported. Since this mismatch can mislead, in our view it is better to report approximate pvalues, but interpret them with care. In particular, one should remember that p-values slightly below the significance level might not indicate significant effects. More research is needed on the performance of significance tests for GAMs.

In the example, random effects of rows were not included in the model, while overdispersion was accounted for through the use of approximate F-tests. As was noted, the main plot error variance was small compared with the subplot error variance. For this reason, the loss by ignoring random effects of main plots was probably small, even negligible. Generalized additive mixed models, i.e., GAMs with random effects, can be fitted in SAS (procedure glimmix) and R (function gamm). Within these procedures, one can also specify correlation structures for modeling temporal or spatial covariance. However, simple models are recommended, because numerical estimation problems can occur (Zuur et al. 2009, p. 323). Verbyla et al. (1999) showed how to analyze designed experiments using generalized additive mixed models and the software ASReml.

Box 2: For non-statisticians active in biological research, terms used in statistical text books and journals are sometimes difficult to understand. The following list highlights some basic steps and considerations when evaluating biological data with GAM. The list is not complete, but may work as a starting point for the beginner.

The beginner's guide to GAM analyses

Plot your data! As always, this will give you a first idea of what kind of statistical model to use. If possible, and if statistical testing is primary, use polynomials or other linear or generalized linear models. If such models fit the data poorly and the aim is to illustrate a complex biological process, consider using a GAM.

Choose statistical software. Many software programs cannot perform GAM analyses.

Choose a distribution. Continuous variables are usually modelled using the normal distribution, but the gamma distribution is another option when the data can take only positive values. Counts are usually assumed to be Poisson distributed. The binomial distribution is usually assumed for proportions and when two outcomes (such as dead/alive, or yes/no) are possible.

Choose a link. The most common choices are the identity link for the normal distribution, the log link for the Poisson distribution, and the logit link for the binomial distribution.

Choose the type of smoothing function. This choice may be restricted by the options available in the chosen software. Nowadays, smoothing splines based on penalized likelihood are often preferred.

Choose the smoothing parameter. Using smoothing splines, the smoothing parameter is chosen through the choice of effective degrees of freedom. Use a software integrated smoothness estimator, e.g., cross validation, or find an appropriate number through graphical examination.

Check the residuals. This can be done in the same manner as for GLMs. Plots of deviance residuals vs fitted values should be free from patterns.

Check for collinearity and concurvity. Plot the explanatory variables against each other and study the correlation matrix. When some variables are highly correlated (i.e. collinearity), consider using only one of them. If one of the variables can be well described by a smooth function of another (i.e., concurvity occurs), then consider using only the other variable.

Look for overdispersion. Oversdispersion takes place when the variance of the observations is larger than expected according to the assumed distribution. In case of severe overdispersion, consider to use another model or distribution.

Hypothesis testing and inference. A reduced model can be compared with a full model, using an approximate chi-square test (Eq. 14). For overdispersed data, the approximate F-test (Eq. 16) can be used instead. Keep in mind that these tests are approximate.

Data transformations can sometimes linearize relationships. This method should be preferred to GAMs when applicable because it makes inference easier (Yee and Mitchell 1991). For the same reason, we believe that polynomials should be preferred to GAMs when the polynomials fit the data well. However, biological data are often complex, and fitted polynomials often describe them poorly, typically with amplified bends outside the scatter of observations. Despite all the problems connected with GAMs mentioned above, we encourage increased use of these powerful models. We recommend the use of GAMs when the primary aim is to illustrate biological processes while statistical testing is secondary. For the ease of the beginner, Box 2 offers a short practical guide for GAM analyses.

ACKNOWLEDGEMENTS

We thank the editors and the reviewers for suggestions that improved the manuscript.

REFERENCES

1. Andersson L., Boström U., Forkman J., Hakman I., Liew J., Magnuski E. (2013). Sprouting capacity from intact root systems of Cirsium arvense and Sonchus arvensis decrease in autumn. Weed Research 53, 183–191.
2. Brandsæter, L. O., Fogelfors, H., Fykse, H., Graglia, E., Jensen, R. K., Melander, B., Salonen, J., Vanhala, P. (2010). Seasonal restrictions of bud growth on roots of Cirsium arvense and Sonchus arvensis and rhizomes of Elymus repens. Weed Research 50, 102–109.
3. Cleveland W.S. (1979). Robust locally weighted regression and smoothing scatterplots. Journal of the American Statistical Association 74, 829–836.
4. Elith J., Leathwick J.R., Hastie T. (2008). A working guide to boosted regression trees. Journal of Animal Ecology 77, 802–813.
5. Everitt S. E., Hothorn T. (2010). A Handbook of Statistical Analyses Using R. 2nd ed. Chapman & Hall/CRC, Boca Raton, FL.
6. Guisan A., Edwards T.C., Hastie T. (2002). Generalized linear and generalized additive models in studies of species distributions: setting the scene. Ecological Modelling 157, 89– 100.
7. Hastie T.J. (1992). Generalized additive models. In: Chambers J.M., Hastie T.J. (Eds.) Statistical Models in S. Chapman & Hall/CRC, Boca Raton, FL.
8. Hastie T., Tibshirani R. (1986). Generalized additive models (with discussion). Statistical Science 1, 297–318.
9. Hastie T., Tibshirani R. (1990). Generalized Additive Models. Chapman & Hall, London.
10. Hastie T., Tibshirani R. (1993). Varying-coefficient models (with discussion). Journal of the Royal Statistical Society, Series B 55, 757–796.
11. Hilbe J. M. (1993). Generalized additive models software. The American Statistician 47, 59–64.
12. Jones K., Almond S. (1992). Moving out the linear rut: the possibilities of generalized additive models. Transactions of the Institute of British Geographers 17, 434–447.
13. Marra G., Wood S.N. (2012) Coverage properties of confidence intervals for generalized additive model components. Scandinavian Journal of Statistics 39, 53–74.
14. McCullagh P., Nelder J.A. (1989). Generalized Linear Models, 2nd ed. Chapman & Hall, Cambridge, UK.
15. Olsson U. (2002). Generalized Linear Models: An Applied Approach. Studentlitteratur, Lund, Sweden.
16. Quinn G.P., Keough M.J. (2002). Experimental Design and Data Analysis for Biologists. Cambridge University Press, Cambridge.
17. Ramsay T.O., Burnett R.T., Krewski D. (2003). The effect of concurvity in generalized additive models linking mortality to ambient particulate matter. Epidemiology 14, 18–23.
18. R Core Team (2012). R: A language and environment for statistical computing. R Foundation for Statistical Computing, Vienna, Austria. http://www.R-project.org/ SAS Institute (2008). SAS/STAT 9.2 User's Guide. Chapter 36: The GAM Procedure. SAS Institute, Cary, NC.
19. Shipley B., Hunt R. (1996). Regression smoothers for estimating parameters of growth analyses. Annals of Botany 78, 569–576.
20. Silverman B.W. (1985). Some aspects of the spline smoothing approach to non-parametric regression curve fitting. Journal of the Royal Statistical Society, Series B 47, 1–53.
21. Thurston S.W., Wand M.P., Wiencke J.K. (2000). Negative binomial additive models. Biometrics 56, 139–144.
22. Venables W.N., Dichmont C.M. (2004). GLMs, GAMs and GLMMs: an overview of theory for applications in fisheries research. Fisheries Research 70, 319–337.
23. Verbyla A.P., Cullis B.R., Kenward M.G., Welham S.J. (1999). The analysis of designed experiments and longitudinal data by using smoothing splines. Journal of the Royal Statistical Society. Series C 48, 269–311.
24. Wahba G. (1983). Bayesian confidence intervals for the cross validated smoothing spline. Journal of the Royal Statistical Society, Series B 45, 133–150.
25. Wood S.N. (2003). Thin plate regression splines. Journal of the Royal Statistical Society, Series B 65, 95–114.
26. Wood S.N. (2006a). Generalized Additive Models: An Introduction with R. Chapman & Hall/CRC, Boca Raton.
27. Wood S.N. (2006b). Low-rank scale-invariant tensor product smooths for generalized additive mixed models. Biometrics 62, 1025–1036.
28. Yee T.W., Mitchell N.D. (1991). Generalized additive models in plant biology. Journal of Vegetation Science 2, 587–602.
29. Zuur A. F., Ieno E.L., Smith G.M. (2007). Analysing Ecological Data. Springer, New York.
30. Zuur A.F., Ieno E.L., Walker N.J., Saveliev A.A., Smith G.M. (2009). Mixed Effects Models and Extensions in Ecology with R. Springer, New York.

CHAPTER 2 THE CHEMEQLB LNG AND THE POOLINGR LNG MODEL (2 BASIC LEARNING MODELS ON LINGO)

MODEL:

! Two level blending or pooling model(POOLINGR.lg4).

Raw materials are selectively blended into pools,

then pools are blended into finished products.

We want to maximize the profit = sales of finished goods,

minus cost of raw materials purchased,

subject to satisfying the quality requirements

of each finished good;

! Keywords: Blending, Pooling, Petroleum refining;

SETS:

! Each raw material has an availability and cost/unit;

RM : A, COST;

! There are a set of intermediate pools, e.g., storage tanks;

POOL: BP ;

! Each finished good has a min required, max sellable,profit contribution/unit and batch size to be determined;

FG: D, E, PRICE, BF;

! There are a set of quality measures;

QM;

! For each combo of RM and QM there is a quality level;

RXQ(RM, QM): QL;

! For each combo QM, FG there are upper and lower limits on quality, and slack on upper quality to be determined;

QXF(QM, FG): U, L, S;

! For each combination of Pool and quality level, there is a quality level to be determined;

PXQ(POOL, QM): QP;

! Set of which RM can go into which POOL;

RXP(RM, POOL): R2P;

! Set of which POOL can go into which FG;

PXF(POOL, FG): P2F;

```
ENDSETS
DATA:
 RM= RMA  RMB  RMC; ! Set of raw materials;
 A = 9999  9999  9999; ! Raw matl availabilities;
 COST = 6   16   10; ! R. M. costs;
 QM = SULFUR;         ! Set of qualities;
 QL = .03  .01  .02; ! Quality by R.M.;

 FG = FGX  FGY;
 D =  0,  0;     ! Min needed of each F.G.;
 E =  100  200;    ! Max sellable of each F.G;
 PRICE= 9   15;   ! Selling price of each F.G.;
 U =  .025  .015;   ! Upper limits on quality for each FG.;
 L =   0  0 ;   ! Lower limits on quality...;
POOL = POOLAB POOLC; ! The names of the pools;
! Which R.M. can go into which POOLS;
 RXP=
   RMA  POOLAB
   RMB  POOLAB
   RMC  POOLC;
! Which pools go into which F.G.;
 PXF = POOLAB FGX  POOLAB FGY  POOLC FGX
POOLC FGY;
! This data set has several local optima:
 ! Local optimum 1:
   Obj = 0,
   Do not buy or sell anything,
   BP( ) = BF( ) = 0,
  Local optimum 2:
   Obj = 300,
   R2P( RMA, POOLAB) =  50.0000
   R2P( RMB, POOLAB)  = 150.0000
   R2P( RMC, POOLC)   =   0.0000
   P2F( POOLAB, FGX)  =   0.0000
   P2F( POOLAB, FGY)  = 200.0000
   QP( POOLAB, SULFUR)=  0.0150
  Local optimum 3 (Global):
   Obj = 400,
   R2P( RMA, POOLAB)  =   0.0000
   R2P( RMB, POOLAB)  = 100.0000
   R2P( RMC, POOLC)   = 100.0000
   P2F( POOLAB, FGX)  =   0.0000
   P2F( POOLAB, FGY)  = 100.0000
   P2F( POOLC, FGX)   =   0.0000
   P2F( POOLC, FGY)   = 100.0000
   QP( POOLAB, SULFUR)=  0.0100
   QP( POOLC, SULFUR) =  0.0200
;

ENDDATA
!--------------------------------------------------------------;
! Variables:
  R2P(r,p) = amount of raw material r transferred to pool p,
  P2F(p,f) = amount of material transferred
       from pool p to finished good f,
  QP(p,q) = level of quality q (e.g., fraction sulfur)
       in pool p,
  BP(p) = batch size (or amount in) of pool p,
  BF(f) = batch size (or amount in) of finished good f;

 ! The model;
  ! Max revenues - cost of raw materials;
    [PROFIT] MAX = @SUM( FG(f): PRICE(f) * BF(f)) -
      @SUM( RXP(r, p): COST( r)*R2P( r, p));
  ! Raw materials stage: raw material availabilities;
   @FOR( RM( r):
     [RMLIM] @SUM( RXP(r, f): R2P( r, f)) <= A( r);
    );
  ! Pools stage;
    @FOR( POOL(p):
  ! How much ( the batch size) is in each pool;
    [INPOOLI] BP(p) = @SUM( RXP(r,p): R2P(r,p));
    [INPOOLO] BP(p) = @SUM( PXF(p,f): P2F(p,f));
  ! The quality level, QP(p,q), of each pool.
    The model is nonlinear and nonconvex because
    of the product terms QP( p, q) * BP( p);
```

```
@FOR( QM( q):
   [QPOOL] QP( p, q)* BP( p) = @SUM( RXP(r,p): QL(r,q)*R2P(r,p));
   );
  );
! Finished goods stage;
 @FOR( FG( f):
 ! Batch size computation;
   [BDEFO] BF( f) = @SUM( POOL( p): P2F( p, f));
 ! Batch size limits;
   [BLO] BF( f) >= D( f);
   [BHI] BF( f) <= E( f);
 ! Quality restrictions for each quality q,
  The model is nonlinear and nonconvex because
  of the product terms QP( p, q) * P2F( p, f);
  @FOR( QM( q):
    [QUP] @SUM( POOL( p): QP( p, q) * P2F( p, f))
        + S( q, f) = U( q, f) * BF( f);
    [QDN] S( q, f) <= ( U( q, f) - L( q, f)) * BF( f);
    );
  );
END
```

CHAPTER 3

AN INTRODUCTION TO THE "SPECIAL VOLUME SPECTROSCOPY AND CHEMOMETRICS IN R"

Katharine M. Mullen and Ivo H. M. van Stokkum

Keywords: Spectroscopy, chemometrics, R.

INTRODUCTION

Statistical computing has assumed a central role in spectroscopy and chemometrics research. The R language and environment for statistical computing (R Development Core Team 2006) is at the forefront of the development of software for statistical computing, and has become a powerful tool for research in spectroscopy and chemometrics. The present special volume gives a flavour of the state of the art of the use of R in these areas.

The summer of 2006 was a watershed moment in the history of the interaction of R and the spectroscopy and chemometrics fields. Prior to then, researchers in these fields had found ample application for the extremely wide range of general-purpose statistical techniques available in R to their research, and a variety of R packages (i.e., modular software components) specifically targeted to the analysis of spectroscopy and chemometrics data had been released. However, no overview of the use of R in these application areas yet existed. Jan de Leeuw took a step toward changing this at the useR! Conference in Vienna in June by suggesting the compilation of this special volume. After the conference we invited contributions from researchers in the community and the outlines of the present volume began to take shape.

Then, in August, a special issue of R News concentrating on applications in chemistry was released, guest edited by Ron Wehrens. This was the first such compilation in the literature, and began with the observation by Wehrens that, "R has become the standard for statistical analysis in biology and bioinformatics, but it is also gaining popularity in other fields of natural sciences" (Wehrens 2006). Three of the articles in the R News special issue have been expanded into contributions here.

For the uninitiated, some explanation may be required as to why an overview of spectroscopy and chemometrics tools and techniques in R in particular is necessary. After all, it may be argued that anything possible in one (Turing-complete) programming language can surely be done in another, and that one's choice of language is therefore not of much interest. However, the features of R in comparison to other currently available systems for statistical computing uniquely advantage spectroscopy and chemometrics researchers working in R. The hundreds of available packages and the fact that R is the lingua franca of computational statisticians means that an implementation of even very new statistical techniques is likely to be found in R. Furthermore, R is available under the

Citation: Katharine M. Mullen and Ivo H. M. van Stokkum, An Introduction to the "Special Volume Spectroscopy and Chemometrics in R", Doi:http://dx.doi.org/10.18637/jss.v018.i01

terms of the GNU General Public License in source code form, and compiles and runs on all major operating systems. This means that R need not be treated as a black box: if necessary, researchers can examine or modify the techniques it implements at a very low level. It also means that without cost and without regard to choice of operating system, researchers using R can reproduce each other's results. Moreover, R (and its package system in particular) facilitates the standardization and distribution of software for spectroscopy and chemometrics, which up until now has too often remained proprietary to the particular department or laboratory in which it was developed, impeding the verification of research results and slowing progress in the dissemination of analysis methodology. The useR! conference during the summer of 2006 not only gave birth to this special volume. It also brought home the point that Journal of Statistical Software (JSS) is an extremely suitable forum for publications describing statistical software based in R. The conference concluded with a panel discussion by representatives of well-established journals in computational statistics, discussing how "Getting recognition for excellence in computational statistics" might be accomplished. In the course of this discussion Jan de Leeuw made a compelling case for the advantages of publishing R-based research in JSS: like R, JSS is free, so that all interested may access it (assuming an internet connection), it supports reproducible research and the validation of software tools by both reviewing and publishing of the software associated with a manuscript, and the electronic format does away with archaic limitations of print journals like page and color restrictions. We are especially pleased that this compendium describing spectroscopy and chemometrics in R appears in JSS.

AN OUTLINE OF THE CONTRIBUTIONS

The volume begins with a contribution from Mevik and Wehrens (2007) on the pls package for principal component and partial least squares regression. These multivariate regression methods are widely applied in spectroscopy and many other fields. The user interface of pls is modeled after the traditional formula interface used in R functions, as exemplified by the lm function. The authors provide an introduction to the algorithms as well as several case studies. Next, Mullen and van Stokkum (2007) present the TIMP package for modeling multi-way spectroscopic measurements. The package employs a partitioned variable projection algorithm for fitting the separable nonlinear models that are common in spectroscopy data analysis. Models are constructed in TIMP with a rich variety of options, e.g., for compartmental models, spectral constraints, and multiexperiment parameterizations. Kirchner, Saussen, Steen, Steen, and Hamprecht (2007) follow with a description of the amsrpm package for preprocessing mass spectra measured in-line with liquid chromatography systems (LC/MS). Using a robust point matching algorithm the package compensates for instrumental distortions, and provides retention time alignment of the LC/MS data. Guha (2007) then introduces the rcdk package that provides R user's easy access to the Chemistry Development Kit CDK, a Java framework for chemical informatics, as well as the rpubchem package that allows access to the data in PubChem, a public repository of molecular structures and associated assay data. The next contribution from Binsl, Mullen, van Beek, and Heringa (2007) describes the FluxSimulator package that is able to simulate isotopomer distributions given a specification of a metabolic network of arbitrary complexity. It is intended for use with isomer labeled substrates measured by mass or NMR spectroscopy to quantify metabolic networks. The sixth contribution by Laptenok, Mullen, Borst, van Stokkum, Apanasovich, and Visser (2007) shows the utility of the TIMP package in fluorescence lifetime imaging microscopy (FLIM), an important technique for the in situ and in vivo study of physico-chemical processes in biological objects. Case studies on real and simulated data evidence the applicability of the package for parameter estimation and also the limitations imposed by practical signal to noise ratios. Next, Fraley and Raftery (2007) review clustering techniques for chemometrics data analysis applications, illustrating their mclust package. They present case studies of density estimation and discriminant analysis. Ritter and Gilliard (2007) then describe ImpuR, a collection of diagnostic tools to explore time-intensity matrices with respect to their bilinear structure and departures from it. It is applied to chromatography combined with (UV) absorption spectroscopy. This is followed by a description by Krastev (2007) of spectrino, an R-based Windows application to organize and pre-process spectra. spectrino also offers a set of features to create data structures and visually manipulate/ compare spectra, in particular from mass spectroscopy. The volume concludes with a contribution by Babu and Mahabal (2007) describing VOStat, an R-based online software suite written mainly for the Virtual Observatory, in which astronomers share large scale spectroscopic data.

CONCLUSIONS AND OUTLOOK

The present volume showcases some of the breadth and depth of software tools for spectroscopy and chemometrics applications for the R language and environment. We hope the collection is inspiring to both computational statisticians interested in applications of their methodologies and to spectroscopists or chemometricians in need of solutions to their data analysis problems. Feedback to the authors may provide an incentive to further develop their methods and software for a broader applicability.

REFERENCES

1. Babu GJ, Mahabal A (2007). "Using R-based VOStat as a Low-Resolution Spectrum Analysis Tool." Journal of Statistical Software, 18(11). URL http://www.jstatsoft.org/v18/ i11/
2. Binsl TW, Mullen KM, van Beek JHGM, Heringa J (2007). "pkgFluxSimulator: An R Package to Simulate Isotopomer Distributions in Metabolic Networks." Journal of Statistical Software, 18(7). URL http://www.jstatsoft.org/v18/i07/.
3. Fraley C, Raftery A (2007). "Model-based Methods of Classification: Using the mclust Software in Chemometrics." Journal of Statistical Software, 18(6). URL http://www. jstatsoft.org/v18/i06/.
4. Guha R (2007). "Chemical Informatics Functionality in R." Journal of Statistical Software, 18(5). URL http://www.jstatsoft.org/v18/i05/.
5. Kirchner M, Saussen B, Steen H, Steen JA, Hamprecht FA (2007). "amsrpm: Robust Point Matching for

Retention Time Alignment of LC/MS Data with R." Journal of Statistical Software, 18(4). URL http://www.jstatsoft.org/v18/i04/.

6. Krastev T (2007). "Spectrino Software: Spectra Visualization and Preparation for R." Journal of Statistical Software, 18(10). URL http://www.jstatsoft.org/v18/i10/.
7. Laptenok S, Mullen KM, Borst JW, van Stokkum IHM, Apanasovich VV, Visser AJWG (2007). "Fluorescence Lifetime Imaging Microscopy (FLIM) Data Analysis with TIMP." Journal of Statistical Software, 18(8). URL http://www.jstatsoft.org/v18/i08/.
8. Mevik BH, Wehrens R (2007). "The pls Package: Principal Component and Partial Least Squares Regression in R." Journal of Statistical Software, 18(2). URL http://www. jstatsoft.org/v18/i02/.
9. Mullen KM, van Stokkum IHM (2007). "TIMP: An R Package for Modeling Multi-way Spectroscopic Measurements." Journal of Statistical Software, 18(3). URL http://www. jstatsoft.org/v18/i03/.
10. R Development Core Team (2006). R: A Language and Environment for Statistical Computing. R Foundation for Statistical Computing, Vienna, Austria. ISBN 3-900051-07-0, URL http: //www.R-project.org/.
11. Ritter C, Gilliard J (2007). "ImpuR: A Collection of Diagnostic Tools Developed in R in the Context of Peak Impurity Detection in HPLC-DAD but Potentially Useful with Other Types of Time–Intensity Matrices." Journal of Statistical Software, 18(9). URL http: //www.jstatsoft.org/v18/i09/.
12. Wehrens R (2006). "Editorial." R News, 6(3), 1–2.

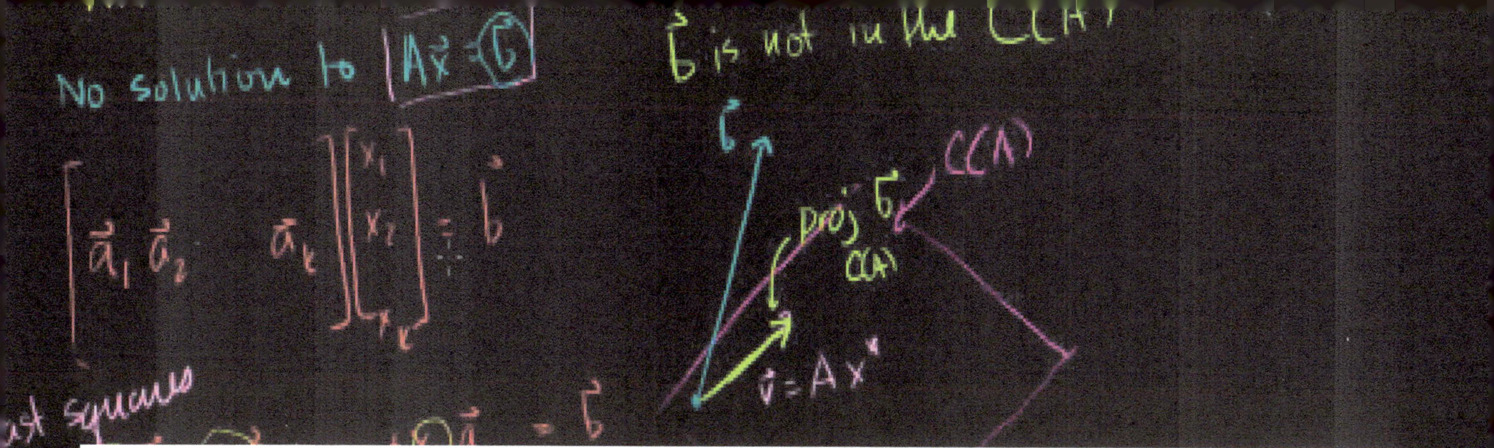

CHAPTER 4

PRINCIPAL COMPONENT AND PARTIAL LEAST SQUARES REGRESSION IN R

Bjørn-Helge Mevik and Ron Wehrens

Keywords: principal component regression, PCR, partial least squares regression, PLSR, R.

INTRODUCTION

Multivariate regression methods like principal component regression (PCR) and partial least squares regression (PLSR) enjoy large popularity in a wide range of fields, including the natural sciences. The main reason is that they have been designed to confront the situation that there are many, possibly correlated, predictor variables, and relatively few samples—a situation that is common, especially in chemistry where developments in spectroscopy since the seventies have revolutionised chemical analysis. In fact, the origin of PLSR lies in chemistry (see, e.g., Wold 2001; Martens 2001). The field of near-infrared (NIR) spectroscopy, with its highly overlapping lines and difficult to interpret overtones, would not have existed but for a method to obtain quantitative information from the spectra.

Also other fields have benefited greatly from multivariate regression methods like PLSR and PCR. In medicinal chemistry, for example, one likes to derive molecular properties from the molecular structure. Most of these quantitative structure–activity relations (QSAR, and also quantitative structure–property relations, QSPR), and in particular, comparative molecular field analysis (ComFA) (Cramer, Patterson, and Bunce 1988), use PLSR. Other applications range from statistical process control (Kresta, MacGregor, and Marlin 1991) to tumour classification (Nguyen and Rocke 2002) to spatial analysis in brain images (McIntosh, Bookstein, Haxby, and Grady 1996) to marketing (Fornell and Bookstein 1982).

In the usual multiple linear regression (MLR) context, the least-squares solution for

$$Y = XB + \mathcal{E} \tag{1}$$

is given by

$$B = (X^\top X)^{-1} X^\top Y \tag{2}$$

The problem often is that $X^T X$ is singular, either because the number of variables (columns) in X exceeds the number of objects (rows), or because of collinearities. Both PCR and PLSR circumvent this by decomposing X into orthogonal scores T and loadings P

$$X = T P \tag{3}$$

and regressing Y not on X itself but on the first a columns of the scores T . In PCR, the scores are given by the left singular vectors of X, multiplied with the corresponding singular

Citation: Bjørn-Helge Mevik, Ron Wehrens, Principal Component and Partial Least Squares Regression in R, Volume 18, Issue 2. Doi:http://dx.doi.org/10.18637/jss.v018.i02

values, and the loadings are the right singular vectors of X. This, however, only takes into account information about X, and therefore may be suboptimal for prediction purposes. PLSR aims to incorporate information on both X and Y in the definition of the scores and loadings. In fact, for one specific version of PLSR, called SIMPLS (de Jong 1993), it can be shown that the scores and loadings are chosen in such a way to describe as much as possible of the covariance between X and Y, where PCR concentrates on the variance of X. Other PLSR algorithms give identical results to SIMPLS in the case of one Y variable, but deviate slightly for the multivariate Y case; the differences are not likely to be important in practice

Algorithms

In PCR, we approximate the X matrix by the first a principal components (PCs), usually obtained from the singular value decomposition (SVD):

$$\boldsymbol{X} = \tilde{\boldsymbol{X}}_{(a)} + \mathcal{E}_X = (\boldsymbol{U}_{(a)}\boldsymbol{D}_{(a)})\boldsymbol{V}_{(a)}^\top + \mathcal{E}_X = \boldsymbol{T}_{(a)}\boldsymbol{P}_{(a)}^\top + \mathcal{E}_X$$

Next, we regress Y on the scores, which leads to regression coefficients

$$\boldsymbol{B} = \boldsymbol{P}(\boldsymbol{T}^\top\boldsymbol{T})^{-1}\boldsymbol{T}^\top\boldsymbol{Y} = \boldsymbol{V}\boldsymbol{D}^{-1}\boldsymbol{U}^\top\boldsymbol{Y}$$

where the subscripts a have been dropped for clarity.

For PLSR, the components, called Latent Variables (LVs) in this context, are obtained iteratively. One starts with the SVD of the crossproduct matrix $S = X^T Y$, thereby including information on variation in both X and Y, and on the correlation between them. The first left and right singular vectors, w and q, are used as weight vectors for X and Y, respectively, to obtain scores t and u:

$$t = \boldsymbol{X}w = \boldsymbol{E}w \quad (4)$$

$$u = \boldsymbol{Y}q = \boldsymbol{F}q \quad (5)$$

where E and F are initialised as X and Y , respectively. The X scores t are often normalised:

$$t = t/\sqrt{t^\top t} \quad (6)$$

The Y scores u are not actually necessary in the regression but are often saved for interpretation purposes. Next, X and Y loadings are obtained by regressing against the same vector t:

$$p = \boldsymbol{E}^\top t \quad (7)$$

$$q = \boldsymbol{F}^\top t \quad (8)$$

Finally, the data matrices are 'deflated': the information related to this latent variable, in the form of the outer products tp^T and tq^T, is subtracted from the (current) data matrices E and F.

$$\boldsymbol{E}_{n+1} = \boldsymbol{E}_n - tp^\top \quad (9)$$

$$\boldsymbol{F}_{n+1} = \boldsymbol{F}_n - tq^\top \quad (10)$$

The estimation of the next component then can start from the SVD of the crossproduct matrix $\boldsymbol{E}_{n+1}^\top \boldsymbol{F}_{n+1}$. After every iteration, vectors w, t, p and q are saved as columns in matrices W, T, P and Q, respectively. One complication is that columns of matrix W cannot be compared directly: they are derived from successively deflated matrices E and F. It has been shown that an alternative way to represent the weights, in such a way that all columns relate to the original X matrix, is given by

$$\boldsymbol{R} = \boldsymbol{W}(\boldsymbol{P}^\top\boldsymbol{W})^{-1} \quad (11)$$

Again, here, only the first a components are used. How many components are optimal has to be determined, usually by cross-validation.

Many alternative formulations can be found in literature. It has been shown, for instance, that only one of X and Y needs to be deflated; alternatively, one can directly deflate the crossproduct matrix S (as is done in SIMPLS, for example). Moreover, there are many equivalent ways of scaling. In the example above, the scores t have been normalised, but one can also choose to introduce normalisation at another point in the algorithm. Unfortunately, this can make it difficult to directly compare the scores and loadings of different PLSR implementations.

On the use of PLSR and PCR

In theory, PLSR should have an advantage over PCR. One could imagine a situation where a minor component in X is highly correlated with Y; not selecting enough components would then lead to very bad predictions. In PLSR, such a component would be automatically present in the first LV. In practice, however, there is hardly any difference between the use of PLSR and PCR; in most situations, the methods achieve similar prediction accuracies, although PLSR usually needs fewer latent variables than PCR. Put the other way around: with the same number of latent variables, PLSR will cover more of the variation in Y and PCR will cover more of X. In turn, both behave very similar to ridge regression (Frank and Friedman 1993)

It can also be shown that both PCR and PLSR behave as shrinkage methods (Hastie, Tibshirani, and Friedman 2001), although in some cases PLSR seems to increase the variance of individual regression coefficients, one possible explanation of why PLSR is not always better than PCR.

Outline of the paper

Section 2 presents an example session, to get an overview of the package. In Section 3 we describe formulas and data frames (as they are used in pls). Users familiar with formulas and data frames in R can skip this section on first reading. Fitting of models is described in Section 4, and cross-validatory choice of components is discussed in Section 5. Next, inspecting and plotting models is described (Section 6), followed by a section on predicting future observations (Section 7). Finally, Section 8 covers more advanced topics such as setting options, using the underlying functions directly, and implementation details.

EXAMPLE SESSION

In this section we will walk through an example session, to get an overview of the package. To be able to use the package, one first has to load it:

R> library("pls")

This prints a message telling that the package has been attached, and that the package implements a function loadings that masks a function of the same name in package stats. (The output of the commands have in some cases been suppressed to save space.) Three example data sets are included in pls:

yarn A data set with 28 near-infrared spectra (NIR) of PET yarns, measured at 268 wavelengths, as predictors, and density as response (density) (Swierenga, de Weijer, van Wijk, and Buydens 1999). The data set also includes a logical variable train which can be used to split the data into a training data set ooliveoil A data set with 5 quality measurements (chemical) and 6 panel sensory panel variables (sensory) made on 16 olive oil samples (Massart, Vandeginste, Buydens, de Jong, Lewi, and Smeyers-Verbeke 1998). See ?oliveoil for details.f size 21 and test data set of size 7. See ?yarn for details.

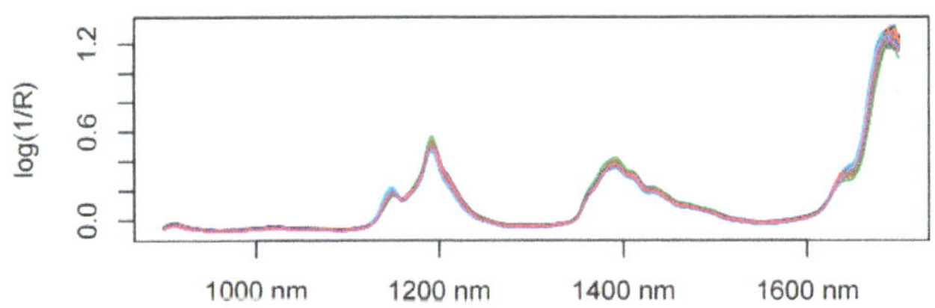

Figure 1: Gasoline NIR spectra

gasoline A data set consisting of octane number (octane) and NIR spectra (NIR) of 60 gasoline samples (Kalivas 1997). Each NIR spectrum consists of 401 diffuse reflectance measurements from 900 to 1700 nm. See? gasoline for details.

These will be used in the examples that follow. To use the data sets, they must first be loaded:

R> data("yarn")

R> data("oliveoil")

R> data("gasoline")

For the rest of the paper, it will be assumed that the package and the data sets have been loaded as above. Also, all examples are run with options(digits = 4)

In this section, we will do a PLSR on the gasoline data to illustrate the use of pls. The spectra are shown in Figure 1. We first divide the data set into train and test data sets:

R> gasTrain <- gasoline[1:50,]

R> gasTest <- gasoline[51:60,]

A typical way of fitting a PLSR model is

R> gas1 <- plsr(octane ~ NIR, ncomp = 10, data = gasTrain, validation = "LOO")

This fits a model with 10 components, and includes leave-one-out (LOO) cross-validated predictions (Lachenbruch and Mickey 1968). We can get an overview of the fit and validation results with the summary method:

R> summary(gas1)

Data: X dimension: 50 401

Y dimension: 50 1

Fit method: kernelpls

Number of components considered: 10

VALIDATION: RMSEP

Cross-validated using 50 leave-one-out segments.

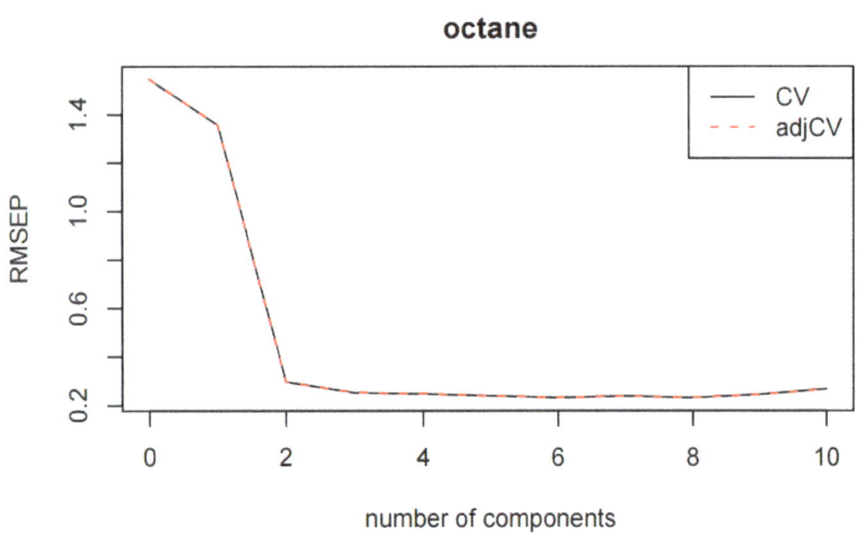

Figure 2: Cross-validated RMSEP curves for the gasoline data

```
       (Intercept)  1 comps  2 comps  3 comps  4 comps  5 comps  6 comps
CV           1.545    1.357   0.2966   0.2524   0.2476   0.2398   0.2319
adjCV        1.545    1.356   0.2947   0.2521   0.2478   0.2388   0.2313
       7 comps  8 comps  9 comps  10 comps
CV      0.2386   0.2316   0.2449    0.2673
adjCV   0.2377   0.2308   0.2438    0.2657

TRAINING: % variance explained
        1 comps  2 comps  3 comps  4 comps  5 comps  6 comps  7 comps  8 comps
X         78.17    85.58     93.4    96.06    96.94    97.89    98.38    98.85
octane    29.39    96.85     97.9    98.26    98.86    98.96    99.09    99.16
        9 comps  10 comps
X         99.02      99.2
octane    99.28      99.4
```

The validation results here are root mean squared error of prediction (RMSEP). There are two cross-validation estimates: CV is the ordinary CV estimate, and adjCV is a bias-corrected CV estimate (Mevik and Cederkvist 2004) (For a LOO CV, there is virtually no difference).

It is often simpler to judge the RMSEPs by plotting them:

R> plot(RMSEP(gas1), legendpos = “topright”)

This plots the estimated RMSEPs as functions of the number of components (Figure 2). The legendpos argument adds a legend at the indicated position. Two components seem to be enough. This gives an RMSEP of 0.297. As mentioned in the introduction, the main practical difference between PCR and PLSR is that PCR often needs more components than PLSR to achieve the same prediction error. On this data set, PCR would need three components to achieve the same RMSEP

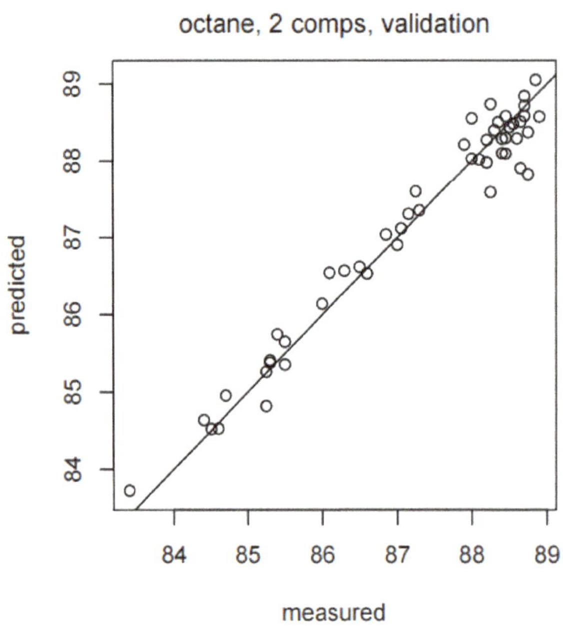

Figure 3: Cross-validated predictions for the gasoline data

Once the number of components has been chosen, one can inspect different aspects of the fit by plotting predictions, scores, loadings, etc. The default plot is a prediction plot:

R> plot(gas1, ncomp = 2, asp = 1, line = TRUE)

This shows the cross-validated predictions with two components versus measured values (Figure 3). We have chosen an aspect ratio of 1, and to draw a target line. The points follow the target line quite nicely, and there is no indication of a curvature or other anomalies. Other plots can be selected with the argument plottype:

R> plot(gas1, plottype = “scores”, comps = 1:3)

This gives a pairwise plot of the score values for the three first components (Figure 4). Score plots are often used to look for patterns, groups or outliers in the data. (For instance, plotting the two first components for a model built on the yarn dataset clearly indicates the experimental design of that data.) In this example, there is no clear indication of grouping or outliers. The numbers in parentheses after the component labels are the relative amount of X variance explained by each component. The explained variances can be extracted explicitly with

R> explvar(gas1)

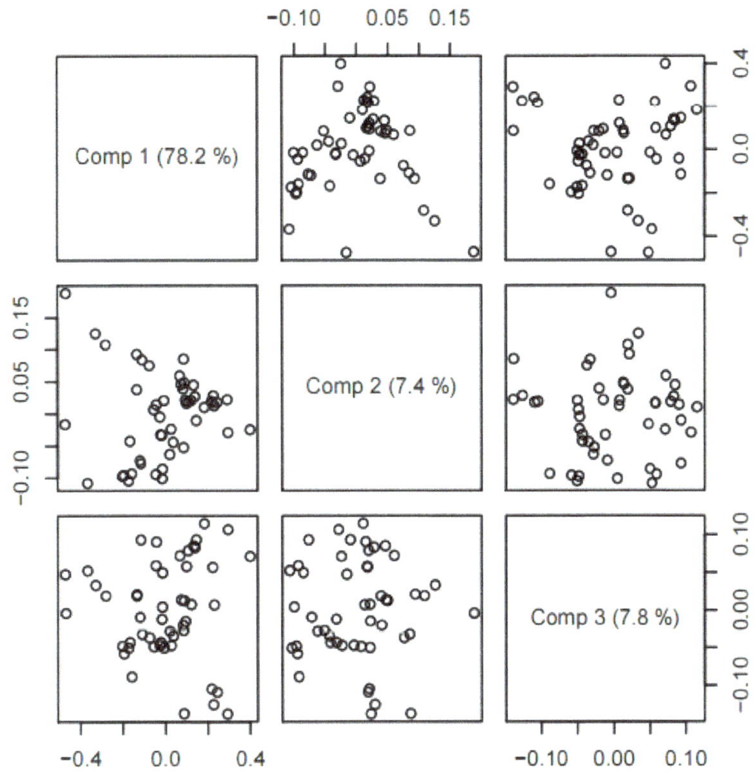

Figure 4: Score plot for the gasoline data

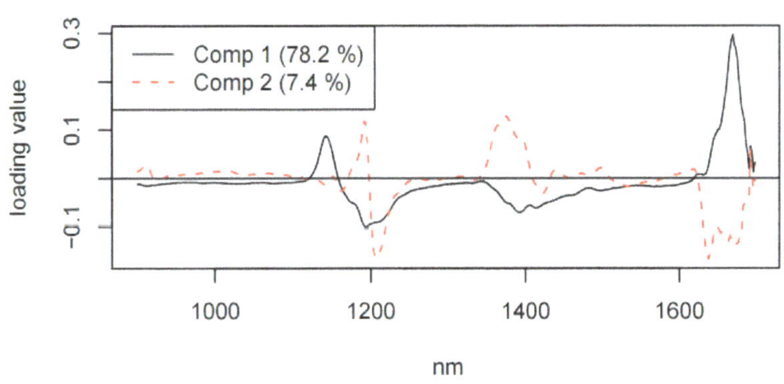

Figure 5: Loading plot for the gasoline data

```
 Comp 1  Comp 2  Comp 3  Comp 4  Comp 5  Comp 6  Comp 7  Comp 8  Comp 9 Comp 10
78.1708  7.4122  7.8242  2.6578  0.8768  0.9466  0.4922  0.4723  0.1688  0.1694
```

The loading plot (Figure 5) is much used for interpretation purposes, for instance to look for known spectral peaks or profiles:

```
R> plot(gas1, "loadings", comps = 1:2, legendpos = "topleft",
+      labels = "numbers", xlab = "nm")
R> abline(h = 0)
```

The labels = "numbers" argument makes the plot function try to interpret the variable names as numbers, and use them as x axis labels.

A fitted model is often used to predict the response values of new observations. The following predicts the responses for the ten observations in gasTest, using two components:

```
R> predict(gas1, ncomp = 2, newdata = gasTest)

, , 2 comps

   octane
51  87.94
52  87.25
53  88.16
54  84.97
55  85.15
56  84.51
57  87.56
58  86.85
59  89.19
60  87.09
```

Because we know the true response values for these samples, we can calculate the test set RMSEP:

```
R> RMSEP(gas1, newdata = gasTest)

(Intercept)     1 comps     2 comps     3 comps     4 comps     5 comps
     1.5114      1.1696      0.2445      0.2341      0.3287      0.2780
    6 comps     7 comps     8 comps     9 comps    10 comps
     0.2703      0.3301      0.3571      0.4090      0.6116
```

For two components, we get 0.244, which is quite close to the cross-validated estimate above (0.297)

FORMULAS AND DATA FRAMES

The pls package has a formula interface that works like the formula interface in R's standard lm functions, in most ways. This section gives a short description of formulas and data frames as they apply to pls. More information on formulas can be found in the lm help file, in Chapter 11 of "An Introduction to R" (R Development Core Team 2006a), and in Chapter 2 of "The White Book" (Chambers and Hastie 1992). These are good reads for anyone wanting to understand how R works with formulas, and the user is strongly advised to read them.

Formulas

A formula consists of a left hand side (lhs), a tilde (~), and a right hand side (rhs). The lhs consists of a single term, representing the response(s). The rhs consists of one or more terms separated by +, representing the regressor(s). For instance, in the formula a ~ b + c + d, a is the response, and b, c, and d are the regressors. The intercept is handled automatically, and need not be specified in the formula

Each term represents a matrix, a numeric vector or a factor (a factor should not be used as the response). If the response term is a matrix, a multi-response model is fit. In pls, the right hand side quite often consists of a single term, representing a matrix regressor: y ~ X. It is also possible to specify transformations of the variables. For instance, log(y) ~ msc(Z) specifies a regression of the logarithm of y onto Z after Z has been transformed by multiplicative scatter (or signal) correction (MSC) (Geladi, MacDougall, and Martens 1985), a pre-treatment that is very common in infrared spectroscopy. If the transformations contain symbols that are interpreted in the formula handling, e.g., +, * or ^, the terms should be protected with the I() function, like this: y ~ x1 + I(x2 + x3). This specifies two regressors: x1, and the sum of x2 and x3.

Data frames

The fit functions first look for the specified variables in a supplied data frame, and it is advisable to collect all variables there. This makes it easier to know what data has been used for fitting, to keep different variants of the data around, and to predict new data. To create a data frame, one can use the data.frame function: if v1, v2 and v3 are factors or numeric vectors, mydata <- data.frame(y = v1, a = v2, b = v3) will result in a data frame with variables named y, a and b.

PLSR and PCR are often used with a matrix as the single predictor term (especially when one is working with spectroscopic data). Also, multi-response models require a matrix as the response term. If Z is a matrix, it has to be protected by the 'protect function' I() in calls to data.frame: mydata <- data.frame(..., Z = I(Z)). Otherwise, it will be split into separate variables for each column, and there will be no variable called Z in the data frame, so we cannot use Z in the formula. One can also add the matrix to an existing data frame:

```
R> mydata <- data.frame(...)

R> mydata$Z <- Z
```

This will also prevent Z from being split into separate variables. Finally, one can use cbind to combine vectors and matrices into matrices on the fly in the formula. This is most useful for the response, e.g., cbind(y1, y2) ~ X.

Variables in a data frame can be accessed with the $ operator, e.g., mydata$y. However, the pls functions access the variables automatically, so the user should never use $ in formulas.

FITTING MODELS

The main functions for fitting models are pcr and plsr. (They are simply wrappers for the function mvr, selecting the appropriate fit algorithm). We will use plsr in the examples in this section, but everything could have been done with pcr (or mvr).

In its simplest form, the function call for fitting models is plsr(formula, ncomp, data) (where plsr can be substituted with pcr or mvr). The argument formula is a formula as described above, ncomp is the number of components one wishes to fit, and data is the data frame containing the variables to use in the model. The function returns a fitted model (an object of class "mvr") which can be inspected (Section 6) or used for predicting new observations (Section 7). For instance:

```
R> dens1 <- plsr(density ~ NIR, ncomp = 5, data = yarn)
```

If the response term of the formula is a matrix, a multi-response model is fit, e.g.,

```
R> dim(oliveoil$sensory)

[1] 16 6

R> plsr(sensory ~ chemical, data = oliveoil)

Partial least squares regression, fitted with the kernel algorithm.

Call:

plsr(sensory ~ chemical, data = oliveoil)
```

(As we see, the print method simply tells us what type of model this is, and how the fit function was called.)

The argument ncomp is optional. If it is missing, the maximal possible number of components are used. Also data is optional, and if it is missing, the variables specified in the formula is searched for in the global environment (the user's workspace). Usually, it is preferable to keep the variables in data frames, but it can sometimes be convenient to have them in the global environment. If the variables reside in a data frame, e.g. yarn, do not be tempted to use formulas like yarn$density ~ yarn$NIR! Use density ~ NIR and specify the data frame with data = yarn as above.

There are facilities for working interactively with models. To use only part of the samples in a data set, for instance the first 20, one can use arguments subset = 1:20 or data = yarn[1:20,]. Also, if one wants to try different alternatives of the model, one can use the function update. For instance

```
R> trainind <- which(yarn$train == TRUE)

R> dens2 <- update(dens1, subset = trainind)
```

will refit the model dens1 using only the observations which are marked as TRUE in yarn$train, and

```
R> dens3 <- update(dens1, ncomp = 10)
```

will change the number of components to 10. Other arguments, such as formula, can also be changed with update. This can save a bit of typing when working interactively with models (but it doesn't save computing time; the model is refitted each time). In general, the reader is referred to "The White Book" (Chambers and Hastie 1992) or 'An Introduction to R" (R Development Core Team 2006a) for more information about fitting and working with models in R.

Missing data can sometimes be a problem. The PLSR and PCR algorithms currently implemented in pls do not handle missing values intrinsically, so observations with missing values must be removed. This can be done with the na.action argument. With na.action = na.omit (the default), any observation with missing values will be removed from the model completely. With na.action = na.exclude, they will be removed from the fitting process, but included as NAs in the residuals and fitted values. If you want an explicit error when there are missing values in the data, use na.action = na.fail. The default na.action can be set with options(), e.g., options(na.action = quote(na.fail)).

Standardisation and other pre-treatments of predictor variables are often called for. In pls, the predictor variables are always centered, as a part of the fit algorithm. Scaling can be requested with the scale argument. If scale is TRUE, each variable is standardised by dividing it by its standard deviation, and if scale is a numeric vector, each variable is divided by the corresponding number. For instance, this will fit a model with standardised chemical measurements:

```
R> olive1 <- plsr(sensory ~ chemical, scale = TRUE, data = oliveoil)
```

As mentioned earlier, MSC (Geladi et al. 1985) is implemented in pls as a function msc that can be used in formulas:

```
R> gas2 <- plsr(octane ~ msc(NIR), ncomp = 10, data = gasTrain)
```

This scatter corrects NIR prior to the fitting, and arranges for new spectra to be automatically scatter corrected (using the same reference spectrum as when fitting) in predict:

```
R> predict(gas2, ncomp = 3, newdata = gasTest)
```

There are other arguments that can be given in the fit call: validation is for selecting validation, and ... is for sending arguments to the underlying functions, notably the crossvalidation function mvrCv. For the other arguments, see ?mvr.

CHOOSING THE NUMBER OF COMPONENTS WITH CROSS-VALIDATION

Cross-validation, commonly used to determine the optimal number of components to take into account, is controlled by the validation argument in the modelling functions (mvr, plsr and pcr). The default value is "none". Supplying a value of "CV" or "LOO" will cause the modelling procedure to call mvrCv to perform cross-validation; "LOO" provides leaveone-out cross-validation, whereas "CV" divides the data into segments. Default is to use ten segments, randomly selected, but also segments of consecutive objects or interleaved segments (sometimes also referred to as 'Venetian blinds') are possible through the use of the argument segment.type. One can also specify the segments explicitly with the argument segments; see ?mvrCv for details

When validation is performed in this way, the model will contain an element comprising information on the out-of-bag predictions (in the form of predicted values, as well as MSEP and R2 values). As a reference, the MSEP error using no components at all is calculated as well. The validation results can be visualised using the plottype = "validation" argument of the standard plotting function. An example is shown in Figure 2 for the gasoline data; typically, one would select a number of components after which the cross-validation error does not show a significant decrease. Unfortunately, no generally applicable tests are available, so the decision on how many components to retain will always be subjective to some extent.

When a pre-treatment that depends on the composition of the training set is applied, the crossvalidation procedure as described above is not optimal, in the sense that the cross-validation errors are biased downward. As long as the only purpose is to select the optimal number of components, this bias may not be very important, but it is not too difficult to avoid it. The modelling functions have an argument scale that can be used for auto-scaling per segment. However, more elaborate methods such as MSC need explicit handling per segment. For this, function crossval is available. It takes an mvr object and performs the cross-validation as it should be done: applying the pre-treatment for each segment. The results can be shown in a plot (which looks very similar to Figure 2) or summarised in numbers.

```
R> gas2.cv <- crossval(gas2, segments = 10)
R> plot(MSEP(gas2.cv), legendpos = "topright")
R> summary(gas2.cv, what = "validation")

Data:        X dimension: 50 401
        Y dimension: 50 1
Fit method: kernelpls
Number of components considered: 10

VALIDATION: RMSEP
Cross-validated using 10 random segments.
       (Intercept)  1 comps  2 comps  3 comps  4 comps  5 comps  6 comps
CV           1.545    1.346   0.2872   0.2612   0.2479   0.2390   0.2503
adjCV        1.545    1.342   0.2841   0.2600   0.2435   0.2353   0.2441
       7 comps  8 comps  9 comps  10 comps
CV      0.2438   0.2653   0.2791    0.2885
adjCV   0.2383   0.2576   0.2697    0.2777
```

Applying MSC in this case leads to nearly identical cross-validation estimates of prediction error

When the scaling does not depend on the division of the data into segments (e.g., log-scaling), functions crossval and mvrCv give the same results; however, crossval is much slower.

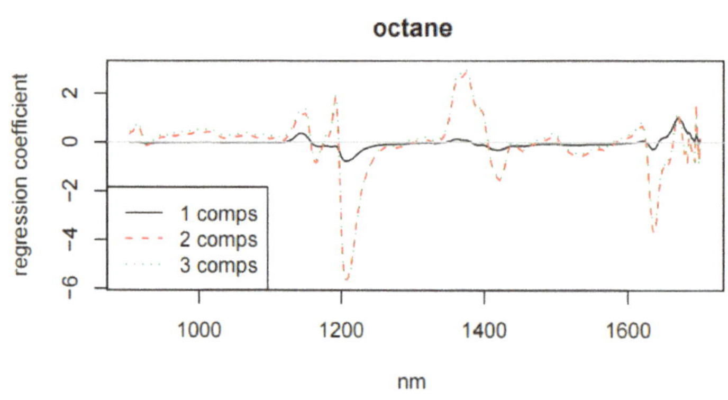

Figure 6: Regression coefficients for the gasoline data

INSPECTING FITTED MODELS

A closer look at the fitted model may reveal interesting agreements or disagreements with what is known about the relations between X and Y. Several functions are implemented in pls for plotting, extracting and summarising model components.

Plotting

One can access all plotting functions through the "plottype" argument of the plot method for mvr objects. This is simply a wrapper function calling the actual plot functions; the latter are available to the user as well.

The default plot is a prediction plot (predplot), showing predicted versus measured values. Test set predictions are used if a test set is supplied with the newdata argument. Otherwise, if the model was built using cross-validation, the cross-validated predictions are used, otherwise the predictions for the training set. This can be overridden with the which argument. An example of this type of plot can be seen in Figure 3. An optional argument can be used to indicate how many components should be included in the prediction. To assess how many components are optimal, a validation plot (validationplot) can be used such as the one shown in Figure 2; this shows a measure of prediction performance (either RMSEP, MSEP, or R^2) against the number of components. Usually, one takes the first local minimum rather than the absolute minimum in the curve, to avoid over-fitting. The regression coefficients can be visualised using plottype = "coef" in the plot method, or directly through function coefplot. This allows simultaneous plotting of the regression vectors for several different numbers of components at once. The regression vectors for the gasoline data set using MSC are shown in Figure 6 using the command

```
R> plot(gas1, plottype = "coef", ncomp = 1:3, legendpos = "bottomleft",
+     labels = "numbers", xlab = "nm")
```

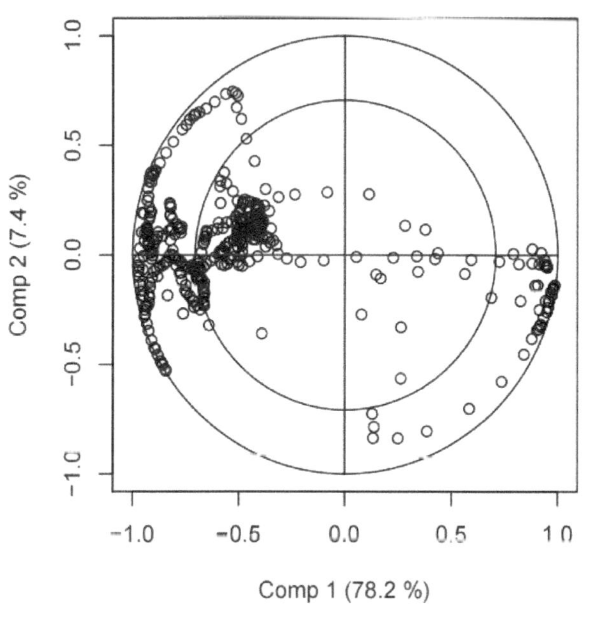

Figure 7: Correlation loadings plot for the gasoline data

Note that the coefficients for two components and three components are similar. This is because the third component contributes little to the predictions. The RMSEPs (see Figure 2) and predictions (see Section 7) for two and three components are quite similar. Scores and loadings can be plotted using functions scoreplot (an example is shown in Figure 4) and loadingplot (in Figure 5), respectively. One can indicate the number of components with the comps argument; if more than two components are given, plotting the scores will give a pairs plot, otherwise a scatter plot. For loadingplot, the default is to use line plots. Finally, a 'correlation loadings' plot (function corrplot, or plottype = "correlation" in plot) shows the correlations between each variable and the selected components (see Figure 7). These plots are scatter plots of two sets of scores with concentric circles of radii given by radii. Each point corresponds to an X variable. The squared distance between the point and the origin equals the fraction of the variance of the variable explained by the components in the panel. The default values for radii correspond to 50% and 100% explained variance, respectively. The plot functions accept most of the ordinary plot parameters, such as col and pch. If the model has several responses or one selects more than one model size, e.g. ncomp = 4:6, in some plot functions (notably prediction plots (see below), validation plots and coefficient plots) the plotting window will be divided and one plot will be shown for each combination of response and model size. The number of rows and columns are chosen automatically, but can be specified explicitly with arguments nRows and nCols. If there are more plots than fit the plot window, one will be asked to press return to see the rest of the plots.

Extraction

Regression coefficients can be extracted using the generic function coef; the function takes several arguments, indicating the number of components to take into account, and whether the intercept is needed (default is FALSE). Scores and loadings can be extracted using functions scores and loadings for X, and Yscores and Yloadings for Y. These also return the percentage of variance explained as attributes. In PLSR, weights can be extracted using the function loading.weights. When applied to a PCR model, the function returns NULL. Note that commands like plot(scores(gas1)) are perfectly correct, and lead to exactly the same plots as using scoreplot.

Summaries

The print method for an object of class "mvr" shows the regression type used, perhaps indicating the form of validation employed, and shows the function call. The summary method gives more information: it also shows the amount of variance explained by the model (for all choices of a, the number of latent variables). The summary method has an additional argument (what) to be able to focus on the training phase or validation phase, respectively. Default is to print both types of information.

PREDICTING NEW OBSERVATIONS

Fitted models are often used to predict future observations, and pls implements a predict method for PLSR and PCR models. The most common way of calling this function is predict(mymod, ncomp = myncomp, newdata = mynewdata), where mymod is a fitted model, myncomp specifies the model size(s) to use, and mynewdata is a data frame with new X observations. The data frame can also contain response measurements for the new observations, which can be used to compare the predicted values to the measured ones, or to estimate the overall prediction ability of the model. If newdata is missing, predict uses the data used to fit the model, i.e., it returns fitted values.

If the argument ncomp is missing, predict returns predictions for models with 1 component, 2 components, . . ., A components, where A is the number of components used when fitting the model. Otherwise, the model size(s) listed in ncomp are used. For instance, to get predictions from the model built in Section 2, with two and three components, one would use

```
R> predict(gas1, ncomp = 2:3, newdata = gasTest[1:5, ])

, , 2 comps

   octane
51  87.94
52  87.25
53  88.16
54  84.97
55  85.15

, , 3 comps

   octane
51  87.95
52  87.30
53  88.21
54  84.87
55  85.24
```

(We predict only the five first test observations, to save space.) The predictions with two and three components are quite similar. This could be expected, given that the regression vectors (Figure 6) as well as the estimated RMSEPs for the two model sizes were similar. One can also specify explicitly which components to use when predicting. This is done by specifying the components in the argument comps. (If both ncomp and comps are specified, comps takes precedence over ncomp.) For instance, to get predictions from a model with only component 2, one can use

```
R> predict(gas1, comps = 2, newdata = gasTest[1:5, ])

   octane
51  87.53
52  86.30
53  87.35
54  85.82
55  85.32
```

The results are different from the predictions with two components (i.e., components one and two) above. (The intercept is always included in the predictions. It can be removed by subtracting mymod$Ymeans from the predicted values.) The predict method returns a three-dimensional array, in which the entry (i, j, k) is the predicted value for observation i, response j and model size k. Note that singleton dimensions are not dropped, so predicting five observations for a uni-response model with ncomp = 3 gives an 5 × 1 × 1 array, not a vector of length five. This is to make it easier to distinguish between predictions from models with one response and predictions with one model size. (When using the comps argument, the last dimension is dropped, because the predictions are always from a single model.) One can drop the singleton dimensions explicitly by using drop(predict(...)):

```
R> drop(predict(gas1, ncomp = 2:3, newdata = gasTest[1:5, ]))

   2 comps 3 comps
51   87.94   87.95
52   87.25   87.30
53   88.16   88.21
54   84.97   84.87
55   85.15   85.24
```

Missing values in newdata are propagated to NAs in the predicted response, by default. This can be changed with the na.action argument. See ?na.omit for details. The newdata does not have to be a data frame. Recognising the fact that the right hand side of PLSR and PCR formulas very often are a single matrix term, the predict method allows one to use a matrix as newdata, so instead of

```
newdataframe <- data.frame(X = newmatrix)

predict(..., newdata = newdataframe)
```

one can simply say

```
predict(..., newdata = newmatrix)
```

However, there are a couple of caveats: First, this only works in predict. Other functions that take a newdata argument (such as RMSEP) must have a data frame (because they also need the response values). Second, when newdata is a data frame, predict is able to perform more tests on the supplied data, such as the dimensions and types of variables. Third, with the exception of scaling (specified with the scale argument when fitting the model), any transformations or coding of factors and interactions have to be performed manually if newdata is a matrix. It is often interesting to predict scores from new observations, instead of response values. This can be done by specifying the argument type = "scores" in predict. One will then get a matrix with the scores corresponding to the components specified in comps (ncomp is accepted as a synonym for comps when predicting scores). Predictions can be plotted with the function predplot. This function is generic, and can also be used for plotting predictions from other types of models, such as lm. Typically, predplot is called like this:

```
R> predplot(gas1, ncomp = 2, newdata = gasTest, asp = 1, line = TRUE)
```

This plots predicted (with 2 components) versus measured response values. (Note: newdata must be a data frame with both X and Y variables.)

FURTHER TOPICS

This section presents a couple of slightly technical topics for more advanced use of the package.

Selecting fit algorithms

There are several PLSR algorithms, and the pls package currently implements three of them: the kernel algorithm for tall matrices (many observations, few variables) (Dayal and MacGregor 1997), the classic orthogonal scores algorithm (A.K.A. NIPALS algorithm) (Martens and Næs 1989) and the SIMPLS algorithm (de Jong 1993). The kernel and orthogonal scores algorithms produce the same results (the kernel algorithm being the fastest of them for most problems). SIMPLS produces the same fit for single-response models, but slightly different results for multi-response models. It is also usually faster than the NIPALS algorithm.

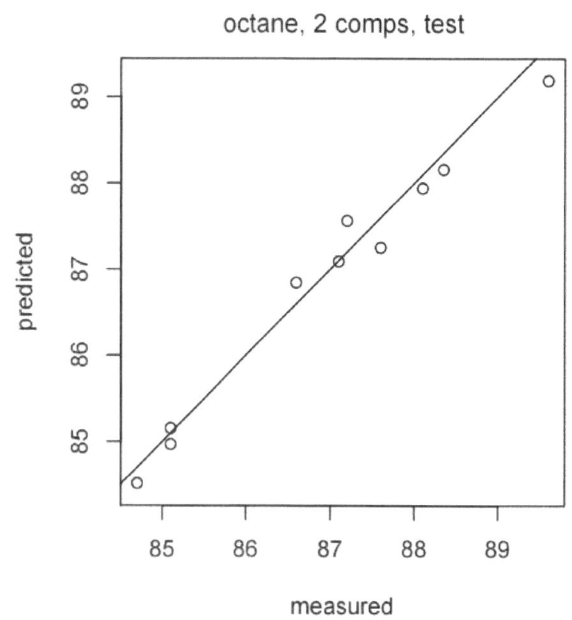

Figure 8: Test set predictions

The factory default is to use the kernel algorithm. One can specify a different algorithm with the method argument; i.e., method = "oscorespls"

If one's personal taste of algorithms does not coincide with the defaults in pls, it can be quite tedious (and error prone) having to write e.g. method = "oscorespls" every time (even though it can be shortened to e.g. me = "o" due to partial matching). Therefore, the defaults can be changed, with the function pls.options. Called without arguments, it returns the current settings as a named list:

R> pls.options()

$mvralg

[1] "kernelpls"

$plsralg

[1] "kernelpls"

$pcralg

[1] "svdpc"

The options specify the default fit algorithm of mvr, plsr, and pcr. To return only a specific option, one can use pls. options("plsralg"). To change the default PLSR algorithm for the rest of the session (and for future settings, if the global environment is saved and restored), one can use, e.g.

R> pls.options(plsralg = "oscorespls")

When changed, the options are stored in a variable .pls.Options in the global environment.

A quick-n-dirty way to get back to the 'factory defaults' is to remove this variable:

R> pls.options("plsralg")

$plsralg

[1] "oscorespls"

R> rm(.pls.Options)

R> pls.options("plsralg")

$plsralg

[1] "kernelpls"

Package design

The pls package is designed such that an interface function mvr handles the formula and data, and calls an underlying fit function (and possibly a cross-validation function) to do the real work. There are several reasons for this design: it makes it easier to implement new algorithms, one can easily skip the time-consuming formula and data handling in computingintensive applications (simulations, etc.), and it makes it easier to use the pls package as a building block in other packages.

The plotting facilities are implemented similarly: the plot method simply calls the correct plot function based on the plottype argument. Here, however, the separate plot functions are meant to be callable interactively, because some people like to use the generic plot function, while others like to use separate functions for each plot type. There are also plot methods for some of the components of fitted models that can be extracted with extract functions, like score and loading matrices. Thus there are several ways to get some plots, e.g.:

plot(mymod, plottype = "scores", ...)

scoreplot(mymod, ...)

plot(scores(mymod), ...)

One example of a package that uses pls is lspls, available on CRAN. In that package LS is combined with PLS in a regression procedure. It calls the fit functions of pls directly, and also uses the plot functions to construct score and loading plots. There is also the plsgenomics package, which includes a modified version of (an earlier version of) the SIMPLS fit function simpls.fit.

Calling fit functions directly

The underlying fit functions are called kernelpls.fit, oscorespls. fit, and simpls.fit for the PLSR methods, and svdpc.fit for the PCR method. They all take arguments X, Y, ncomp, and stripped. Arguments X, Y, and ncomp specify X and Y (as matrices, not data frames), and the number of components to fit, respectively. The argument stripped defaults to FALSE. When it is TRUE, the calculations are stripped down to the bare minimum required for returning the X means, Y means, and the regression coefficients. This is used to speed up cross-validation procedures

The fit functions can be called directly, for instance when one wants to avoid the overhead of formula and data handling in repeated fits. As an example, this is how a simple leave-one-out cross-validation for a uni-response-model could be implemented, using the SIMPLS:

R> X <- gasTrain$NIR

```
R> Y <- gasTrain$octane
R> ncomp <- 5
R> cvPreds <- matrix(nrow = nrow(X), ncol = ncomp)
R> for (i in 1:nrow(X)) {
+ fit <- simpls.fit(X[-i, ], Y[-i], ncomp = ncomp, stripped = TRUE)
+cvPreds[i,]<-(X[i,]-fit$Xmeans)%*%drop(fit$coefficients) +
+ fit$Ymeans
+ }
```

The RMSEP of the cross-validated predictions are

```
R> sqrt(colMeans((cvPreds - Y)^2))
[1] 1.3570 0.2966 0.2524 0.2476 0.2398
```

which can be seen to be the same as the (unadjusted) CV results for the gas1 model in Section 2.

Formula handling in more detail

The handling of formulas and variables in the model fitting is very similar to what happens in the function lm: The variables specified in the formula are looked up in the data frame given in the data argument of the fit function (plsr, pcr or mvr), or in the calling environment if not found in the data frame. Factors are coded into one or more of columns, depending on the number of levels, and on the contrasts option. All (possibly coded) variables are then collected in a numerical model matrix. This matrix is then handed to the underlying fit or cross-validation functions. A similar handling is used in the predict method. The intercept is treated specially in pls. After the model matrix has been constructed, the intercept column is removed. This ensures that any factors are coded as if the intercept was present. The underlying fit functions then center the rest of the variables as a part of the fitting process. (This is intrinsic to the PLSR and PCR algorithms.) The intercept is handled separately. A consequence of this is that explicitly specifying formulas without the intercept (e.g., y ~ a + b - 1) will only result in the coding of any factors to change; the intercept will still be fitted.

REFERENCES

1. Chambers JM, Hastie TJ (1992). Statistical Models in S. Chapman & Hall, London.
2. Cramer R, Patterson D, Bunce J (1988). "Comparative Molecular-Field Analysis (ComFA). 1. Effect of Shape on Binding of Steroids to Carrier Proteins." Journal of the American Chemical Society, 110(18), 5959–5967.
3. Dayal BS, MacGregor JF (1997). "Improved PLS Algorithms." Journal of Chemometrics, 11(1), 73–85.
4. de Jong S (1993). "SIMPLS: An Alternative Approach to Partial Least Squares Regression." Chemometrics and Intelligent Laboratory Systems, 18, 251–263.
5. Fornell C, Bookstein F (1982). "Two Structural Equation Models – LISREL and PLS Applied to Consumer-Exit Voice Theory." Journal of Marketing Research, 19(4), 440–452.
6. Frank IE, Friedman JH (1993). "A Statistical View of Some Chemometrics Regression Tools (with Discussion)." Technometrics, 35(2), 109–148.
7. Geladi P, MacDougall D, Martens H (1985). "Linearization and Scatter-Correction for NIR Reflectance Spectra of Meat." Applied Spectroscopy, 39, 491–500.
8. Hastie T, Tibshirani R, Friedman J (2001). The Elements of Statistical Learning. SpringerVerlag, New York.
9. Kalivas JH (1997). "Two Data Sets of Near Infrared Spectra." Chemometrics and Intelligent Laboratory Systems, 37, 255–259.
10. Kresta J, MacGregor J, Marlin T (1991). "Multivariate Statistical Monitoring of Process Operating Performance." The Canadian Journal of Chemical Engineering, 69(1), 35–47.
11. Lachenbruch PA, Mickey MR (1968). "Estimation of Error Rates in Discriminant Analysis." Technometrics, 10(1), 1–11.
12. Martens H (2001). "Reliable and Relevant Modelling of Real World Data: A Personal Account of the Development of PLS Regression." Chemometrics and Intelligent Laboratory Systems, 58(2), 85–95.
13. Martens H, Næs T (1989). Multivariate Calibration. Wiley, Chichester.
14. Massart DL, Vandeginste BGM, Buydens LMC, de Jong S, Lewi PJ, Smeyers-Verbeke J (1998). Handbook of Chemometrics and Qualimetrics: Part B. Elsevier.
15. McIntosh A, Bookstein F, Haxby J, Grady C (1996). "Spatial Pattern Analysis of Functional Brain Images Using Partial Least Squares." Neuroimage, 3(3), 143–157.
16. Mevik BH, Cederkvist HR (2004). "Mean Squared Error of Prediction (MSEP) Estimates for Principal Component Regression (PCR) and Partial Least Squares Regression (PLSR)." Journal of Chemometrics, 18(9), 422–429.
17. Nguyen D, Rocke D (2002). "Tumor Classification by Partial Least Squares Using Microarray Gene Expression Data." Bioinformatics, 18(1), 39–50.
18. R Development Core Team (2006a). An Introduction to R. R Foundation for Statistical Computing, Vienna, Austria. ISBN 3-900051-12-7, URL http://www.R-project.org/.
19. R Development Core Team (2006b). R: A Language and Environment for Statistical Computing. R Foundation for Statistical Computing, Vienna, Austria. ISBN 3-900051-07-0, URL http://www.R project.org/.
20. Swierenga H, de Weijer AP, van Wijk RJ, Buydens LMC (1999). "Strategy for Constructing Robust Multivariate Calibration Models." Chemometrics and Intelligent Laboratory Systems, 49(1), 1–17.
21. Wold S (2001). "Personal Memories of the Early PLS Development." Chemometrics and Intelligent Laboratory Systems, 58(2), 83–84.

CHAPTER 5

TIMP: AN R PACKAGE FOR MODELING MULTI-WAY SPECTROSCOPIC MEASUREMENTS

Katharine M. Mullen and Ivo H. M. van Stokkum

Keywords: spectroscopy, separable nonlinear least squares, superposition model, compartmental model, global analysis, target analysis.

INTRODUCTION

TIMP[1] is an R package for modeling multiway spectroscopic measurements. It is a fully crossplatform problem-solving environment for fitting a wide range of models to spectroscopy data, and is freely available under the GNU General Public License (GPL)

This paper outlines the capabilities, structure, and application of the package. The introduction gives an overview of the problems in scientific model discovery that the package has been designed to address. Section 2 describes some aspects of the implementation, and includes a description of the partitioned variable projection algorithm that allows application of the variable projection functional to parameter estimation problems in the absence of large memory resources. Section 3 describes in brief user-accessible functions. Section 4 describes general model options. Section 5 contains descriptions of kinetic models and their specification, fitting and validation with the package, and includes a case study in the application of a kinetic model to the simultaneous analysis of two datasets. Section 6 describes spectral models and their specification, fitting and validation in TIMP, and includes a case study in the application of a spectral model to the description of time-dependent change in spectral bandshapes. Section 7 discusses in brief the extension of TIMP to new model types. Conclusions are contained in Section 8.

Interactive scientific model-discovery

We term scientific model discovery the identification of a statistical model able to reproduce experimentally collected measurements to a satisfactory degree of accuracy, with the additional constraint that the model be well-interpretable according to physico-chemical theory. Scientific model discovery very often requires iterating the steps of formulation of a candidate model, model fitting, and model validation. Postulation of a candidate model is guided by a priori knowledge of the system underlying the data as well as by exploratory analysis of the dataset (e.g., with decomposition techniques like the singular value decomposition). Fitting provides estimates for free parameters that are more statistically likely than the starting estimates provided during postulation of the candidate model. Validation considers whether the fitted parameter values are precise and likely to be correct according to physico-chemical theory, whether the residuals are sufficiently small and unstructured, and whether adjustment of the candidate

Citation: Katharine M. Mullen and Ivo H. M. van Stokkum, TIMP: An R Package for Modeling Multi-way Spectroscopic Measurements, https://www.jstatsoft.org/article/view/v018i03/v18i03.pdf

model is desirable. The cycle of model formulation, fitting and validation is often iterative because validation often results in the identification of a new candidate model.

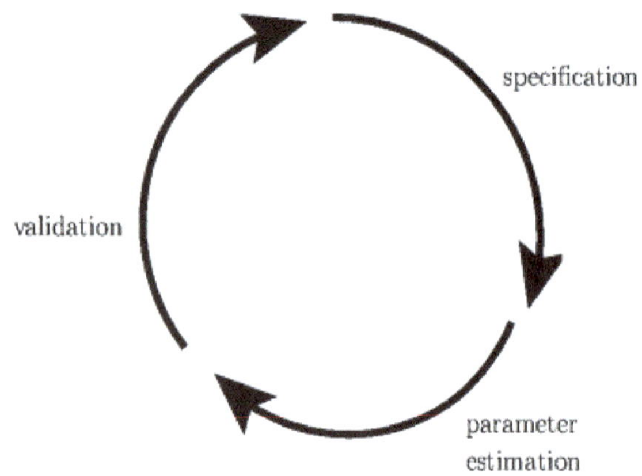

Figure 1: Scientific model discovery is often an iterative process of model specification, parameter estimation and validation.

Scientific model discovery is interactive in the case that the time to complete the model formulation, fitting, and validation cycle is determined primarily by the ability of the researcher to postulate and validate candidate models. To allow for interactive scientific model discovery the applied computer hardware and software must enable the researcher to quickly specify and fit a model, and must provide information for model validation that allows for efficient evaluation of model fit and physical feasibility. The primary goal of the TIMP package for the R system for statistical computing (R Development Core Team 2006) is to provide software for interactive scientific model discovery in the multiway spectroscopy data modeling problem domain.

Multiway spectroscopy data and models

Let Ψ_q denote a spectroscopic dataset arising under experimental conditions q. Ψ_q may be represented as the matrix

$$\Psi_q = \left[\begin{array}{c|cccc} & \lambda_1 & \lambda_2 & \dots & \lambda_n \\ \hline t_1 & \psi(t_1,\lambda_1) & \psi(t_1,\lambda_2) & \dots & \psi(t_1,\lambda_n) \\ t_2 & \psi(t_2,\lambda_1) & \psi(t_2,\lambda_2) & \dots & \psi(t_2,\lambda_n) \\ \vdots & \vdots & \vdots & \ddots & \vdots \\ t_m & \psi(t_m,\lambda_1) & \psi(t_m,\lambda_2) & \dots & \psi(t_m,\lambda_n) \end{array}\right] \tag{1}$$

Each row of Ψ_q is a spectrum in the spectral variable λ (which is often wavelength, but may be wavenumber or magnetic field strength; Laptenok, Mullen, Borst, van Stokkum, Apanasovich, and Visser (2007) describe the case that the variable is location). Spectra are represented at m instances of independent experimental variables t such as time, pH, pD, temperature, excitation wavelength or quencher concentration. The independent experimental variables are chosen so as to monitor spectral change in a manner that provides information on the dynamics of the underlying system

Ψ_q represents a contribution from nncomp spectrally distinct components. The concentration and spectral property of each component may be represented as column l of matrices C and E, respectively, in the superposition model

$$\Psi_q = \left[\begin{array}{c|ccc} & 1 & \dots & n_{\text{ncomp}} \\ \hline t_1 & c(t_1,1) & \dots & c(t_1,n_{\text{ncomp}}) \\ t_2 & c(t_2,1) & \dots & c(t_2,n_{\text{ncomp}}) \\ \vdots & \vdots & \ddots & \vdots \\ t_m & c(t_m,1) & \dots & c(t_m,n_{\text{ncomp}}) \end{array}\right] \left[\begin{array}{c|ccc} & 1 & \dots & n_{\text{ncomp}} \\ \hline \lambda_1 & \epsilon(\lambda_1,1) & \dots & \epsilon(\lambda_1,n_{\text{ncomp}}) \\ \lambda_2 & \epsilon(\lambda_2,1) & \dots & \epsilon(\lambda_2,n_{\text{ncomp}}) \\ \vdots & \vdots & \ddots & \vdots \\ \lambda_n & \epsilon(\lambda_n,1) & \dots & \epsilon(\lambda_n,n_{\text{ncomp}}) \end{array}\right]^\top \tag{2}$$

$$= CE^\top \tag{3}$$

Each column of C represents a concentration profile of a component in the independent variable t in which spectra are resolved. Likewise, each column of the matrix E represents a spectrum of a component.

In modeling multiway spectroscopy data Ψ_q, the inverse problem of recovery of the entries of C or E in terms of physically significant parameters (descriptive of, e.g., the decay rate of a component, or the location of the maximum of a spectrum) using Equation 3 is often of interest. Adequate parameterizations of either C or E are nonlinear, and are usually comprised of many submodels to represent various model aspects. Such parameterizations have been reviewed for the case of time-resolved spectroscopy data by van Stokkum, Larsen, and van Grondelle (2004). Very often, a model-based description is possible for either C or E, but not both matrices. Then the variable projection algorithm (Golub and LeVeque 1979; Golub and Pereyra 2003) allows the estimation of the entries of the matrix for which a model-based description is unavailable as conditionally linear parameters (clp). The resulting reduction in the size of the nonlinear parameter search space is very significant for problems arising in typical spectroscopy data modeling applications, as is discussed further in Section 2.3. TIMP includes an implementation of a refinement of the standard variable projection implementation, which we call partitioned variable projection that allows the application of the variable projection functional to large estimation problems in the absence of large memory resources. Partitioned variable projection is discussed further in Section 2.3. Is is compared to the standard variable projection implementation (implemented in R in the plinear option of the nls function) in detail in Appendix A.

Multiway spectroscopy experiments often result in measured datasets $\Psi_1, \dots, \Psi_x$ collected under similar but not identical experimental conditions 1, . . . , x. $\Psi_1, \dots, \Psi_x$ may vary with one experimental condition, in which case the data is three-way, or with multiple experimental conditions, in which case the data is four-or-higher-way. In the latter case, distinct groups of datasets in $\Psi 1, \dots, \Psi_x$ represent variance with respect to each experimental condition representing a dimension of the data Ψ as a whole, (for an example of five-way data Ψ resolved with respect to time, wavelength, temperature, pH, and polarization see e.g., van Stokkum and Lozier (2002)). It may be desirable to parameterize a model for all datasets Ψ = $\{\Psi_1, \ldots, \Psi_x\}$ so as to extract information regarding some model parameters from all available data, while fitting other model parameters on a per-dataset basis. Parameters fit per-dataset may then be used to describe the effects of variances in experimental conditions on the underlying system.

TIMP is designed to efficiently specify, fit and validate models for multiway spectroscopy data Ψ from possibly many experiments simultaneously. The package currently implements a wide variety of model options for the parameterization of kinetic and spectral models. Extension to other model types requires a minimum of additional code. An overview of the goals, structure and capabilities of the package is contained in this paper, along with two case studies its application.

An introduction to modeling with TIMP

The application of TIMP to a very simple parameter estimation task on a single simulated spectroscopy dataset will introduce the use of the package. As discussed in the previous section, a spectroscopy dataset Ψ_q can be considered to represent a superposition of the concentration profiles C and spectral properties E of components, so that $\Psi_q = CE^T$ (Equation 3).

We will simulate a dataset in which two spectrally distinct components contribute to Ψ_q, where C represents concentration in time and E represents spectra resolved with respect to wavenumber. The simplest realistic model for C lets the time-profile of each component cl be described by an exponential decay with decay rate parameter k_l , so that

$$c_l(t) = \exp(-k_l t), \tag{4}$$

where t is time. (for elaboration on the use of exponential models for kinetic processes, see e.g., the review by Istratov and Vyvenko (1999)). We let the two components contributing to Ψ_q have associated decay rate parameters .5 and 1, and let the concentrations be measured at 51 timepoints equidistant in the interval 0-2 ns. Then the following R commands calculate C.

```
R> C <- matrix(nrow = 51, ncol = 2)
R> k <- c(.5, 1)
R> t <- seq(0, 2, by = 2/50)
R> C[, 1] <- exp( - k[1] * t)
R> C[, 2] <- exp( - k[2] * t)
```

The most basic model for the spectrum el associated with a single component in wavenumber $\bar{\nu}$ is a Gaussian with parameters $\mu_{\bar{\nu}}$, $\Delta_{\bar{\nu}}$ and al , for the location, full width at half maximum (FWHM), and amplitude, respectively, so that

$$e_l(\bar{\nu}) = a_l \exp\left(-\ln(2)\left(2\frac{(\bar{\nu}-\mu_{\bar{\nu}})}{\Delta_{\bar{\nu}}}\right)^2\right), \tag{5}$$

(see e.g., van Stokkum (1997) and references therein regarding the ubiquity of Gaussian models for spectra). Let us consider spectra represented by 51 wavenumbers equidistant in the interval 18000 - 28000 cm^{-1} , with locations 25000 and 20000, FWHMs 5000 and 7000, and amplitudes 1 and 2, respectively. In R we can then calculate E as

```
R> E <- matrix(nrow = 51, ncol = 2)
R> wavenum <- seq(18000, 28000, by=200)
R> location <- c(25000, 20000)
R> delta <- c(5000, 7000)
R> amp <- c(1, 2)
R> E[, 1] <- amp[1] * exp( - log(2) * (2 * (wavenum - location[1])/delta[1])^2)
R> E[, 2] <- amp[2] * exp( - log(2) * (2 * (wavenum - location[2])/delta[2])^2)
```

Given these R expressions for C and E, a dataset Ψ_q with Gaussian noise with zero mean and width σ = .001 may be simulated as

```
R> sigma <- .001
R> Psi_q <- C %*% t(E) + sigma * rnorm(dim(C)[1] * dim(E)[1])
```

Plots of C and E are shown in Figure 2. The simulated dataset Ψ_q is shown in Figure 3

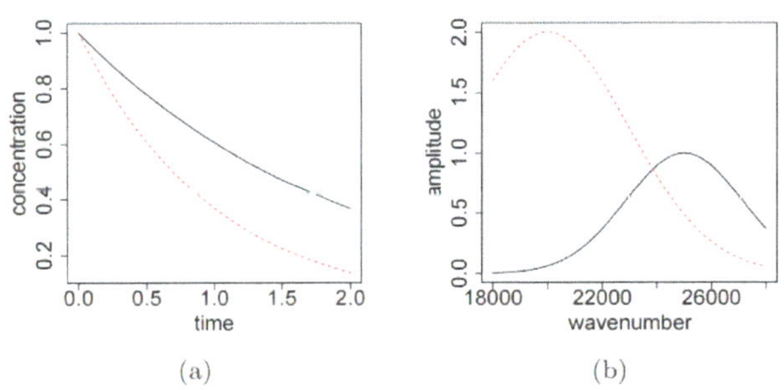

Figure 2: (a) Simulated concentrations (b) simulated spectra. Component 1 is in black; component 2 is in red.

Given a dataset such as the simulated Psi_q shown in Figure 3 it is often the case that either a kinetic model for C or a spectral model for E is known to apply whose parameters are nonlinear. Given estimates for the nonlinear model parameters that determine one of the two matrices, the entries of the other matrix may be solved for as clp. Let us assume that the

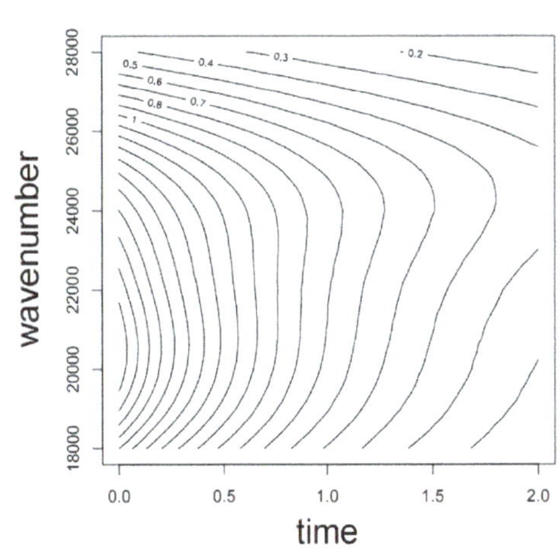

Figure 3: Simulated data; model fitting will resolve the two contributing components

kinetic model used in simulating the data is known to describe Psi_q, and that approximate starting values for the two rate constants are known. Then the following R commands can be used to estimate the rate constants k and clp E.

TIMP is loaded with

R> library("TIMP")

The simulated dataset is placed into an instance of the class "dat" which is used to store data and model objects in TIMP. The "dat" object contains not only the data but also some information like its dimensions.

R> Psi_q_data <- dat(psi.df = Psi_q, x = t, nt = length(t),

+ x2 = wavenum, nl = length(wavenum))

A model to be applied to the data is initialized with the initModel function. The seqmod=FALSE option indicates that the components decay in parallel; the starting values for the decay rates are given in the vector kinpar.

R> kinetic_model <- initModel(mod_type = "kin", seqmod = FALSE, + kinpar = c(.1, 2))

With the next command model parameters are optimized over the course of four iterations using the fitModel function of TIMP.

R> kinetic_fit <- fitModel(data = list(Psi_q_data), model = kinetic_model, + opt = list(iter=4, paropt=list(mar=c(2,2,2,2),mgp=c(1,.2,0))))

The call to the fitModel function results in a composite plot displaying the fit of the model to the data at each wavenumber, a window showing parameter estimates for the two rate constants and information regarding the residuals, and a summary figure showing the estimated spectra and rates constants. For the case considered here the estimated spectra and concentration profiles well-approximate the entries of E and C used in simulation. TIMP employed the nls function in estimating the nonlinear parameters $k = \{k_1, k_2\}$; examination of the summary object returned by nls, as below, verifies that the standard errors of estimated parameters are low

```
R> kinetic_fit$sumonls

Formula: 0 ~ rescomp(t, d)

Parameters:
    Estimate Std. Error t value Pr(>|t|)
t1 0.4997107  0.0007933   629.9   <2e-16 ***
t2 1.0002352  0.0005581  1792.1   <2e-16 ***
---
Signif. codes:  0 '***' 0.001 '**' 0.01 '*' 0.05 '.' 0.1 ' ' 1

Residual standard error: 0.000999 on 2497 degrees of freedom

Number of iterations till stop: 4
Achieved convergence tolerance: 0.1151
Reason stopped: number of iterations exceeded maximum of 4
```

This introduction to the application of TIMP shows how, given a single dataset $_{\Psi q}$, a parametric description of the concentration profiles of contributing components can be fit, while the spectra E are solved for as clp. Subsequent sections will describe the application of TIMP to fitting more complex models to possibly many datasets.

Hierarchical models for possibly many datasets

Let $\Psi = \{\Psi_1 . . . \Psi_x\}$ denote multiway spectroscopy data collected over the course of possibly many experimental conditions indexed 1. . . x. Let Θ denote the nonlinear parameters of the multidataset model applied to Ψ. The goal of model fitting is to solve $\min \|\hat{\Psi}(\Theta) - \Psi\|_F^2$, e.g., to minimize the residuals associated with the fit of the model to the data. The model specifies a prescription to determine residual matrices $Z_1(\Theta), . . . , Z_x(\Theta)$, where $Z_q(\Theta)$ are those residuals associated with dataset Ψ_q. $Z_1(\Theta), . . . , Z_x(\Theta)$ are concatenated together in vectorized form to form the residual vector to be minimized with respect to Θ by a nonlinear optimization routine, (such as the nls function from R, which is employed by TIMP). The multidataset model may be such that residual matrix Z_q depends on only a subset of Θ. Parameters in Θ that determine the residuals associated with multiple datasets are said to be linked. Parameters that determine the residuals of only one dataset are said to be unlinked.

Models applied to data Ψ may be comprised of many submodels, each of which describe a distinct aspect of the underlying system giving rise to the measurements. Examples of submodels include a parametric description of an instrument response function (IRF), a coherent artifact, or the shape of a spectrum. Submodels may also include a prescription for the transformation of parameters in order to enforce constraints or apply relations between parameters, as illustrated in the case study contained in Section 5.6.

Submodels may themselves be comprised of submodels. For instance, a multidataset model is comprised of a model for each dataset, which is in turn comprised of submodels, for, e.g., an IRF, a kinetic model, etc. Description in terms of a tree is often well-representative of such parameterizations. As an example, a tree representation for a model describing two datasets $\Psi = \{\Psi_1, \Psi_2\}$ is given in Figure 4. The parameters associated with models for the individual datasets Ψ_1 and Ψ_2 are themselves comprised of submodels for various aspects of the underlying system. Note that distinct submodels may depend on the same parameter $\theta_i \in \Theta$, so that parameters are linked between datasets, as Figure 5 illustrates diagrammatically.

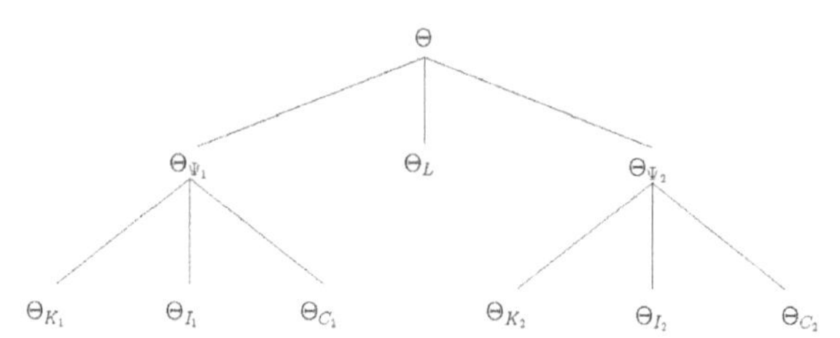

Figure 4: A hierarchical model Θ for datasets $\Psi = \{\Psi_1, \Psi_2\}$. The model is comprised of a submodel for each dataset with associated nonlinear parameters $\Theta\Psi_1$ and $\Theta\Psi_2$. These submodels are each comprised of a submodel for the kinetic decay rates parameterized with ΘK_1 and ΘK_2, a submodel for the IRF, parameterized with Θ_{I1} and Θ_{I2}, and a submodel for a coherent artifact parameterized with Θ_{C1} and Θ_{C2}.

THE IMPLEMENTATION OF TIMP

TIMP has been designed to facilitate interactive scientific model discovery of the multidataset hierarchical models introduced in Section 1.4 and reviewed in van Stokkum et al. (2004). In this section we attempt to give a brief overview of the design of the package.

The role of S4 classes and methods in TIMP

S4 classes and methods are the preferred means by which object-oriented programming is implemented in R. See e.g., Chambers (1998) for a review of the S4 classes and methods system, and e.g., Bates and DebRoy (2003) for a description of their utility within a large R package.

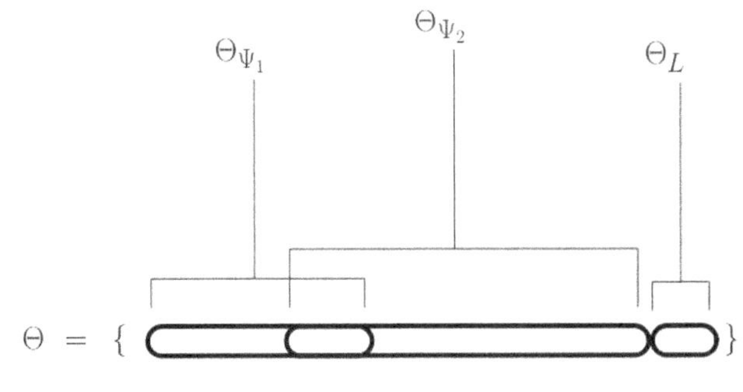

Figure 5: The vector of nonlinear model parameters Θ may parameterize a model for multiple datasets. Some parameters $\theta \in \Theta$ may be used to model more than one dataset, so that θ is linked between datasets, while other parameters may be used to model only one dataset. Still other parameters may be used to model a relationship between datasets, such as a linear scaling. In the above figure, the vectors of parameters $\Theta_{\Psi1}$ and $\Theta_{\Psi}2$ parameterizing the models for datasets Ψ_1 and Ψ_2, respectively, include some of the same elements from the full vector of nonlinear model parameters Θ. Parameters to determine a relationship between datasets are written as Θ_L, and apply to how the model functions associated with datasets are scaled.

S4 classes and methods are central to the implementation of TIMP. Slots of S4 classes are used to store each component of a hierarchical model. In the course of optimization (initiated with the fitModel function), an object is associated with the parameterization of the multidataset model and the current values of the parameters associated with this model, respectively. The object containing a prescription for the multidataset model is initialized to the (hidden) global variable .currModel, and is of class multimodel. The object used to store the current parameter estimates associated with this model is initialized to the (hidden) global variable .currTheta, and is of class multitheta.

During each iteration of parameter estimation, TIMP updates the object representing the current nonlinear parameter estimates .currTheta using the updated vector of non-linear parameter estimates and the model prescription object .currModel. The update is performed by the function getThetaCl. The object .currTheta is subsequently used to form the residual vector to be minimized with respect to the nonlinear parameter vector. By determining the residuals as a function of .currTheta as opposed to as a direct function of the "raw" vector of nonlinear parameter estimates Θ (which is typically of length $10^1 - 10^2$), the residual function implementation is relatively concise and readable. All bookkeeping necessary to account for fixed parameters, parameter relations, and parameter constraints is performed in getThetaCl, not the residual function, further facilitating its implementation and readability. This also allows for the rapid implementation of support for model types associated with new residual functions, as elaborated in Section 7.

Table 1: Classes currently used in TIMP. These classes are initialized in the package by functions prefaced by init., followed by the name of the class in the table above.

S4 classes defined by **TIMP**	
class	represents
`dat`	single dataset Ψ_q and an associated model
`kin`	kinetic model specification, inherits from `dat`
`spec`	spectral model specification, inherits from `dat`
`theta`	parameter estimates associated with model for single dataset Ψ_q
`multitheta`	parameter estimates associated with (possibly) multiple dataset model
`multimodel`	all information associated with multiple dataset model and fit
`res`	results of fitting (possibly) many datasets

As elaborated in Section 4, aspects of model parameterization not specific to model type (e.g., a weighting specification, constraints, or a specification of fixed parameter values) are specified in the slots of the class dat. Aspects of model parameterization specific to a given sort of model (e.g., a model in which the nonlinear parameters apply to the kinetics) are specified in the slots of classes inheriting from dat. S4 methods switch the definition of the function that determines the residuals and the output/plotting based on the class of the model type.

Model specification

Model specification in TIMP has possibly two steps. First, a model applicable to one dataset in $\Psi = \{\Psi_1, \ldots, \Psi_x\}$ is specified via the function initModel. If no per-dataset model differences are required, the model and the dataset list become arguments to the function fitModel, the call of which fits the model to the data, and outputs this fit and information for model validation. In the case that the model varies per-dataset, a second specification step describing per-dataset model differences from the model defined in initModel is required. The description of per-dataset model differences takes the form of a list input to fitModel. An example of this two-step process is found in Section 5.6.

Note that when the fitModel function is called, its arguments are used to initialize a list of objects of class dat, stored in the slot modellist of the (hidden) global variable .currModel. Element i of modellist stores dataset i and the model associated with dataset i. Many TIMP functions index into modellist.

Breaking the model specification into two steps allows the datasets associated with a given model to be specified just prior to fitting, as an argument to the fitModel function. This is convenient for applying the same model to many different combinations of datasets, but is not convenient when there are many differences between datasets, as the differences must be respecified with each call to fitModel. In the next version of TIMP a function to allow multidataset model specification to be performed in a single step will be included to avoid this inconvenience. This function will also facilitate the specification of model differences between groups of datasets.

Parameter estimation

The variable projection algorithm that allows for the solution of separable nonlinear parameter estimation problems is central to parameter estimation with TIMP. Separable nonlinear parameter estimation problems often arise in spectroscopy data modeling due to the bilinear form of the superposition model $\Psi = CE^T$, and the fact that it is often possible to welldescribe either the concentration profiles C or the spectra E, but not both matrices, with a parameterized model. By separating the parameters into intrinsically nonlinear parameters and clp descriptive of the entries of the matrix that is not described in terms of a parametric model, the nonlinear search space is often very significantly reduced, since datasets Ψ are often large (of dimension on the order of $10^3 \times 10^3$) with on the order of 101 contributing components. Furthermore, a parametric description of C or E is also often of interest in its own right, insofar as the model and parameter estimates may be assigned physical significance, and thereby serve as a simplified system description.

Let X denote the matrix (either C or E) directly determined by nonlinear parameters Θ, and let β denote the matrix determined as conditionally linear on X. The estimation problem associated with fitting Θ to $\Psi = X(\Theta)\beta$ such that the residuals k $\Psi - X(\Theta)\beta$ k_2 are minimal may be formulated as

$$\text{Minimize} \parallel (I - X(\Theta)X^+(\Theta))\Psi \parallel_2 . \tag{6}$$

This is the variable projection functional for which Golub and Pereyra (1972, 1973) determined an analytical gradient by deriving the derivative of the pseudoinverse X^+.

Given starting estimates for Θ, a solution to Equation 6 may be approached iteratively using gradient-based techniques. TIMP uses a finite difference approximation for the gradient.

TIMP's parameter estimation technique to solve Problem 6 can be summarized as follows, where "convergence" is some appropriate stopping criterion

Algorithm Finite Difference VarPro:

1. Choose starting Θ approximately
2. for s: = 1, 2 . . . until convergence do

determine gradient in Θ-space with finite diff. approx. of $\frac{d(I-XX^+)}{d\Theta}\Psi$

$\Theta_{s+1:}$ = step (Θ_s, gradient . . .)

Implementation in TIMP of the finite difference variable projection algorithm described above is via the nls function of R. nls is given as input starting values for Θ and a residual function that returns the vectorized version of (I − X(Θ) X+(Θ))Ψ. Then the numericDeriv function is used by nls to determine the finite difference approximation of the gradient of this residual vector in Θ-space. The determination of the step-size to move Θ in the gradient direction each iteration (the algorithm step) is also performed by nls. Section 2.4 describes the extension of this methodology to the case in which X is dependent on conditions such as the wavelength or timepoint of the data, and to multiple datasets.

Partitioned variable projection

For application to multiway spectroscopy modeling problems domain we seek to address, the algorithm for finite difference variable projection given in the previous section must be generalized to the case of simultaneously modeling multiple datasets, and to the case where the model for the matrix X is dependent on condition, where a condition is an instance of a variable with respect to which the data is resolved, such as wavelength, time, temperature, etc. The desired extension must operate on (i.e., perform QR-decomposition on, for instance) small matrices, so as to allow parameter estimation to be performed on computer systems lacking large memory resources.

Let X_q be the matrix determined by nonlinear parameters Θ that is associated with dataset Ψ_q. The model for X_q may be a function of Θ that is different for distinct datasets in Ψ_1, . . . , Ψ_x. The clp may be equated between datasets, so that $\beta_1 = \beta_2 = \ldots = \beta_x$ (the "link clp" case in the following description of PartitionedVarPro) or the clp may be estimated per-dataset, or per-group of datasets, so that $\beta_p \neq \beta_q$ for datasets Ψ_q and Ψ_p in Ψ. These two possibilities for the clp correspond to the equations

$$\begin{pmatrix} \Psi_1 \\ \vdots \\ \Psi_x \end{pmatrix} = \begin{matrix} X_1 \\ \vdots \\ X_x \end{matrix} \beta = X_{\text{super}} \beta \tag{7}$$

and

$$\begin{pmatrix} \Psi_1 \\ \vdots \\ \Psi_x \end{pmatrix} = \begin{matrix} X_1\beta_1 \\ \vdots \\ X_x\beta_x \end{matrix} \tag{8}$$

In the case that the matrix X_q is dependent on a data condition (as in wavelength-dependent kinetic models, for instance, or in time-dependent spectral models), X_q must be re-calculated for every condition with which it varies, so that Models 7 and 8 are

$$\psi_{\text{super } p} = \begin{pmatrix} \psi_{1p} \\ \vdots \\ \psi_{xp} \end{pmatrix} = \begin{matrix} X_{1p} \\ \vdots \\ X_{1p} \end{matrix} \beta_p = X_{\text{super}} \beta_p \tag{9}$$

and

$$\begin{pmatrix} \psi_{1p} \\ \vdots \\ \psi_{xp} \end{pmatrix} = \begin{matrix} X_{1p}\beta_{1p} \\ \vdots \\ X_{xp}\beta_{xp} \end{matrix} \tag{10}$$

where p is the condition index, ψ_{qp} is a vector comprising the pth row or column of dataset Ψ_q, β_{qp} is the pth row of βq, and X_{qp} is associated with dataset q at condition p.

In general p need not represent a single condition, but may represent a collection of conditions (e.g., wavelengths or times) for which the matrix X is constant.

The residuals associated with Equations 9 and 10 may be calculated using the variable projection function under the partitioned variable projection algorithm. This algorithm is presented for the general case in which p may represent a single condition or a group of conditions associated with the same prescription for X. A group of conditions associated with the same prescription for X is termed in the algorithm description a part. If Equation 7 or 9 are desired, then (link clp) is true. If Equation 8 or 10 are desired, then (link clp) is false

Note that the variable projection functional is $(I-CC^{+})\Psi = Q_2 Q_2^{\top}\Psi$ using the QR-decomposition, and that these residuals are formulated in Q-space as $Q_2^{\top}\Psi$.

```
PARTITIONEDVARPRO(MODEL, Θ)
  for parts p in 1:n:
      if(link clp)
            get(X_super p( MODEL, Θ))
            QR(X_super p)
            append Q2ᵀ ψ_super p to Z
      else
            for q in 1:x
                get(X_pq( MODEL, Θ))
                QR(X_pq)
                append Q2ᵀ ψ_pq to Z
  return: Z
```

After formation of the residual vector Z, a finite difference method may be applied to determine an update of Θ as described in Section 2.3 for Finite Difference VarPro. The ability to apply the variable projection functional without operating on large matrices is the main motivation for introduction of PartitionedVarPro. PartitionedVarPro results in the same residuals as under the standard variable projection algorithm as implemented in, e.g., the plinear function of nls. The advantage of PartitionedVarPro is that significantly less memory resources are required, allowing application of the algorithm on large datasets on a personal computer. The memory resources required by PartitionedVarPro and the standard variable projection implementation are considered in detail in Appendix A. To examine the implementation of partitioned variable projection in TIMP, see the rescomp function which collects the residual vector Z and one of the S4 methods for residPart, which returns the residuals (with a contribution from possibly many datasets) associated with a single part p.

Validation

Model validation in TIMP may be performed via a variety of means. Nonlinear parameter estimates as returned by nls may be validated via the linear approximation standard error estimates returned. Relatively large standard errors indicate an over-parameterization of the model.

For each model type, an S4 method plotter is implemented to output model type-specific results, both in the form of plots and in the form of ASCII files representing the model fit and other estimates.

Analysis of the residuals is an important aspect of model validation. This analysis is often facilitated by taking a singular value decomposition (SVD) of the residual matrix, which allows structure in the misfit of the model to the data to be readily observed. Also important to an evaluation of the model fit are plots of the fit of the model to the data for each row or column of the data (possibly for each of multiple datasets).

It is often desirable in scientific modeling applications to perform model validation after the application of the fitting algorithm (in the case of TIMP, nls) results in the satisfaction of stopping criteria, as opposed to convergence criteria. For example, it is often desirable to validate the model fit after a set number of iterations, or at the starting values of parameters to be optimized. In order to allow validation to be performed after a set number of iterations, the nls function of R was extended with new options, which are described in Appendix B. These options allow the fitModel of TIMP to return information validating model fit after any desired number of iterations.

USER-ACCESSIBLE FUNCTIONS

TIMP is currently structured around five core user-accessible functions, described in this section in turn. More complete information regarding function arguments and output is found in the help functions of the package; here a higher-level description of the purpose and structure of the arguments and output is given.

readData

readData takes as an argument a string containing the path to an ASCII-file containing data and reads the data into R. There are currently three supported data formats, which are described in Appendix C. The data (and inferred attributes, such as the number of wavelengths by which spectra are represented, etc.) are returned as an object of class dat.

preprocess

preProcess takes an argument of class dat and a specification of the desired data sampling, selection, baseline correction, or axis scaling, and the dimension in which to perform the preprocessing (which may be the spectral dimension or the dimension in which spectra are resolved, e.g., time). The preProcess function returns an object of class dat.

initModel

The initModel function is used to specify a model. A string mod_type giving the class of the model being specified, ("kin" for kinetic models, "spec" for spectral models, and so forth) is a mandatory argument. Additional arguments may be any model options described in the help page for the dat class, plus options described on the help page of the desired model class given as mod_type. Output is an object of the desired model class, which inherits from dat

fitModel

fitModel performs optimization of a model to an arbitrary number of datasets. Arguments to fitModel include a list of datasets to be fit, a model that is applied to all datasets, a list specifying any model differences to apply per-datasets, and a list of control and printing options. A call to this function results in optimization of free parameters. By default results are then printed to the screen, and optionally to text files and/or postscript. Returned is a list whose elements include the output of the call to the function nls that is used to iteratively improve the starting estimates for nonlinear parameter values.

examineFit

examineFit takes as input the list returned by fitModel and re-calls the plotting and functions to write output. This function is useful for the comparison of fit of several models. An output object is not returned.

GENERAL MODEL OPTIONS

This section seeks to outline at a higher level than that found in the package's help pages model parameterization options that may be applied to all model types. The specification of each such option becomes a slot in the class dat, possibly after processing (within the initModel or getModel functions). Options that are specific to a given model type that inherits from dat (e.g., the class kin for kinetic model options or the class spec for spectral model options) are described in later sections.

The function initModel takes as arguments a specification of model options. Note that for multidataset models, any aspect of a parameterization of a model possible to specify in initModel may be modified, removed or added to the prescription of per-dataset model differences given as the modeldiffs argument to the fitting function fitModel.

Data weighting

A weighting scheme W has the form of an m by n matrix (where m by n is the dimension of the data matrix Ψ). W may lessen the weight of portions of the data known to contain less information regarding the model parameters. Such data may result from noisy experimental conditions, for instance. The application of a weighting scheme may also be desirable from first principles, e.g., in the case that the data are known to have a variance related to their magnitude as in single photon counting (SPC) experiments, in which the data are Poisson distributed, with

$$\hat{\sigma}_{\Psi(t_i,\lambda_j)} = \sqrt{\Psi(t_i,\lambda_j)}. \tag{11}$$

For the case of Poisson distributed count data, W(i, j) is

$$W(i,j) = \frac{1}{\hat{\sigma}_{\Psi(t_i,\lambda_j)}}. \tag{12}$$

Once W has been determined, the Hadamard product (element-by-element product) of W and Ψ is taken, so that

$$\Psi^W(t_i,\lambda_j) = \Psi(t_i,\lambda_j) * W(t_i,\lambda_j) \tag{13}$$

which we write as

$$\Psi^W = \Psi \circ W \tag{14}$$

In the case that Ψ is transformed by a weight matrix W into Ψ^W, the associated concentration matrix becomes C^W where

$$C^W_{\lambda_j}(t_i,l) = C_{\lambda_j}(t_i,l)W(t_i,\lambda_j) = C \circ W \tag{15}$$

Ψ^W and C^W may then be used in place of Ψ and C in parameter estimation.

Specification in TIMP: Weighting

The list argument weightpar specifies a prescription for the matrix of weights to be applied to the dataset. weightpar is a list of vectors. The vectors have form

c(first_x, last_x, first_x2, last_x2, weight). first_x and last_x are the least and greatest timepoints (or other variable with which spectra are resolved) having weight weight; first_x2 and last_x2 are the least and greatest values of the spectral variable having weight weight.

Note that if vector elements 1-4 are NA, the first-most point of the data is taken for elements 1 and 3, and the last-most points are taken for 2 and 4. For example, for a dataset in which spectra are measured in wavelength at many different times in picoseconds, the specifi- cation weightpar = list(c(40, 1500, 400, 600, .9), c(NA, NA, 700, 800, .1)) will weight data between 40-1500 ps and 400 and 600 nm by .9, and will weight data at all times between 700 and 800 nm by .1.

For single photon counting data or other types of count datasets, weightpar = list(poisson = TRUE) will apply Poisson weighting to all non-zero elements of the data.

Fixed parameters

It is often of interest to set nonlinear model parameters to fixed values. Fixing model parameters may make use of a priori knowledge of true parameter values, or may be performed to decrease the number of free parameters of the model.

Specification in TIMP: Fixed parameters

Every model parameterization option with an associated list or vector of nonlinear parameter starting values is named. For instance, the name of the starting values for kinetic decay rates is "kinpar". In order to fix nonlinear parameters, the name of their list or vector of starting values is given, along with the indices into the list or vector at which parameter values should be fixed. This specification is contained in a list fixed.

For instance fixed = list(kinpar = c(1,3,5), parmu = list(c(1,1), c(1,2), c(1,3))) will fix the 1st, 3rd, and 5th elements of the kinpar vector of starting values for kinetic decay rates, and the 1st, 2nd, and 3rd elements of the 1st list of parameters in the parmu list of starting values for parameters describing wavelength-dependence of the IRF.

Constraint of clp

The basic superposition model $\Psi = CE^T$ (Equation 3) is of bilinear form, where the matrix C describes concentrations and the matrix E describes spectra. A nonlinear model may be used to describe either C or E, and the entries of the remaining matrix estimated as clp, as described in Section 2.3.

It is often desirable to constrain clp to account for a priori knowledge or to reduce the number of free parameters in the model. TIMP currently allows clp to be constrained to a linear relationship with a scaling parameter (that may be fixed at 1 to equate clp), or to be constrained to zero.

It is often useful to name the clp to be constrained in terms of the component they represent, (i.e., the column of C or E they are contained in).

Specification in TIMP: Constraint of clp to zero

The list clp0 contains lists that specify clp to constraint to zero. The elements of these lists are named low, high, comp, specifying the least and greatest absolute values of the clp dimension to constrain to zero, and the component to which to apply the zero constraint, respectively. For example, where clp represent spectra in wavelength, clp0 = list(list(low=400, high = 600, comp=2), list(low = 600, high = 650, comp=4)) applies zero constraints to the spectra associated with component 2 between 400 and 600 nm, and to the spectra associated with component 4 between 600 and 650 nm.

Specification in TIMP: Constraint of clp to a linear relationship

The list clpequspec contains lists that specify collections of clp to relate. The elements of these lists are named to, from, low, high. An optional element named dataset specifies the dataset from which to get the reference clp determining the relationship. to is the component from which clp are to be fixed in relation to clp from some other component; from is the reference component. low and high are the least and greatest absolute values of the clp dimension to constrain. For example, where clp represent spectra in wavelength, clpequspec = list(list(low = 400, high = 600, to = 1, from = 2)) will constrain the spectra associated with the first and the second components to equality between 400 and 600 nm according to $\epsilon_1 \leftarrow \epsilon_2\theta_\epsilon$ where ϵ_1 and ϵ_2 are the spectra associated with components 1 and 2, ← indicates that ϵ_1 is dependent on ϵ_2, and θ_ϵ parameterizes the linear relation.

The vector clpequ contains length(clpequspec) numerics, where the ith numeric is a starting value θi parameterizing the linear relation specified between clp by the ith list in clpequspec. Fixing the starting value of an element of clpequ at 1 constrains the associated clp to equality.

Relations between nonlinear parameters

It may be desirable to enforce a relationship between two nonlinear parameters θ_1 and θ_2 so that $\theta_2 = f(\theta_1)$. The relationship of nonlinear parameters is usually performed in order to take into account a priori knowledge of the system being modeled.

A linear relationship between parameters is currently implemented, (and other functional relationships will be added).

Specification in TIMP: Nonlinear parameter relations

As in the case of fixing parameters, in order to relate nonlinear parameters, the name of the associated list or vector of starting values is given, along with the indices into the list or vector at which parameter values are related. This specification is contained in a list prelspec.

Each element of prelspec is a list having elements named what1 (a character string describing the parameter type to relate, e.g., "kinpar"), what2 the parameter type on which the relation is based; usually the same as what1), ind1 (an index into what1) and ind2 (an index into what2), and the optional argument rel (a character string to specify the functional relation type, by default "linear"). For examples, prelspec = list(list(what1 = "kinpar", what2 = "kinpar", ind1 = 1, ind2 = 5)) relates the 1st element of kinpar to the 5th element of kinpar according to kinpar[2] ← kinpar[1] where ← indicates that kinpar[2] is dependent on kinpar[1].

The vector prel is of length length(prelspec) and contains numeric starting values parameterizing the linear relationships described in prelspec.

Constraint of nonlinear parameters to positivity

It may be known a priori that the nonlinear parameters associated with a submodel should be positive. Then optimizing on the log of these parameters and transforming the results by exponentiation of the estimated parameters is a means of enforcing the desired constraint.

Specification in TIMP: Constraint of parameters to positivity

The vector positivepar includes a character string containing the name of the vector or list of starting parameters that should be constrained to positivity, e.g., positivepar = c("kinpar").

KINETIC MODELS

Kinetic models describe the concentrations of components in time. For multiway spectroscopy data modeling applications the basic model for the kinetics of each component (i.e., each column of the matrix C) is an exponential decay in time t, so that

$$\Psi = CE^\top = \sum_{l=1}^{n_{\text{ncomp}}} c_l \epsilon_l^\top = \sum_{l=1}^{n_{\text{ncomp}}} (\exp(-\phi_l t) \star i(t))\epsilon_l^\top \tag{16}$$

here i is the instrument response function (IRF) and ? is convolution. The parameter estimation problem of optimal values for the amplitudes ϵ_l of the exponential decays (i.e., the spectra E) along with the decay rates φl under least squares criteria is called the multiexponential analysis problem, and is ubiquitous in physics applications in which data is modeled by the solution of first-order differential equations, as Istratov and Vyvenko (1999) review. Kinetic models are represented in TIMP with the class kin. Options for the parameterization of kinetic models in TIMP are here outlined.

Model for the decay of components

The basic model for the concentration matrix C is a sum of n_{ncomp} exponential decays parameterized as ΘK = (k1, . . . ,

k_{nncomp}), so that the entries of the concentration matrix C(i, l) are given as

$$C(i,l) = e^{-k_l t_i}. \tag{17}$$

Specification in TIMP: Kinetic decay rates

The vector kinpar contains numeric starting values for the kinetic decay rates, which in the absence of a compartmental scheme parameterize exponential decays. The number of values given determines n_{ncomp}, the number of components dedicated to modeling kinetic decays.

Instrument response models

Multiway spectroscopy experiments often employ a short laser pulse to excite the system under study and measure the resulting spectra in time. The convolution of the shape of this exciting pulse and the detector response is the IRF. With pump-probe spectroscopy the IRF is given by the convolution a pump and a probe pulse. With Gaussian-shaped pump and probe pulses, the convolution of the two will again be Gaussian-shaped, but with an increased width.

An IRF may either be parameterized in the model, or measured. A parametric model for the IRF i as a Gaussian in the time dimension t gives the following model for the concentration matrix C

$$C(t,k_l,\mu,\Delta) = e^{-k_l t} \star i(t) = \frac{e^{-k_l t}}{2} e^{k_l(\mu + k_l \tilde{\Delta}^2/2)} \left\{ 1 + erf\left[\frac{t - (\mu + k_l \tilde{\Delta}^2)}{\sqrt{2}\tilde{\Delta}} \right] \right\} \tag{18}$$

where t is time, kl is the lth kinetic component, μ and Δ are the location and full width half maximum (FWHM) parameters of the Gaussian distribution, respectively, Δ is the Gaussian $\tilde{\Delta}$ width parameter (such that $\tilde{\Delta} = \Delta/2\sqrt{2\ln 2}$), and ? is convolution. Recall that the FWHM of the Gaussian distribution is related to the standard deviation σ as $FMWM = 2\sqrt{2\ln 2}\sigma$.

In the case that a measured IRF is used as opposed to a model for the IRF with parameters to be estimated, its numerical convolution with the exponential decay function yielding the kinetic decays is required. TIMP allows for either methods of including the effects of the IRF to be applied.

Specification in TIMP: Gaussian IRF

The vector irfpar contains starting values for the parametric description of the IRF in terms of a Gaussian. The vector is ordered irfpar = c(μ, Δ). For example, irfpar = c(-2, .05) specifies a starting location parameter μ as -2 and a width Δ as .05, where the starting values are in the unit of time of the timepoints in the data, e.g., picoseconds.

Specification in TIMP: Measured IRF

The vector measured_irf is of length equal to the number of timepoints in the data, and contains the measured IRF; if specified, it will be applied to the data.

The integer convalg is between 1-4 and determines the numerical convolution algorithm used. convalg=1 is the default and is recommended.

Models for dependence of IRF parameters on spectral variable

In the case that an IRF is included in the model using a parametric description, it is often desirable that the parameters involved are dependent on spectral variable (e.g., wavelength). The dependence of IRF on the spectral variable is termed dispersion.

Time-gated spectra are measurements of an optical property (e.g., emission or absorption) as a function of a spectral variable like wavelength taken simultaneously across some range, repeated at many distinct times. In kinetic models of time-gated spectra, dispersion is often well-modeled as a smooth function of the spectral variable. Polynomial functions of degree nparmu and npartau are often used to describe the variance in the spectral variable of the IRF location parameter μ and FWHM parameter Δ, the coefficients of which become additional parameters of Θ_I. Then the IRF μ and Δ are calculated per wavelength as

$$\mu(\lambda) = \mu_0 + \sum_{i=1}^{nparmu} \mu_i \left(\frac{\lambda - \lambda_c}{100} \right)^i, \tag{19}$$

$$\Delta(\lambda) = \Delta_0 + \sum_{i=1}^{npartau} \Delta_i \left(\frac{\lambda - \lambda_c}{100} \right)^i \tag{20}$$

where μ is a function that takes a wavelength or wavenumber and gives the location of the IRF, μ_0 is the location of the IRF at λ_c, Δ is a function that takes a wavelength or wavenumber and gives the scaled width of the IRF, and Δ_0 is the scaled width of the IRF at λ_c.

For data collected by measuring decay traces at many different wavelengths dispersion is often well-modeled by shift parameters for μ and Δ per-wavelength. Where nl is the number of points whereby spectra are represented, this results in nl shift parameters parameterizing the dispersion of μ and nl shift parameters parameterizing the dispersion of Δ. We refer to models of dispersion employing a shift parameter per-wavelength as discrete.

Specification in TIMP: Dispersion models

The character strings dispmufun and disptaufun determine the functional form of the dispersion of the IRF location parameter. The default is a polynomial description. If dispmufun or disptaufun is equal to "discrete" a discrete model for dispersion of the corresponding variable is applied. For the discrete case parmu or partau should contain a starting value for the shift for every point by which spectra are represented (e.g., for each wavelength). The numeric lambdac is supplied if a polynomial description of either dispersion of location or FWHM is applied. Then lambdac is the center-wavelength in this description (λc in Equations 19 and 20).

The list parmu contains starting values for the parameters of the model for dispersion of the IRF location. The vector partau contains starting values for the parameters of the model for dispersion of the IRF FWHM.

Coherent artifact/scatter models

Should the measurements contain an instantaneous response or coherent artifact due to Raman scatter, it may be desirable to include in the model for the concentrations C its description in time. This is done by adding components (which are columns of the C matrix) to represent its contribution.

A commonly used model for coherent artifact/scatter has the time characteristics of the IRF, in which case a single column is appended to the concentration matrix with the IRF time profile. A variation of this model maintains separate coherent artifact spectra for each of x datasets Ψ_q.

Another commonly used model type employs a sequential scheme with, e.g., femtosecond lifetimes, in which the signs of the amplitude of consecutive components alternates. This model type often well-describes an oscillatory coherent artifact. In the case that the instantaneous response of the exciting pulse has both scatter and coherent response components, a linear superposition of the model that follows the IRF time profile and a model based on a sequential scheme may be applied.

An aside on the implementation of ultra-fast coherent artifact lifetimes

Models for the coherent artifact model type that employ a sequential kinetic scheme are often well-fit with ultra-fast lifetimes, and hence large exponential decay rates. For very large decay rates under Equation 18, it may be desirable to employ the exponentially scaled error function $\exp(x^2)\mathrm{erfc}(x)$, as in the decayirf function of TIMP. An implementation of the exponentially scaled error function (erfce) is currently unavailable within R. Its use in TIMP is made possible by porting the implementation found in the Cephes Mathematical Library (Moshier 1992) to an R shared library in C.

Specification in TIMP: Coherent artifact/scatter models

The list cohspec describes the model for coherent artifact/scatter component(s). If cohspec$type is "irf", the coherent artifact/scatter has the time profile of the IRF. If cohspec$type is "freeirfdisp" the coherent artifact/scatter has a Gaussian time profile whose location and width are parameterized in the vector coh (independent from the IRF parameters). If cohspec$type is "irfmulti" the time profile of the IRF is used for the coherent artifact/scatter model, but the IRF parameters are taken per dataset (for the multidataset case), and cohspec$numdatasets must be equal to the number of datasets modeled. If cohspec$type is "seq" a sequential exponential decay model is applied, whose parameters are contained in coh. If cohspec$type is "mix" a sequential exponential decay model is applied along with a model that follows the time profile of the IRF; the coherent artifact/scatter contribution is then a linear superposition of these two models.

The vector coh contains starting values for the parameterization of a coherent artifact/scatter model.

Compartmental models

A linear time-invariant compartmental model may be used to describe allowed transitions between components, as Godfrey (1983) reviews. Transitions between compartments are described by microscopic rate constants which constitute the off-diagonal elements of the transfer matrix K. The diagonal elements of K contain the total decay rates of each compartment. The concentrations of the compartments in continuous time are described by a vector $c(t) = [c_1(t), \dots, c_{nncomp}(t)]^T$. Thus, a linear compartmental model with nncomp compartments is described by a differential equation for these concentrations

$$\frac{d}{dt}c(t) = Kc(t) + j(t) \tag{21}$$

where the input to the system is described by a vector $j = i(t)[j_1\ j_2\ \dots\ j_{n_{\mathrm{ncomp}}}]$ such that Pnncomp $\sum_{l=1}^{n_{\mathrm{ncomp}}} j_l = 1$, and $j_{n_{\mathrm{ncomp}}} \equiv 1 - \sum_{l=1}^{n_{\mathrm{ncomp}}-1} j_l$. Also, generally $j_l \geq 0$. The IRF i(t) describes the time-profile of the inputs. Under the assumption that the eigenvalues of K are different, and that $c(-\infty) = 0$, Equation 21 is solved as

$$c(t) = e^{Kt} \star j(t). \tag{22}$$

Equation 22 is used as the prescription for the concentrations of the components in discrete time. Note that Equation 22 requires evaluation of the exponential of the K matrix. This exponentiation is implemented in TIMP as described in van Stokkum (2005).

Analysis in the absence of a compartmental model is equivalent to the application of a K matrix with non-zero elements only on the diagonal (Figure 6). A commonly applied compartmental model is termed a sequential model, in which $n_{\mathrm{ncomp}} - 1$ components decay to a single other component, one component decays directly to the ground state and no loops are present (Figure 7).

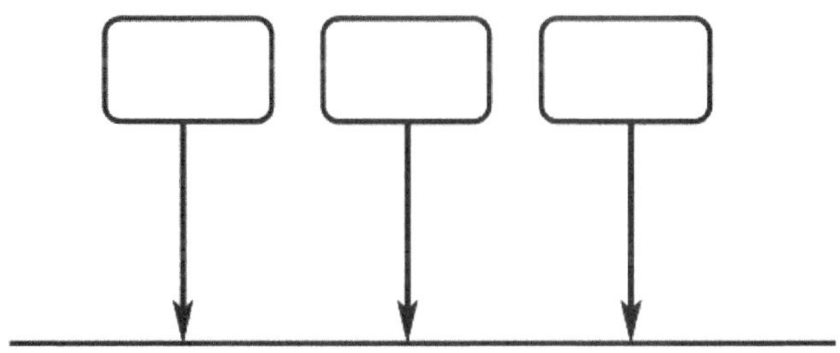

Figure 6: Diagram of a compartmental model of three components, all of which decay in parallel to the ground state.

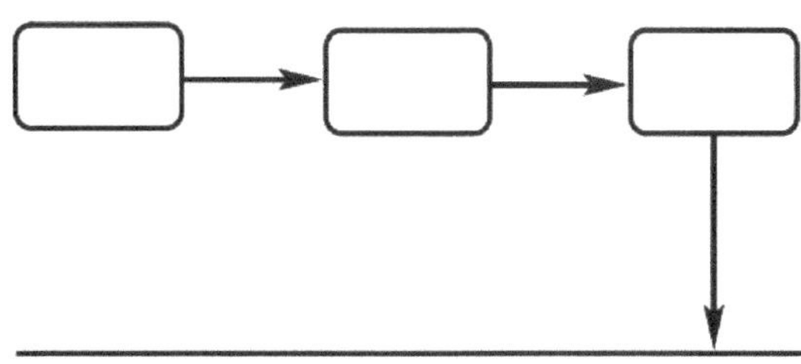

Figure 7: Diagram of a compartmental model of three components sequentially decaying to the ground state.

Specification in TIMP: Parallel/sequential compartmental model

The logical seqmod determines whether a sequential compartmental model is applied. If seqmod=FALSE then a parallel compartmental model is applied. The default is seqmod=TRUE.

Specification in TIMP: Full compartmental model

The array kmat contains an array with dimension attribute c(2, n_ncomp, n_ncomp). That is, kmat contains two matrices of dimension $n_{ncomp} \times n_{ncomp}$. The matrix in kmat[1, 1 : n_ncomp, 1 : n_ncomp] contains i at position kmat[1, j, i] if a transition from component i to component j is allowed, and else 0.

The matrix in kmat[2, 1 : n_ncomp, 1 : n_ncomp] contains k at position kmat[2, j, i] if the transition from component i to component j is parameterized by a branching parameter from kinscal with index k, and else 0.

The vector jvec contains the j vector descriptive of the inputs to the transfer matrix K.

The vector kinscal is descriptive of starting values for branching parameters of K.

Case study: Multiexperiment analysis

An example of multiexperiment kinetic modeling of data $\Psi = \{\Psi_1, \Psi_2\}$ is considered in this section. Datasets Ψ_1 and Ψ_2 were collected under the same experimental conditions except the excitation laser intensity was doubled during collection of Ψ1 relative to the laser intensity used during collection of Ψ2. The challenge in modeling is to obtain a parametric description of the kinetics of the underlying system as evidenced by both datasets, as well as a parametric description of how the difference in laser intensity affects the system. The system underlying the data and physical evidence informing formulation of the initial model and interpretation of model fit will be described elsewhere (manuscript in preparation).

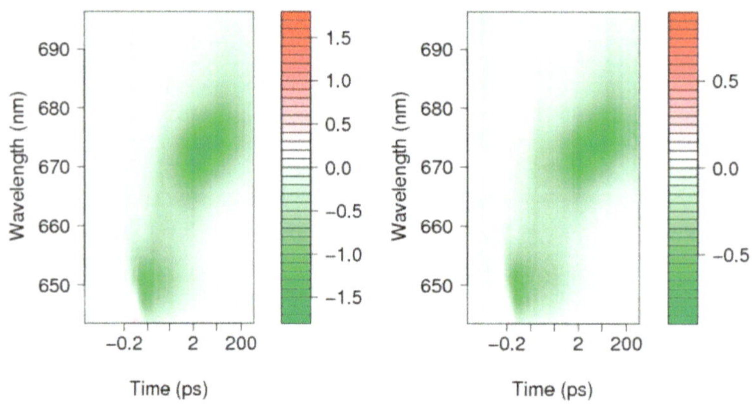

Figure 8: (Left) Dataset Ψ_1 measured using twice the laser intensity used to measure (Right) dataset Ψ_2. The color palette used to display these datasets is generated with the diverge_hsv function of the vcd package (Meyer et al. 2006).

Data input

The data $\Psi = \{\Psi_1, \Psi_2\}$ is read into TIMP in the time explicit format described in Appendix C via the commands

```
R> psi_1 <- readData("psi_1.txt")
Read 1 item
Read 2385 items
R> psi_2 <- readData("psi_2.txt")
Read 1 item
Read 2385 items
```

where Ψ1 is stored in the file "psi_1.txt" and Ψ2 is stored in the file "psi_2.txt" (distributed with this paper).

Data preprocessing

The wavelength axis of the data is scaled via the prescription x = 3.78x + 643.5.

```
R> psi_1 <- preProcess(data = psi_1, scalx2 = c(3.78, 643.5))
R> psi_2 <- preProcess(data = psi_2, scalx2 = c(3.78, 643.5))
```

An initial model

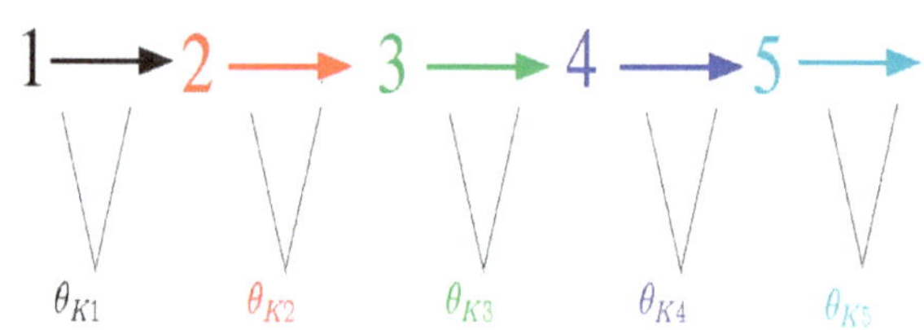

Figure 9: Kinetics are described by a five-component sequential compartmental model with identical decay rate parameters for both datasets.

It is known a priori that the underlying system is likely to contain five components decaying sequentially, that a Gaussian IRF model is likely to be appropriate, and that a contribution from a coherent artifact must be accounted for. Approximate starting estimates for parameter values are also known.

This leads to the following model specification in TIMP.

```
R> model1 <- initModel(mod_type = "kin",
+ kinpar=c(7.9, 1.08, 0.129, .0225, .00156),
+ irfpar=c( -.1018, 0.0434),
+ parmu = list(c(.230)),
+ lambdac = 650, seqmod=TRUE,
+ positivepar = c("kinpar"), title="model 1",
+ cohspec = list( type = "irf"))
```

The vector kinpar contains the starting values for five kinetic components. These parameters are constrained to positive values during fitting by including "kinpar" in the vector argument positivepar. The seqmod argument determines that the kinetics are described with a sequential compartmental model. The irfpar argument gives starting values for the parameters of the default Gaussian IRF model. The parmu argument gives starting values for the default model for dispersion in terms of a polynomial, in this case of first-order. The lambdac argument gives the center wavelength for the polynomial description of dispersion. The cohspec argument determines that a model for a coherent artifact/scatter component is to be added with the time-profile of the IRF.

Fitting and validating the initial model

For simultaneous analysis a list is initialized to group dataset Ψ_1 and dataset Ψ_2 together.

R> psi <- list(psi_1, psi_2)

In the absence of information regarding the effect of laser intensity on the underlying system, the model initialized in Section 5.6.3 may be fit to both dataset Ψ_1 and dataset Ψ_2 simultaneously, with the addition of a dataset scaling parameter Θ_L to account for the difference in amplitude between datasets due to laser intensity.

A call to the fitting function fitModel fits the initial model model1 to both datasets, with the argument modeldiffs specifying per-dataset differences in the applied model. By visual inspection of the data it is clear that the intensity of dataset Ψ_2 is approximately half that of dataset Ψ_1. It is not obvious from visual inspection what other per-dataset differences to include in the model. Therefore modeldiffs only specifies that the second dataset is scaled to .5 times the first dataset (via the argument dscal = list(list(to=2,from=1,value=.5))). The input argument opt specifies plotting and output options.

```
R> res_model1 <- fitModel(psi, model1,
+ modeldiffs = list(dscal = list(list(to=2,from=1,value=.5))),
+ opt=list(iter=5, superimpose = c(1,2), divdrel = TRUE, linrange = .2,
+ makeps = "den1", selectedtraces = c(1,5,10), plotkinspec =TRUE,
+ xlabel = "time (ps)", ylabel = "wavelength",
+ paropt=list(cex.main=1.2,cex.lab=1.2,cex.axis=1.2)))
```

Figure 10 shows the fit of selected traces after fitting for five iterations. Large misfits are present. Misfits around time zero are likely to be due to an insufficient IRF model, whereas misfits at later times are likely to be due to differences in the kinetics of the two measured datasets. The root mean square (RMS) error associated with this fit is .040. Note that the datasets appear to have the same amplitudes due to the argument divrel to opt in the call to fitModel, which divides Ψ_2 and and the fit of the model to Ψ_2 by the estimate for the dataset scaling parameter.

Model refinement and re-validation

Fitting the IRF location parameters and the slowest two kinetic decay rate parameters perdataset attempts to address the inadequacies of the model identified in validation.

```
R> res_model1_refined <- fitModel(psi, model1,
+ modeldiffs = list(dscal = list(list(to=2,from=1,value=.5)),
+ free = list(
+ list(what = "irfpar", ind = 1, dataset = 2, start=-.1932),
+ list(what = "kinpar", ind = 5, dataset = 2, start=.0004),
+ list(what = "kinpar", ind = 4, dataset = 2, start= .0159)
+ )),
+ opt=list(iter=5,superimpose = c(1,2), divdrel = TRUE, linrange = .2,
+ makeps = "den2", selectedtraces = c(1,5,10),
+ paropt=list(cex.main=1.2,cex.lab=1.2,cex.axis=1.2)))
```

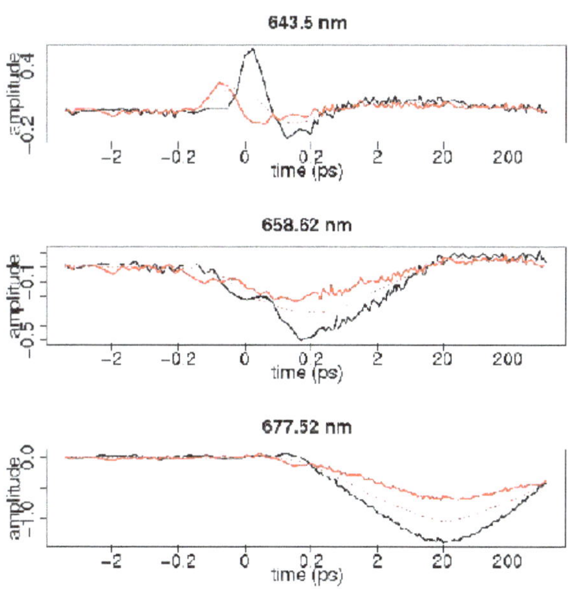

Figure 10: A plot of three selected traces resulting from the fit of the initial model to the data $\Psi = \{\Psi_1, \Psi_2\}$. Black represents data Ψ_1 (solid) and the fit of the initial model to Ψ_1 (dotted); Red represents data Ψ_2 (solid) and the fit of the initial model to Ψ_2 (dotted). The RMS error associated with this fit is .040.

The resulting fit is improved over the initial model, though a misfit remains evident at early times. The data has an oscillatory nature in some traces indicative of a contribution from a coherent artifact, as evident in Figure 11. The RMS error associated with this fit is .027.

A satisfactory model

The model for the coherent artifact has followed the time profile of the IRF, which is not suffi- cient to account for the oscillatory nature of its apparent contribution to the data. Therefore the coherent artifact model is replaced via a re-definition of the model (which could also be performed by specifying a different model prescription for both datasets in the model differences list):

```
R> model2 <- initModel(mod_type = "kin",
+ kinpar=c(7.9, 1.08, 0.129, .0225, .00156),
+ irfpar=c( -.1018, 0.0434),
+ parmu = list(c(.230)),
+ lambdac = 650,
+ seqmod=TRUE,
+ positivepar = c("kinpar", "coh"),
+ title="Model 2",
+ cohspec = list( type = "seq", start = c(8000, 1800)))
```

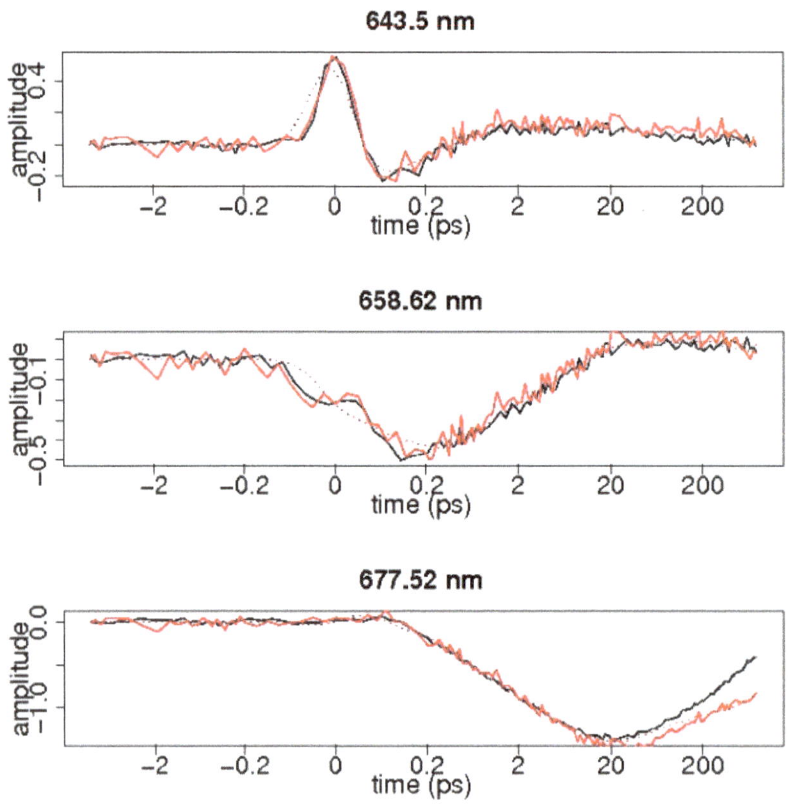

Figure 11: A plot of three selected traces resulting from the fit of the refined model to the data $\Psi = \{\Psi_1, \Psi_2\}$, with the IRF location parameters and the two slowest decay rates fit per-dataset. Black represents data Ψ_1 (solid) and the fit of the initial model to Ψ_1 (dotted); Red represents data Ψ_2 (solid) and the fit of the initial model to Ψ_2 (dotted). The RMS error associated with this fit is .027.

The new model is fit to the datasets. Again the IRF location and the two slowest data rates are fit per-dataset. The estimated value of the data scaling parameter between Ψ_1 and Ψ_2 obtained in fitting the refinement of the initial model is used as a starting value.

```
R> res_model2 <- fitModel(psi, model2,
+ modeldiffs = list(dscal = list(list(to=2,from=1,value=.457)),
+ free = list(
+ list(what = "irfpar", ind = 1, dataset = 2, start=-.1932),
+ list(what = "kinpar", ind = 5, dataset = 2, start=.0004),
+ list(what = "kinpar", ind = 4, dataset = 2,
+ start= .0159))),
+ opt=list(iter=1,superimpose = c(1,2), divdrel = TRUE,
linrange = .2,
+ makeps = "den3", selectedtraces = c(1,5,10), plotkinspec =TRUE,
+ xlabel = "time (ps)", ylabel = "wavelength", breakdown = c(1,5,10),
+ paropt=list(cex.main=1.2,cex.lab=1.2,cex.axis=1.2)))
```

The resulting parameter estimates mapped to a hierarchical representation of the model are shown in Figure 15. For clarity the standard errors estimates returned by nls have been omitted. These standard errors are typically 1-5 in the last significant digit reported, except for θC (which is not of interest) where they are huge.

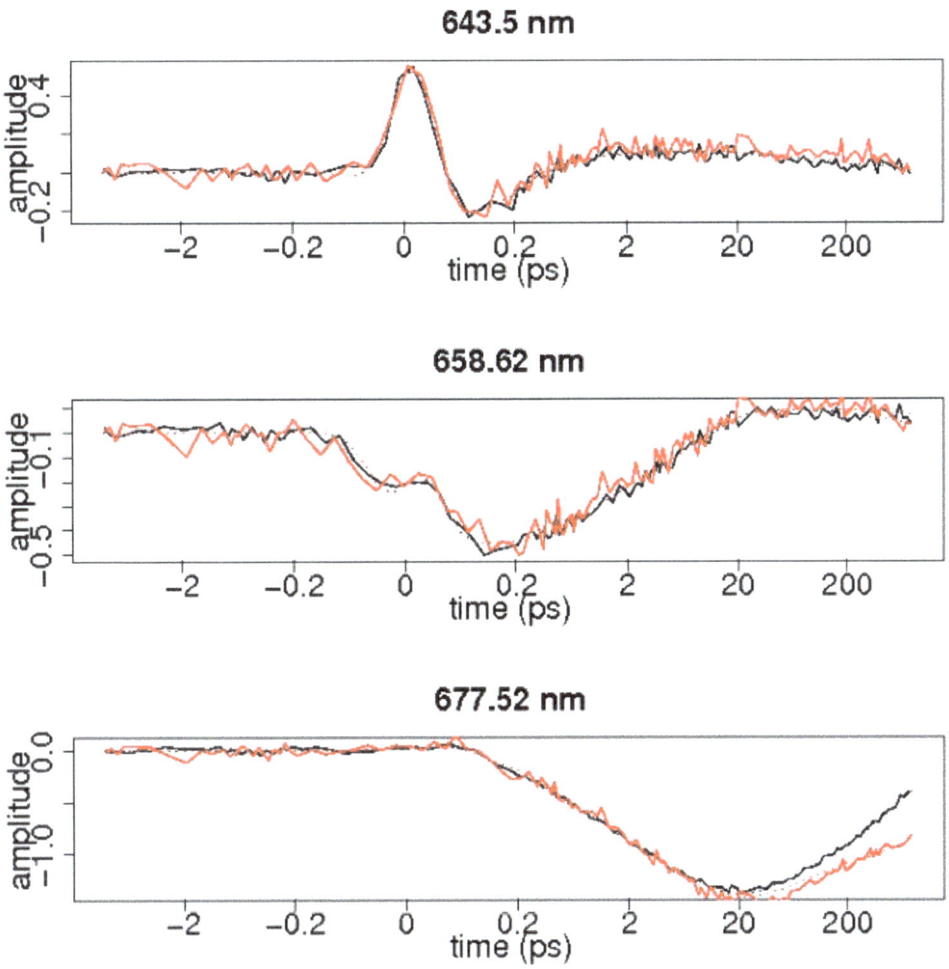

Figure 12: A plot of three selected traces resulting from the fit of a satisfactory model to the data $\Psi = \{\Psi_1, \Psi_2\}$, with the IRF location parameters and the two slowest decay rates fit per-dataset, and under application of an oscillatory coherent artifact model. Black represents data Ψ_1 (solid) and the fit of the initial model to Ψ_1 (dotted); Red represents data Ψ_2 (solid) and the fit of the initial model to Ψ_2 (dotted). The RMS error associated with this fit is .025.

Figure 13 shows just the estimates of the kinetic decay rates, with the color of each component the same as in Figure 16 and Figure 14. Figure 14 shows the contribution to the fit of the kinetic decay components and the coherent artifact (which is in pink) at three different wavelengths. The fit associated with the model after five iterations shown at three selected wavelengths in Figure 12 is deemed acceptable. The RMS error associated with this fit is .025. The spectra associated with the kinetic decay components (Figure 16) have physically plausible shapes. The discovery of an appropriate model for the data allows the differences in the slow rate constant estimates between datasets to be attributed to the effect of a difference in laser power. The parameter Θ_L is also interpretable as quantifying this difference.

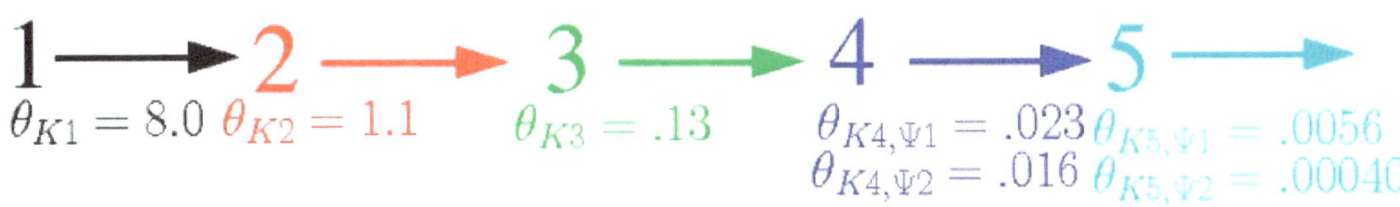

Figure 13: Transitions between the five components of the compartmental model are now fit with decay rates as labeled; the slowest two decays are fit independently for each dataset.

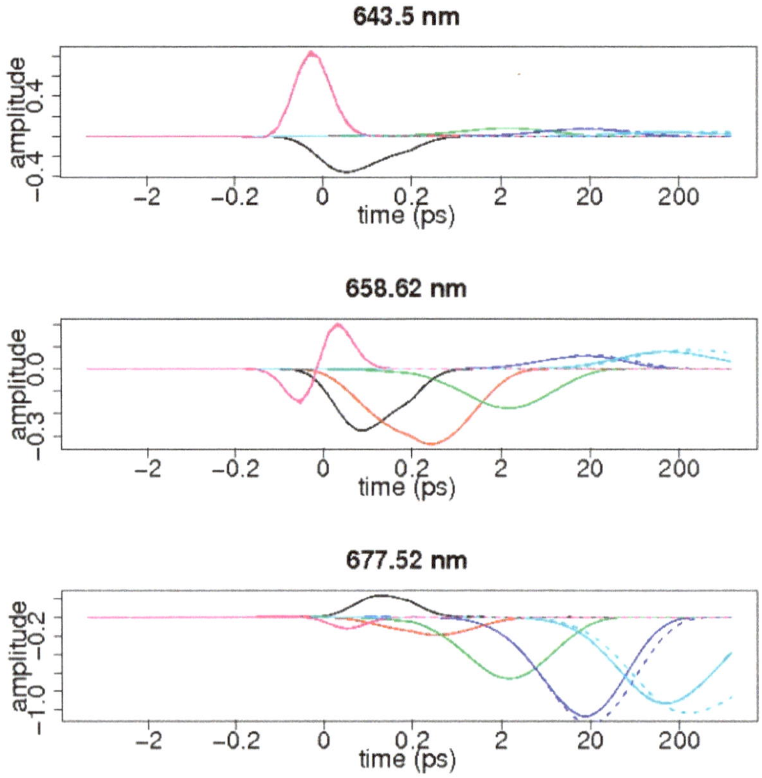

Figure 14: Contributions to fit per component show evolution. Pink represents the coherent artifact component; key to other colors is in Figure 13. Dashed lines indicate the fit of the second dataset Ψ_2, which has slower decay rate estimates.

SPECTRAL MODELS

Spectral models are those models in which the nonlinear parameters determine the matrix of spectra E. The spectral bandshapes are typically described in terms of a linear superposition of standard band shapes (e.g., Gaussian, Lorentzian, Voigt, skewed Gaussian) or in terms of splines. In the case that a superposition of standard band shapes is used, each column of E, ϵ_l, is modeled as

$$\epsilon_l = (\mathrm{amp}_1)(g_{1l}) + \ldots + (\mathrm{amp}_h)(g_{hl}) \tag{23}$$

where $\mathrm{amp}_1, \ldots, \mathrm{amp}_h$ are amplitude parameters, and g is the band shape function. The concentration profiles are then estimated as clp.

Spectral models are represented in TIMP with the class spec. Model parameterization options for spectral models are here outlined.

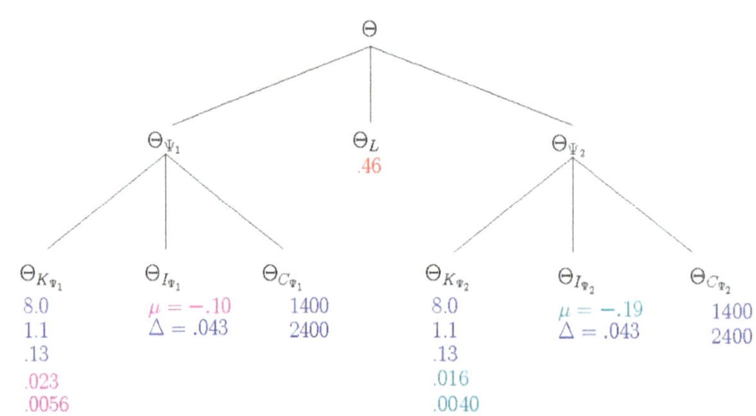

Figure 15: Parameter estimates associated with the fit of the satisfactory model mapped to a hierarchical model representation. Θ_L is a dataset scaling parameter descriptive of the difference in intensity of Ψ_2 as compared to Ψ_1. Parameter estimates linked between datasets Ψ_1 and Ψ_2 are in blue. Parameter estimates fit per-dataset are in magenta and green respectively, for Ψ_1 and Ψ_2.

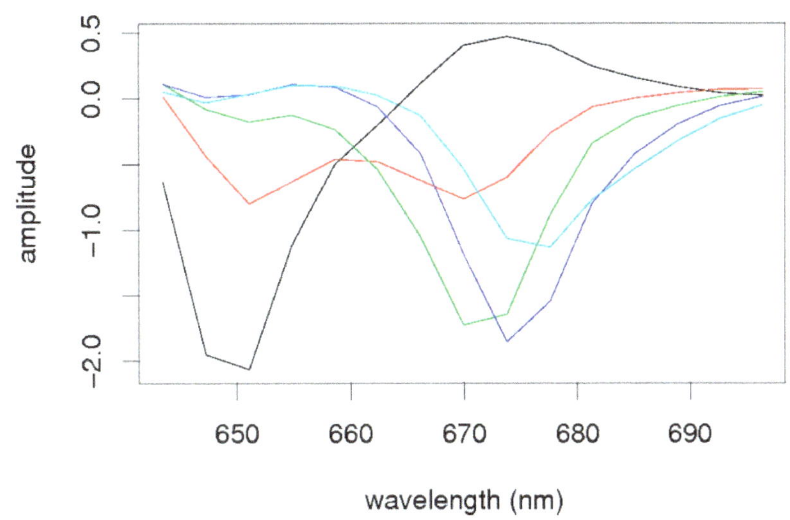

Figure 16: Estimated spectra associated with the five kinetic decays have physically plausible shapes. The mapping of spectra to kinetic decays is given in the color code of Figures 13 and 14.

Bandshape models

The most commonly applied bandshape model is a superposition of skewed Gaussians. This underlying model for E is chosen because it is a simple model capable of representing real spectra in practice and because the use of (skewed) Gaussians to represent spectral shapes is wide-spread, (see, e.g., van Stokkum et al. (2004) and van Stokkum (1997) and references therein).

The model for ϵ_l under a single skewed Gaussian distribution where $\bar{\nu} = 10^7/\lambda$ is

$$\epsilon_l(\bar{\nu}_{max}, \Delta\bar{\nu}, b) \equiv \bar{\nu}^{-n} \exp\left(-\ln(2)\left(\ln\left(1 + \frac{2b(\bar{\nu} - \bar{\nu}_{max})}{\Delta\bar{\nu}}\right)/b\right)^2\right), \tag{24}$$

except if $1 + (2b(\bar{\nu}_i - \bar{\nu}_{max}))/\Delta\bar{\nu} \leq 0$, in which case $\epsilon_l(\bar{\nu}_{max}, \Delta\bar{\nu}, b) \equiv 0$. When skewness b = 0, (and the skewed Gaussian distribution reduces to the Gaussian distribution), $\bar{\nu}_{max}$ corresponds to the maximum of the distribution, and $\Delta\bar{\nu}$ corresponds to the full width at half maximum (FWHM). When $b \neq 0$, $\bar{\nu}_{max}$ and $\Delta\bar{\nu}$ do not have this exact correspondence. The FWHM may then be determined as $\Delta\bar{\nu}\sinh(b)/b$. The average wavenumber of the skewed Gaussian is given by

$$\bar{\nu}_{avg} = \bar{\nu}_{max} + \frac{\Delta\bar{\nu}}{2b}\left(\exp\left(-\frac{3b}{4\ln(2)}\right) - 1\right) \tag{25}$$

A linear superposition of spectra as in Equation 24 is a common model for a single spectrum ϵ_l.

Specification in TIMP: Bandshape model

The list specpar contains vectors of starting values for spectral parameters; the number of vectors gives the number of components in the resulting spectral model. Each vector contains the parameters associated with a component. e.g., specpar = list(c(20000, 3000, .3, 21000, 2000, .4), c(18000, 1000, .2)); the parameters in each vector are grouped c(location_spectra, width_spectra, skew_spectra). The location and width parameters are given in wavenumbers. Note that this means that each component may be modeled with a superposition of many skewed Gaussians.

The character string specfun specifies the model used in the description of bandshapes. Inclusion of the amplitude parameters is not yet implemented, so that all skewed bandshapes contributing to a single component spectrum ϵ_l contribute equally.

Dependence of bandshape parameters on independent variable

It is often desirable to model bandshape parameters of spectra as a function of the independent variable in which spectra are resolved, (e.g., time or temperature). Parameterization of this variation is a means of studying protein conformational stability (for temperature-resolved data), and vibrational relaxation (for time-resolved data).

Dependence on the variable in which spectra are resolved may often be well-described by a polynomial model. Where θ is some parameter contributing to the determination of a bandshape (i.e., either a location, width, skewness or amplitude parameter), t is a value of the independent variable with which spectra are resolved, v is the number of coefficients parameterizing the polynomial (in addition to θ), tref is the center-point of the polynomial, and $\alpha_1, \ldots, \alpha_v$ parameterize the dependence, then

$$\theta(t) = \theta + \sum_{i=1}^{v} \alpha_i((t - t_{\text{ref}})/100)^i. \tag{26}$$

An exponential model of dependence on the variable in which spectra are resolved is also often of interest. Then where t_{ref} is a reference time, and other variables are as in the polynomial description,

$$\theta(t) = \theta + \alpha_1 \exp(-\alpha_2(t - t_{\text{ref}})). \tag{27}$$

A multiexponential model may also be of use in some cases. For the bi-exponential case

$$\theta(t) = \theta + ((\theta_0 - \theta)\frac{\exp(-k_1(t - t_{\text{ref}})) + \alpha_2 \exp(-k_2(t - t_{\text{ref}}))}{1 + \alpha_2}), \quad (28)$$

where $\theta(t_0) = \theta_0$ and $\theta(\infty) = \theta$.

For the general case of $i \geq 1$ multiexponential decays,

$$\theta(t) = \theta + (\theta_0 - \theta)\frac{\sum(\alpha_i \exp(-k_i(t - t_{\text{ref}})))}{\sum \alpha_i} \quad (29)$$

where $\alpha_1 \equiv 1$.

Specification in TIMP: Dependence of bandshape parameters on independent variable

The character string parmufunc determines the function modeling dependence of the bandshape parameters on the independent variable. Options are "poly" for the polynomial case (Equation 26) "exp" for the single exponential description (Equation 27) and "multiexp" for the multiexponential description (Equation 29).

The numeric specref gives the index of the center variable, t_{ref} in Equations 26, 27 and 29). The list specdispindex defines those indices of specpar whose dependence on the variable in which spectra are resolved is to be modeled. For example, specdispindex = list(c(1,1), c(1,2), c(1,3)) indicates that parameters 1-3 of spectra 1 are to be modeled as variable.

The list specdisppar contains vectors of the parameters describing time-dependence. One vector of parameters is given for each vector of indices in specdispindex. These parameters describe a polynomial time-dependence by default. There are three ways to interpret these vectors, depending on the value of parmufunc. If parmufunc is "poly" for the polynomial case, the vectors are of the length of the desired degree of the polynomial parameterization; e.g., use specdisppar=list(c(-2000, 1, .1), c(1, .1, .01), c(.2, .1)) for a 3rd order dependence of two spectral parameters, and a 2nd order dependence on one spectral parameter. If parmufunc is "exp" the first parameter is a linear coefficient and the second is a rate; if all but the first vector have the rate omitted then rates will be linked across all parameters that are time-dependent; otherwise rates will be fit per-parameter that is dependent on the variable with which spectra are resolved. If parmufunc is "multiexp" an arbitrary number of vectors of coefficients and rates in the form c(α_1, k_1, α_2, k_2, . . .) may be specified as elements of the specdisppar list.

Case study: Time-dependence of spectral parameters

A spectral model is fit to the time-resolved dataset shown in Figure 17. A goal is the parametric description of the relaxation of the bandshape in time. As in the kinetic modeling case study, (Section 5.6) this section describes the use of TIMP for interactive model discovery. The system underlying the data and motivation for the initial model as well as physicochemical interpretation of model fit will be described elsewhere (manuscript in preparation)

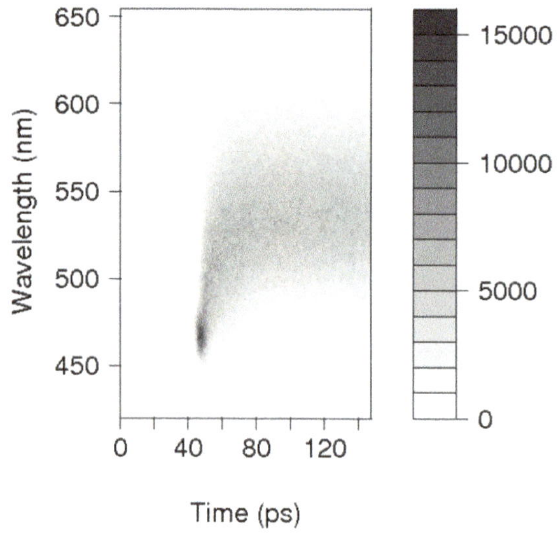

Figure 17: A dataset of time-resolved spectra. We are interested in application of a spectral model to the selection of this data shown in Figure 18.

Data input

The data $\Psi = \{\Psi_1\}$ is read into TIMP in the time explicit format described in Appendix C via the command

R> psi_1 <- readData("psitspec.txt")

Read 1 item

Read 181063 items

where Ψ_1 is stored in the file "psitspec.txt", (distributed with this paper)

Data preprocessing

The dataset was truncated to the wavelength range 440-640 nm, and to times after 53 ps (from time indices 178 to 478), since we are only interested in modeling processes occurring in this range.

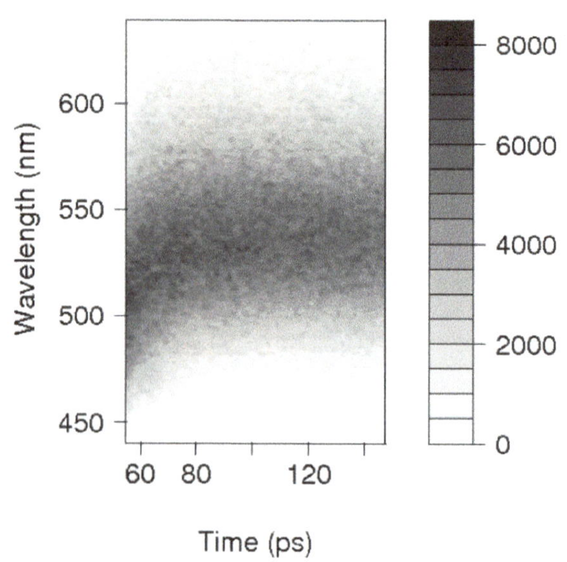

Figure 18: The dataset in Figure 17 selected to include only the range of timepoints and wavelengths of interest for modeling, and sampled to include only every fifth timepoint to expedite parameter estimation.

```
R> psi_1_full <-preProcess(psi_1, sel_time=c(178, 478),
+ sel_lambda_ab = c(440, 640))
```

To expedite parameter estimation only every fifth timepoint is sampled from the selected dataset. When a model likely to be satisfactory is identified on the sampled dataset, it may be applied to the unsampled selected dataset.

```
R> psi_1_sampled <- preProcess(psi_1, sel_time=c(178, 478),
+ sel_lambda_ab = c(440, 640), sample_time=5)
```

Figure 18 shows the selected and sampled dataset resulting from the above calls to the preProcess function.

Initial model: Polynomial dependence of spectral variables on time

A spectra with bandshape model comprised of a single skewed Gaussian is initialized, first with linear polynomial dependence of the spectral variables on time.

```
R> model_polylin <- initModel(mod_type = "spec",
+ specpar=list(c(20000,3100,-.3) ),
+ specdispindex = list(c(1,1), c(1,2), c(1,3)),
+ specdisppar=list(c(-2000),c(1), c(.2)),
+ specref=53, specfun="gaus",
+ iter=1, nupow=5, make_ps="badan_lin",
+ title = "Linear" )
R> res_polylin <- fitModel(data = list(psi_1_sampled),
+ model = model_polylin,
+ opt=list(iter=7, linrange = 20,
+ makeps = "polylin", nospectra = TRUE,
+ selectedspectra = seq(1, psi_1_sampled@nt, by=7),
+ residplot = TRUE,
+ xlabel = "time", ylabel = "wavelength",
+ ))
```

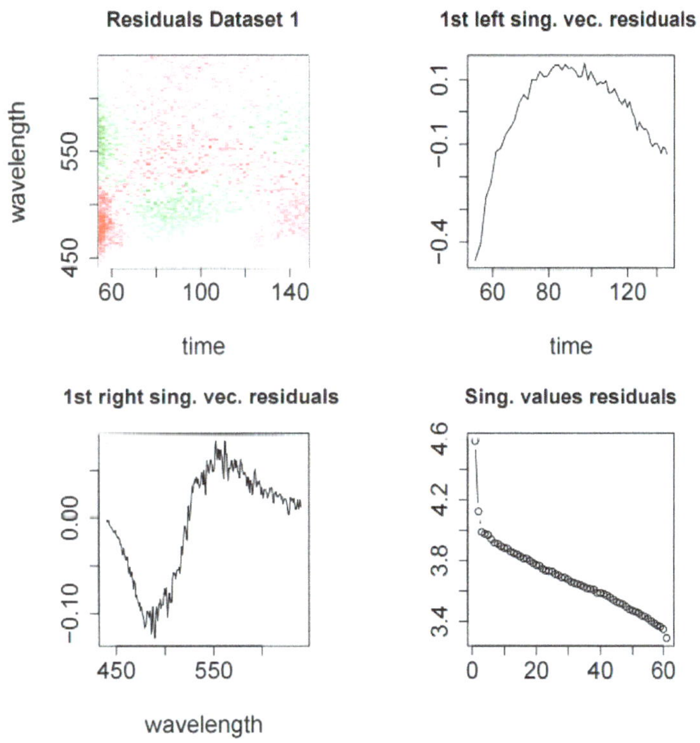

Figure 19: Residuals associated with the fit of the initial spectral model in which dependence of the spectral bandshape on parameters is described with a first-order polynomial. The first left and right singular vector and the singular values associated with their SVD are also plotted. Note the structure in the first left singular vector of the residuals, indicating inadequacy in the model fit. The RMS error associated with this fit is 414.

Structure in the residuals as evidenced by their SVD in Figure 19 indicates an inadequate model. A different parameterization of the dependence of spectral parameters on time will be applied as a model refinement.

Refined model: Exponential dependence of spectral variable on time

The initial model is refined to describe an exponential dependence of spectral variables on time, in an attempt to address the inadequacy in the fit of the initial model. The applied model is such that α_2 from Equation 27 is equated for all spectral bandshape parameters.

```
R> model_exp_linkedrates <- initModel(mod_type = "spec",
+ specpar=list(c(18000, 3200, -.1)),
+ specdispindex = list(c(1,1), c(1,2), c(1,3)),
+ specdisppar=list(c(600,1/20), c(400), c(.1)),
+ specref=53, specfun="gaus", parmufunc = "exp",
+ nupow=5,
+ title = "Exponential parameterization of time dep., linked rates")
```

```
R> res_model_exp_linkedrates <- fitModel(data = list(psi_1_sampled),
+ model = model_exp_linkedrates,
+ opt=list(iter=5, linrange = 20, residplot=TRUE,
+ makeps = "explinked", nospectra = TRUE,
+ selectedspectra = seq(1, psi_1_sampled@nt, by=7),
+ xlabel = "time", ylabel = "wavelength"))
```

The fit of the refined spectral model that employs an exponential description of time-dependence of spectral bandshape parameters is well-fit to the data, as evidenced by plots showing the fit of the model to the data in Figure 20. Further evidence for the satisfactory model fit is contained in plots regarding the residuals in Figure 21. An SVD decomposition of the residuals shows little structure in the first left singular vector. Some structure remains evident in the first right singular vector, indicating that there remains room for improvement in the model. Despite this, we conclude that the model fit is satisfactorily descriptive of the data. The exponential rate estimate $\alpha_1 = .07$ is therefore useful as a descriptor of the relaxation of the spectral bandshape parameters in time.

EXTENSION OF SUPPORTED MODEL TYPES

TIMP has been designed to allow for rapid implementation of new options and new model types. New options can be added by modification of the class definitions associated with existing model types and the class for parameter estimates theta, and modification of the associated residual functions to work with slots describing new options. For instance, imagine that for a model type modelx, a parameterization of an additional sort of submodel is desired.

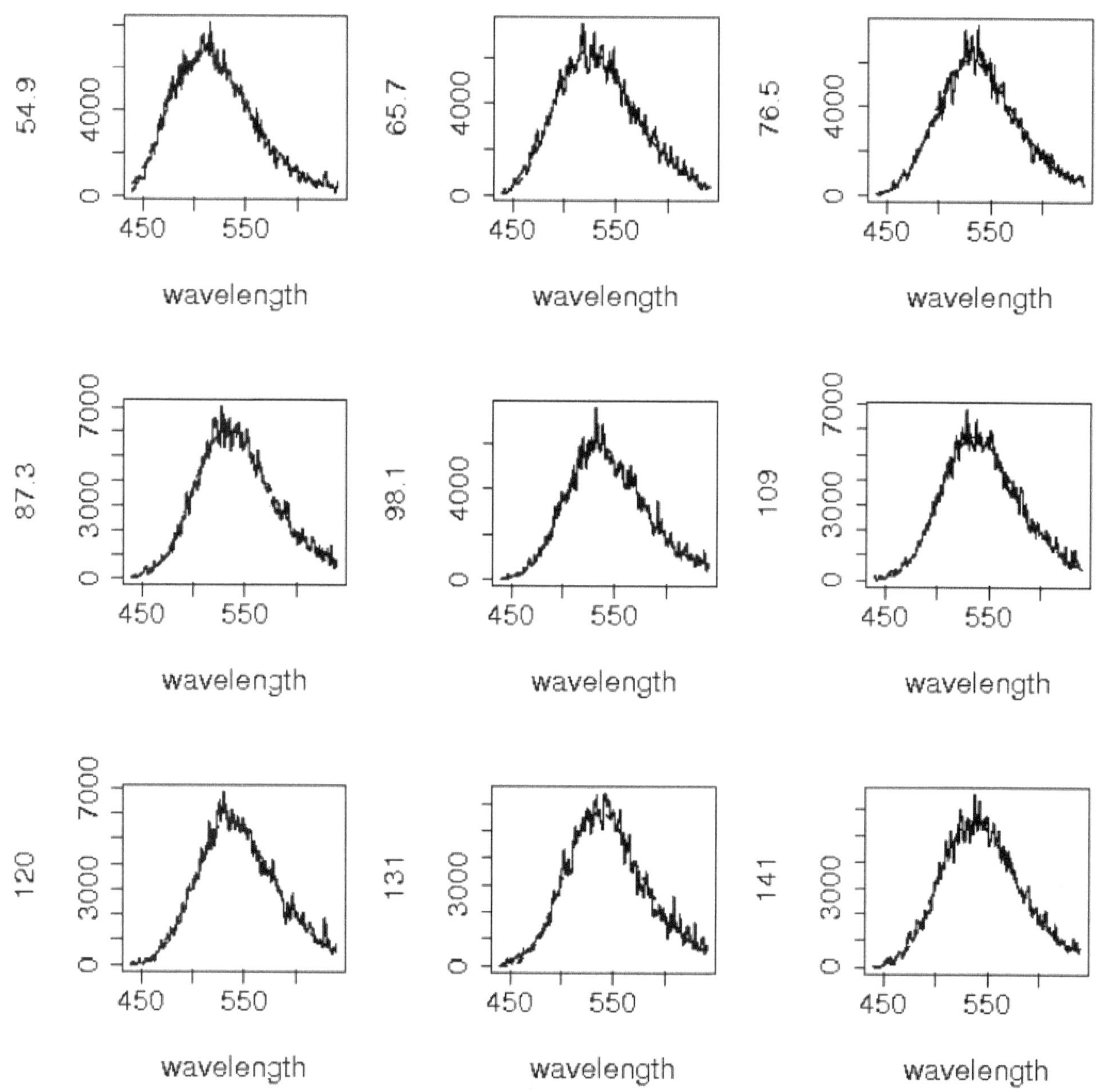

Figure 20: (Left) A plot of selected spectra resulting from the fit (dashed line) of a spectral model in which time-dependence of spectral bandshape parameters is modeled with an exponential function to data $\Psi = \{\Psi 1\}$ (solid line). Each sub-plot represents the data at the timepoint on the left axis. The RMS error associated with this fit is 312.

Then two slots are added to the class definition for model type modelx: a slot to represent options to parameterize the new submodel new, and a slot parnew to represent the starting values of nonlinear parameters associated with the new submodel. A slot named parnew is also added to the class theta. Then new and parnew are given as inputs to the inputModel function. To use the new parameterization to change how the residuals are determined, the S4 methods associated with determining the residuals for class modelx are modified. The S4 method plotter determining plotting/output for modelx is also modified to use the new options as desired. No additional modification of the code is necessary.

The process of adding an entirely new model type newmodel can be described in terms of four steps:

- definition of a new class for the model type newmodel that inherits from dat
- for every slot par representing a list/vector of nonlinear parameter value starting values in the definition of newmodel adding a slot for a list/vector of the same name par to the class theta so that the vector of parameters Θ can be inferred, and so that updated parameter estimates can be plugged back into the model specification each iteration
- definition of methods for residPart (and/or getClpIndepX depending on whether clp are involved) that supply a prescription for the calculation of residuals for a single dataset Ψq given a model
- definition of desired output plots and other information via a method for plotter

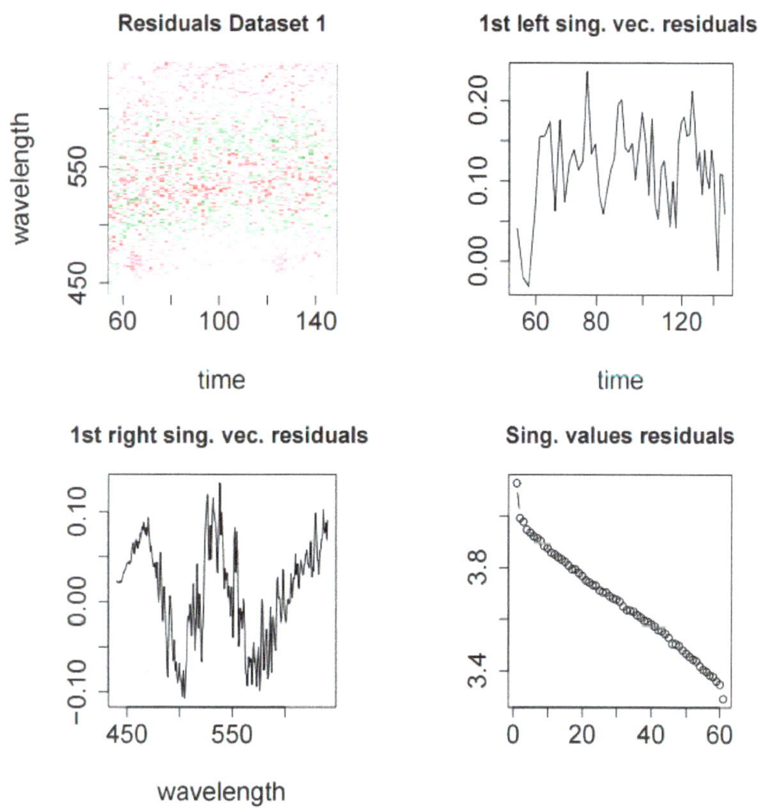

Figure 21: Residuals associated with the fit of the spectral model, and the first left and right singular vector and the singular values associated with their SVD. Note the lack of structure in the first left singular vector of the residuals, a sign that the model fit is satisfactory. There remains some structure in the first right singular vector, but we accept the model as satisfactory nonetheless.

The remaining code of TIMP should not require any modification. Planned extensions to additional model types will allow the application of the package to fitting models for photocycles and polarization-dependent effects.

CONCLUSIONS

TIMP, an R package for interactive scientific model discovery for multiway spectroscopy data has been introduced. The design of the package has been outlined. The partitioned variable projection algorithm that is central to solving separable nonlinear least squares parameter estimation problems with TIMP was presented. General options for models in TIMP were introduced, along with options specific to kinetic and spectral model types. A case study in application of a kinetic model to two datasets simultaneously illustrated many kinetic model options. A case study in application of a spectral model illustrated many spectral model options. TIMP is in active development. Future work includes the development of new model types for data collected in multipulse laser experiments, anisotropy experiments, and experiments designed to extract information on photocycles, as well as the implementation of additional options to support model specification and validation. The development of a GUI to support interactivity is also planned.

ACKNOWLEDGMENTS

This research was funded by Computational Science grant #635.000.014 from the Netherlands Organization for Scientific Research (NWO). Mikas Vengris, Denitsa Grancharova and Rienk van Grondelle provided the data modeled in Section 5.6. Rob Koehorst, Bart van Oort, Sergey Laptenok, Ton Visser and Herbert van Amerongen provided the data modeled in Section 6.3. Joris Snellenburg is thanked for constructive comments on the text. Uwe Ligges and Martin M¨achler collaborated in the implementation of the nls options described in Section B. Achim Zeileis contributed helpful suggestions regarding the figures.

A. PARTITIONED VS. UNPARTITIONED VARIABLE PROJECTION ALGORITHMS

This appendix considers in detail the memory requirements of the partitioned variable projection algorithm introduced in Section 2.4, and compares these requirements to those of the standard (unpartitioned) variable projection algorithm. The large memory requirements of the standard variable projection algorithm complicate its application, and sometimes prohibit its use entirely (as e.g., in the experience of Verveer, Squire, and Bastiaens (2000)). The ability to apply the variable projection functional without operating on large matrices is the main motivation for introduction of the partitioned variant of the algorithm. PartitionedVarPro and the standard implementation of variable projection, as found for instance in the plinear function of nls, return the same results. The algorithms differ only in the memory resources required.

Under PartitionedVarPro with free clp (as defined in Section 2.4) for part p, the residual vector representing dataset Ψ_q is formed by taking QR(X_{pq}). The matrix $Q_2^\top$ associated with this decomposition is of dimension ($m - n_{ncomp}$) × m. Ψ_{pq} is of length m. Hence the residual for part p, dataset q is of length ($m - n_{ncomp}$). Concatenating the results for all x datasets at all n parts p yields a total residual of length nx($m - n_{ncomp}$).

Under unpartitioned variable projection (VarPro), Ψ is not treated by part p. Hence the entire matrix Ψ is vectorized. Recall that where A, B and C are matrices of compatible dimensions, vec(ABC) = (C > ⊗ A)vec(B), where vec is the vector representation, and ⊗ is the Kronecker product. Let Ψ_{super} be the m×n matrix comprised of concatenating datasets $\Psi_1, \ldots, \Psi_x$. Then

$$vec(\Psi_{\text{super}}) = vec(X_{\text{super}}\beta_{\text{super}}^\top I_n) = (I_n \otimes X_{\text{super}})vec(\beta_q^\top), \tag{30}$$

where

$$I_n \otimes X_{\text{super}} = \begin{bmatrix} X_{1\text{super}} & & & & \\ & \ddots & & & \\ & & X_{p\text{super}} & & \\ & & & \ddots & \\ & & & & X_{n\text{super}} \end{bmatrix} \tag{31}$$

$I_n \otimes X_{super}$ has zero entries except as noted, and has dimension $n_{mx} \times n_{nncomp}$.

PartitionedVarPro with linked clp performs n QR decompositions on matrices of dimension mx × nncomp, whereas PartitionedVarPro with free clp performs nx QR decompositions on matrices of dimension m × n_{ncomp}. VarPro performs one QR decomposition on a matrix of dimension nmx × n_{nncomp}. For m = n = 1000, n_{ncomp} = x = 10, the QR decomposition is performed on matrices of dimensions 10000 × 10 and 1000 × 10 in the case of linked and free clp, respectively, whereas VarPro performs the QR decomposition on a matrix of dimension $10^7 \times 10^4$. Avoiding storage and manipulation of a large matrix is a primary motivation to prefer PartitionedVarPro over VarPro. In order to make VarPro usable for realistic problems, storage of only non-zero elements of I_{nl} ⊗QR(X_{super}) and operation on the result with a modified QR-routine could be attempted. However, doing so would provide no advantage over PartitionedVarPro, (and furthermore, requires implementation of a QR routine to operate on the special form of the matrix).

A.1. Empirical comparison of partitioned and unpartitioned variable projection algorithms

The nls function of R allows application of the variable projection algorithm via the plinear option. We show here how the simple kinetic modeling problem considered in Section 1.3 may be fit using nls and the plinear option. We then compare the memory requirements under plinear to those under the partitioned variable projection algorithm implemented in TIMP.

We assume that a dataset Psi_q is simulated in R as described in Section 1.3. Commands to simulate the dataset and set up the workspace are contained in full in the file "memory_prof_plin.R" (distributed with this paper). In order to use nls under the plinear option, the model for concentrations is placed into a function calcC, which is also found in TIMP, as follows.

```
"calcC" <- function (k, t)
{
conc <- matrix(nrow = length(t), ncol = length(k))
for (i in 1:length(k)) {
conc[, i] <- exp(-k[i] * t)
}
conc
}
```

The sum-of-exponentials model can then be fit to the data using calcC, with the plinear option using the standard variable projection functional to determine the residuals, and with the spectra as conditionally linear parameters.

R> psi_q_vector <- as.vector(Psi_q)

R> onls <- nls(psi_q_vector ~ kronecker(diag(length(wavenum)),

+ calcC(k, t)), data.frame(psi_q_vector),

+ start = list(k = c(.1,2)), alg = "plinear", trace = T)

To profile the memory allocated in the course of solving this problem, we apply the gc function, (note that more refined memory profiling is possible with the Rprof and Rprofmem functions under builds of R compiled to enable memory profiling). Before and after the call to nls, the results of a call gc(verbose=TRUE) on on our system are respectively. This shows that under the plinear implementation of variable projection about 12 Mbytes of vector space is allocated in the course of solving this example problem.

Garbage collection 11 = 9+0+2 (level 2) ...

6.4 Mbytes of cons cells used (58%)

0.9 Mbytes of vectors used (14%)

and

Garbage collection 100 = 45+26+29 (level 2) ...

6.6 Mbytes of cons cells used (53%)

13.2 Mbytes of vectors used (32%)

To contrast this with the memory allocated under the partitioned variable projection implementation found in TIMP on the same system, we load the package and initialize a model object as described in in Section 1.3 and in the file "memory_prof_pvarpro.R". A call to gc(verbose=TRUE) before calling the fitModel function that applies the partitioned variable projection algorithm to fitting the sum-of-exponentials model with the spectra as clp has the following result

Garbage collection 35 = 31+2+2 (level 2) ...

7.5 Mbytes of cons cells used (68%)

1.0 Mbytes of vectors used (16%)

Then fitting the model as in Section 1.3 with

R> kinetic_fit <- fitModel(data = list(Psi_q_data), model = kinetic_model,

+ opt = list(iter=4, plot=FALSE))

and subsequently calling gc(verbose=TRUE) results in

Garbage collection 52 = 46+3+3 (level 2) ...

7.8 Mbytes of cons cells used (62%)

1.5 Mbytes of vectors used (25%)

This shows that in the course of applying the partitioned variable projection implementation to the problem, about .5 Mbytes of vector space is allocated, about 20 times less than under plinear on the same problem. The savings in memory allocated via the use of the partitioned variable projection algorithm found in TIMP is very significant for problems of interest in the multiway spectroscopy modeling domain. As larger amounts of data are involved the memory requirements of the standard non-partitioned implementation found in the plinear option grow so large as to prohibit its use on a modern personal computer, while the memory requirements of the partitioned version of the algorithm found in TIMP remain modest

B. NEW NLS OPTIONS

It is often desirable in scientific modeling applications to terminate the iterative optimization of free model parameters when stopping criteria are met, as opposed to when convergence criteria are met. For instance, it is often desirable to evaluate the fit of a model at a given set of starting estimates, or after fitting for a modest number of iterations. Then the stopping criteria is completion of a maximum number of iterations, after which output is desired, even though the fit may be far from satisfying convergence criteria.

It is often also desirable to examine output in the case that the fitting algorithm encountered a problem and terminated fitting with an error. For instance, if the gradient of the residual vector with respect to nonlinear parameter estimates becomes singular, examination of the current parameter estimates may shed light on how the model can be modified to be better determined with respect to the data.

The R function nls is widely applied in scientific model discovery and is used in TIMP to iteratively improve nonlinear parameter estimates. Prior to R version 2.5 nls did not return output in the case that any of the following conditions are met

- the maximum number of iterations x is met (as specified with nls(. . ., control = list(maxiter = x, . . .)))
- the step-size is below the minimum x (as specified with nls(. . ., control = list(minFac = x, . . .)))
- a singular gradient occurs

In all these cases return of the output object may be valuable for scientific model discovery, for the reasons sketched above. We have implemented the option 'warnOnly' to determine if an output object is returned in the case that one of the above conditions is met. The implementation of this option has been incorporated into R version 2.5. A logical slot in the class nls.control is used to toggle the 'warnOnly' option. To output a result object even in the case that an error is triggered, nls is called with nls(. . ., control = list(warnOnly = TRUE, . . .).

A better understanding of the residual surface on which optimization occurs is sometimes gained by knowledge of how many times nls halves the step-size in the gradient direction. We have implemented the option 'printEval' to print the number of evaluations (of the step-size) required each iteration; this option is also included in R version 2.5. A logical slot in the class nls.control is used to toggle the 'printEval' option. To print the number of evaluations required (as well as the achieved convergence tolerance), nls may be called with nls(. . ., control = list(warnOnly = TRUE, . . .).

C. DATA FORMATS FOR INPUT INTO TIMP

Currently TIMP allows the input of data in three formats. More formats will be added in the future.

C.1. TIME EXPLICIT FORMAT

The time explicit format for data input contains 5 lines and then a matrix of data in which each row represents a concentration profile, and each column represents spectra.

Heading line 1

Heading line 2

Time explicit

Intervalnr 5

	t_1	t_2	...	t_{m-1}	t_m
λ_1	$\Psi(t_1, \lambda_1)$	$\Psi(t_2, \lambda_1)$	...	$\Psi(t_{m-1}, \lambda_1)$	$\Psi(t_m, \lambda_1)$
λ_2	$\Psi(t_1, \lambda_2)$	$\Psi(t_2, \lambda_2)$	...	$\Psi(t_{m-1}, \lambda_2)$	$\Psi(t_m, \lambda_2)$
...	...	...	...	...	...
λ_{n-1}	$\Psi(t_1, \lambda_{n-1})$	$\Psi(t_2, \lambda_{n-1})$	...	$\Psi(t_{m-1}, \lambda_{n-1})$	$\Psi(t_m, \lambda_{n-1})$
λ_n	$\Psi(t_1, \lambda_n)$	$\Psi(t_2, \lambda_n)$	...	$\Psi(t_{m-1}, \lambda_n)$	$\Psi(t_m, \lambda_n)$

Data matrix elements are space-delimited. 'Heading line 1' and 'Heading line 2' are two lines that may be filled as desired (e.g., with a data file title). The string 'Time explicit' indicates the data format. The string 'Intervalnr' and a scalar m indicates the number of distinct points m at which spectra were measured, (note that the number of wavelengths n need not be specified). The following line contains the real-valued variable values (such as times) $t_1, \ldots, t_m$ at which measurements were taken. The first value of each of the remaining lines represents the wavelength at which the concentration profile contained on that row was taken. The rest of each remaining row represents a (space-delimited) concentration profile $\Psi(t_1, \lambda), \Psi(t_2, \lambda), \ldots, \Psi(t_m, \lambda)$.

C.2. WAVELENGTH EXPLICIT FORMAT

The wavelength explicit format for data input contains 5 lines and then a matrix of data in which each row represents spectra, and each column represents a concentration profile.

Heading line 1

Heading line 2

Wavelength explicit

Intervalnr 5

	λ_1	λ_2	⋮	λ_{n-1}	λ_n
t_1	$\Psi(t_1, \lambda_1)$	$\Psi(t_1, \lambda_2)$	⋮	$\Psi(t_1, \lambda_{n-1})$	$\Psi(t_1, \lambda_n)$
t_2	$\Psi(t_2, \lambda_1)$	$\Psi(t_2, \lambda_2)$	⋮	$\Psi(t_2, \lambda_{n-1})$	$\Psi(t_2, \lambda_n)$
⋮	⋮	⋮	⋮	⋮	⋮
t_{m-1}	$\Psi(t_{m-1}, \lambda_1)$	$\Psi(t_{m-1}, \lambda_2)$	⋮	$\Psi(t_{m-1}, \lambda_{n-1})$	$\Psi(t_{m-1}, \lambda_n)$
t_m	$\Psi(t_m, \lambda_1)$	$\Psi(t_m, \lambda_2)$	⋮	$\Psi(t_m, \lambda_{n-1})$	$\Psi(t_m, \lambda_n)$

All entries above are space delimited.'Heading line 1' and 'Heading line 2' are two lines that may be filled as desired (e.g., with a data file title). The string 'Wavelength explicit' indicates the format that the input data is to take. The string 'Intervalnr' and a scalar n indicates the number of distinct wavelengths n at which measurements were taken, (note that the number of time points m need not be specified). The following line contains the real-valued wavelengths $\lambda_1, \ldots, \lambda_n$ at which measurements were taken. The first value of each of the remaining lines represent the independent variable value (such as a time) at which the spectrum contained on that row was taken. The rest of each remaining row represents a (space-delimited) spectrum $\Psi(t, \lambda_1), \Psi(t, \lambda_2), \ldots, \Psi(t, \lambda_n)$.

C.3. FLIM format

Fluorescence Lifetime Imaging Microscopy (FLIM) data is read into TIMP in the format described in Laptenok et al. (2007).

REFERENCES

1. Bates DM, DebRoy S (2003). "Converting a Large R Package to S4 Classes and Methods." In K Hornik, F Leisch, A Zeileis (eds.), "Proceedings of the 3rd International Workshop on Distributed Statistical Computing, Vienna, Austria," ISSN 1609-395X, URL http://www. ci.tuwien.ac.at/Conferences/DSC-2003/ Proceedings/.
2. Chambers JM (1998). Programming with Data, A Guide to the S Language. Springer-Verlag, New York.
3. Godfrey K (1983). Compartmental Models and Their Application. Academic Press, London.
4. Golub G, Pereyra V (2003). "Separable Nonlinear Least Squares: the Variable Projection Method and its Applications." Inverse Problems, 19, R1–R26
5. Golub GH, LeVeque RJ (1979). "Extensions and Uses of the Variable Projection Algorithm for Solving Nonlinear Least Squares Problems." In "Proceedings of the 1979 Army Numerical Analysis and Computers Conference," volume ARO Report 79-3, pp. 1–12.

6. Golub GH, Pereyra V (1972). "The Differentiation of Pseudo-inverses and Nonlinear Least Squares Problems whose Variables Separate." Technical report, Stanford University, Department of Computer Science.

7. Golub GH, Pereyra V (1973). "The Differentiation of Pseudoinverses and Nonlinear Least Squares Problems whose Variables Separate." SIAM Journal on Numerical Analysis, 10, 413–432.

8. Istratov AA, Vyvenko OF (1999). "Exponential Analysis in Physical Phenomena." Review of Scientific Instruments, 70(2), 1233–1257.

9. Laptenok S, Mullen KM, Borst JW, van Stokkum IHM, Apanasovich VV, Visser AJWG (2007). "Fluorescence Lifetime Imaging Microscopy (FLIM) data analysis with TIMP." Journal of Statistical Software, 18(8). URL http://www.jstatsoft.org/v18/i08/.

10. Meyer D, Zeileis A, Hornik K (2006). vcd: Visualizing Categorical Data. R package version 1.0-2.

11. Moshier SL (1992). Cephes Mathematical Library. URL http://www.moshier.net/.

12. R Development Core Team (2006). R: A Language and Environment for Statistical Computing. R Foundation for Statistical Computing, Vienna, Austria. ISBN 3-900051-07-0, URL http: //www.R-project.org/.

13. van Stokkum IHM (1997). "Parameter Precision in Global Analysis of Time-Resolved Spectra." IEEE Transactions on Instrumentation and Measurement, 46(4), 764–768.

14. van Stokkum IHM (2005). "Global and Target Analysis of Time-resolved Spectra, Lecture notes for the Troisi`eme Cycle de la Physique en Suisse Romande." Technical report, Department of Physics and Astronomy, Faculty of Sciences, Vrije Universiteit, Amsterdam, The Netherlands. URL http://www.nat.vu.nl/~ivo/lecturenotes.pdf.

15. van Stokkum IHM, Larsen DS, van Grondelle R (2004). "Global and Target Analysis of Time-resolved Spectra." Biochimica et Biophysica Acta, 1657, 82–104, and erratum, 1658, 262.

16. van Stokkum IHM, Lozier RH (2002). "Target Analysis of the Bacteriorhodopsin Photocycle Using a Spectrotemporal Model." Journal of Physical Chemistry B, 106(13), 3477–3485

17. Verveer PJ, Squire A, Bastiaens PIH (2000). "Global Analysis of Fluorescence Lifetime Imaging Microscopy Data." Biophysical Journal, 78(4), 2127–2137.

CHAPTER 6

A MATHEMATICAL MODEL FOR THE INDUSTRIAL HAZARDOUS WASTE LOCATION-ROUTING PROBLEM

Omid Boyer, Tang Sai Hong, Ali Pedram, Rosnah Bt Mohd Yusuff, andNorzima Zulkifli

INTRODUCTION

The harmful wastes are produced from industries, manufacturers, hospitals, and other sectors which have harmful health effect on human and environment. The main sources of producing hazardous wastes are industrial and manufacturing processes; other generators can be household or commercial groups. There is wide and various range of industries to produce hazardous waste, for instance, chemicals manufacturing, oil refining, waste acids, and electroplating companies. Hazardous material is divided into four subgroups based on their features: toxicity (e.g., barium, cadmium, and chloroform); reactivity (e.g., sulfides and peroxide formers); corrosiveness (acids and bases); and ignitability (e.g., flammable liquid, oxidizers). These materials need specific transporting and treating to reduce their effect on environment and human health.

Definition of hazardous waste management is collecting, transporting, treating, recycling, and disposing waste residue under safe, efficient, and cost effective manner [1]. Two important objectives are included in transportation of hazardous waste: cost and risk. For carrier firms the best route is the shortest with minimum cost, but for government the basic problem is risk minimization. Hazardous waste also requires different technologies for treatment such as incineration, chemical, biological, and immobilization. Compatible treatment technology should be chosen based on waste features. In addition, recycling and disposing are two important stages in hazardous waste management.

The proposed framework for management of industrial hazardous waste is shown in Figure 1; it starts with generation nodes, and then wastes are divided into three groups. First group is recyclable waste; it is transferred to the recycling centers. Second group requires treatment facility, so they are routed to the treatment centers with incineration technology. Third group is nonrecyclable and nontreatable waste, therefore it is transported to the disposal centers directly. In the end, residues from treatment and recycling centers are carried to the disposal centers.

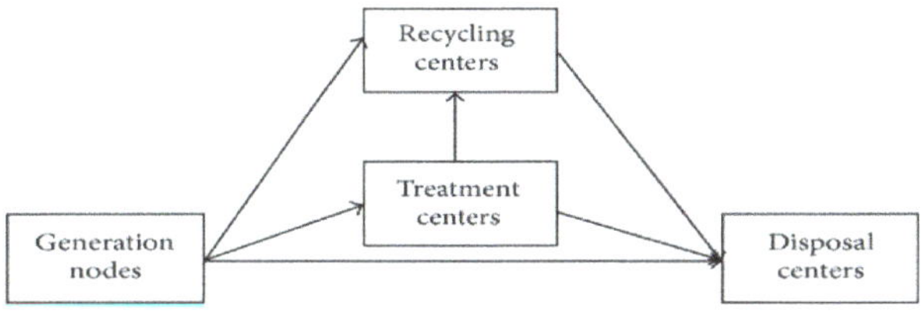

Figure 1: Proposed framework for hazardous waste management.

Citation: Omid Boyer, Tang Sai Hong, Ali Pedram, Rosnah Bt Mohd Yusuff, and Norzima Zulkifli, "A Mathematical Model for the Industrial Hazardous Waste Location-Routing Problem," Journal of Applied Mathematics, vol. 2013, Article ID 435272, 10 pages, 2013. doi:10.1155/2013/435272

At present, there have been little studies on the location of undesirable facilities and routing hazardous waste between these facilities simultaneously. So far, however, no research has been found that presented a comprehensive mathematical model based on proposed frame work (Figure 1). Recycling waste is an important stage in waste management [2], but this stage is often ignored in the literature. Therefore, in this model also locations for recycling centers are considered. In this study, a two-stage work is performed. First, appropriate site for undesirable facilities is selected by GIS. Second, the new mathematical model aids in locating facilities and routing waste. In addition, using total cost and transportation risk as objectives for this model can help decision makers to have a good tradeoff between environmental and economic aspects. In this way, using operational cost parameter for each facility and also cost saving parameter for recycled hazardous waste are suggested to have a more comprehensive cost objective. Applying the proposed model in a real case study showed that this model is practical in real world.

The remainder of this paper is arranged as follows. First a review from related paper is presented in Section 2. Mathematical model for proposed framework is formulated in Section 3. Section 4 gives a solution for multiobjective model. In Section 5, appropriate sites selection via GIS is illustrated and then the model is applied in Markazi province in Iran. A conclusion will be discussed in Section 6.

LITERATURE REVIEW

The initial works in the field of hazardous waste management only focused on locating facilities such as disposal and treatment centers. In this area a simple classification based on related papers considering hazardous waste management aspects will be illustrated as follows :(1)locating undesirable facility,(2)routing hazardous material via risk assessment,(3) locating and routing hazardous material.

Several studies have addressed the location of undesirable facilities such as Ratick and White [3]. They provide a model with three objectives to locating undesirable facilities; the objectives are maximization of facility location cost, maximization of opposition, and equity maximization. Erkut and Neuman [4] also developed a multiobjective model which is the same as that of Ratick and White. They used maximizing of disutility objective instead of minimizing equity. In their model transportation cost is also considered. Wyman and Kuby [5] introduced a frame work and model for locating waste incineration with three conflicting objectives: minimizing cost, risk, and inequity. Rakas et al. [6] developed a multiobjective model to locate undesirable facilities with two objectives. First is to minimize cost including transportation and initial investment cost. Second objective is to minimize population opposition. They also believe that some input data in the model are not deterministic. Therefore, an extension of model has been proposed, considering uncertainty for a parameter. The fuzzy theory is used to show uncertainty for amount of produced waste per person. Locating incineration facilities is a sophisticated problem because of environmental and economic effects. Alçada-Almeida et al. [7] utilized a mixed integer multiobjective programming model to solve this problem. Their model has considered capacity for incineration. They also used a developed interactive decision support system by utilizing geographical information system (GIS) to select suitable option. Eiselt [8] worked on a model that is based on hub location model with cost minimization objective for locating landfills and transfer station. Krarup et al. [9] state the most locational models considering positive or negative effect for locating facilities close to demand center or far away from special places, respectively. They have presented a group of models with push and pull objectives for locating semiobnoxious facilities. In this way some papers have used different methods for locating undesirable facility such as Saameño et al. [10]. In their method, a weighted location problem in polygonal area is utilized.

Some models use assessment risk for routing and transporting hazardous material. Erkut and Verter [11] reviewed previous models to measure hazardous material transportation risk. They utilized one example to compare different types of risk models (traditional risk, population exposure, incident probability, conditional risk, and perceived risk). The results proved that various models have various consequences for choosing suitable transporting route. They believe selecting risk models for transporting HazMats (hazardous materials) requires more attention. Zhang et al. [12] applied Gaussian Plume model to model the number of people imposed by airborne contaminates. In this way, undesirable consequence probability is modeled to assess HazMats transportation as well as GIS to estimation risk. Erkut and Ingolfsson [13] believe that for satisfying risk models three reasonable axioms should be considered: path evaluating via monotonicity axiom, optimality principle for selecting path, and attribute monotonicity axiom. All previous models have violated from one or more axioms. They proposed new set models to fulfil all axioms. These axioms are also mentioned by Erkut and Verter [11]. Bonvicini and Spadoni [14] developed a new OPTIPATH model using TRAT4-GIS software to route HazMats and also decreasing risk. They illustrated that TRAT4-GIS is helpful for selecting suitable routes. Kazantzi et al. [15] proposed a framework for reducing risk and cost in transportation of hazardous material simultaneously. They used Mont Carlo simulation for finding impacts of risk incident and accident rate uncertainty on transportation model.

Several attempts have been made on locating and routing hazardous material concurrently. ReVelle et al. [16] developed mathematical model for locating storage facilities and selecting road for transporting spent fuel rods while minimizing transportation cost and perceived risk. Jacobs and Warmerdam [17] presented a biobjective model for hazardous waste locating and routing with risk and cost objective in 10-month horizon plan. Their model considers hazardous waste risk in transporting, storage and disposal. Wyman and Kuby [5] have used a multiobjective mixed integer programming model for toxic waste. In their model a variable for choosing treatment technology is introduced and the goals of model are minimizing cost, risk and inequity. Giannikos [18] developed a multiobjective model for routing hazardous material and locating related facilities. A goal programming method has been used to solve the problem with four objectives including, operation cost minimization, perceived risk minimization, Minimum of Maximum individual perceived risk, and minimum of maximum disutility. Alumur and Kara [19] proposed a comprehensive new model for locating and routing hazardous waste with two objectives. First objective is to minimize total cost including transportation and fixed annual cost for opening a related facility. Second objective is to minimize transportation risk via population exposure method. In their model transporting different types of hazardous waste, compatibility of treatment technology with type of waste, percentage of recyclable waste, and percentage of mass reduction after using different treatment technology have been considered. Zhao [20] developed a locating and routing model for hazardous waste with the same objectives as those of Alumur and Kara [19] model. He used a goal programing method to solve the problem. In this area, Samanlioglu [2] also developed a multiobjective model based on [19,20] models. Samanlioglu considered some additional features such as considering location for recycling centers, assessing site risk for hazardous waste facilities, and routing nonrecyclable waste from recycling centers to disposal centers. Xie et al. [21] proposed a multimodal hazardous material location-routing problem. In their model transfer yard location and suitable routes are optimized. Multiple modes in routing plans such as highways and railways are used.

Before proposing a model to locate facilities, finding suitable sites based on important criteria is necessary. One of the popular methods to select site in waste management is using GIS. Wang et al. [22] have used qualitative and quantitative factors to select proper site. Their proposed method has used the geographical information system (GIS) to select disposal site based on spatial data with environmental and economic criteria. Gorsevski et al. [23] have used multicriteria decision method based on geographical information system (GIS) for landfill site selection. Mahamid and Thawaba [24] also utilized (GIS) tool to find proper landfill site considering environmental factors. In this research, GIS tool will be utilized to find an appropriate site for treatment, recycling, and disposal centers based on some important criteria.

The complexity in hazardous waste management becomes apparent while solving model with conflict objectives simultaneously. In general one of the most popular approaches to solve such problems is combined multiobjectives to the single objective that is named the scalarization method. In this field the weighted sum method is more popular than other methods. Using the weighted sum method a group of supported efficient solutions from the Pareto frontier are achieved. Alumur and Kara [19] and Rakas et al. [6] are instances that have used this method in waste management. In this paper the weighted sum method will be utilized to find efficient solution from Pareto frontier.

The frame work of our model is close to the Alumur and Kara [19] and Samanlioglu [2]; they proposed a multiobjective model to location-routing hazardous waste. The difference between this paper and that of Alumur and Kara [19] included (1) establishing the locations for recycling centers and routing hazardous waste to these centers, (2) routing a part of generated hazardous waste to disposal centers directly, (3) considering population risk for all routes, (4) considering unit operation cost for treatment centers, disposal centers, and recycling centers in cost objective function, and (5) considering income from selling recycled waste in cost objective as cost saving parameter. Samanlioglu [2] also presented a comprehensive model with three objectives; our model is different from his model in some aspects and input data. We route a piece of generated hazardous waste to disposal centers directly. We take into account operation cost for all facilities and also recycled waste income in cost objective.

This assay attempts to develop a multiobjective model to locating-routing industrial hazardous waste with two objectives; for this reason the appropriate sites are selected to related facilities by GIS tool based on important criteria, then the weighted sum method is utilized to solve model and find efficient solution of Pareto frontier.

MATHEMATICAL MODEL

With respect to formulating location and routing problem for industrial hazardous waste, a multiobjective mixed integer programming model is proposed considering two conflicting objectives. First objective is minimization of total cost and second objective is minimization of transportation risk. The purpose of the model is to locate treatment centers with incineration technology and route hazardous waste to these centers, also to locate recycling centers and route hazardous waste from generation nodes and treatment centers to recycling centers, in addition, location of disposal centers and routing waste residue from other facilities to disposal centers.

Note that the model is an NP-hard problem; even single objective location routing problem is an NP-hard one [25]. A framework of model with input parameters and variables that help to understand the model easier is illustrated in Figure 2. The Objectives and constraints of this problem are as follows.

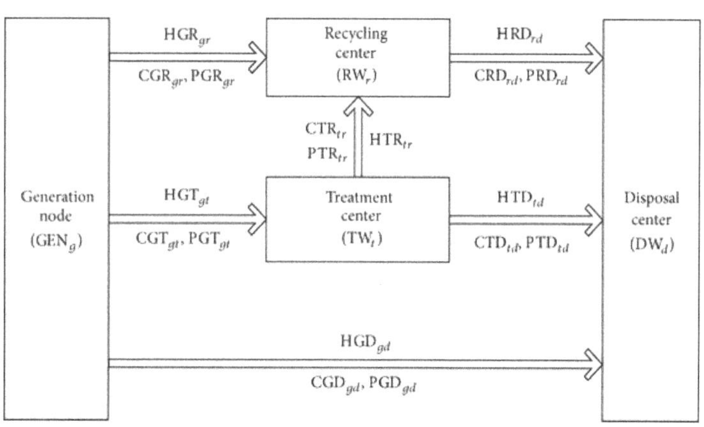

Figure 2: Input parameters and decision variables on proposed framework.

Objectives

Two objectives are considered in the model:

$$\text{Minimize}\,(f_1) = \text{Total cost},$$
$$\text{Minimize}\, f_2\,(x) = \text{Transportation risk.} \tag{1}$$

Cost Objective

The total cost is formulated with sum of the transportation cost, operation cost, and initial investment cost to open facilities minus cost saving or income from recycled waste in recycling centers:

$$\begin{aligned}\text{Minimize}\, f_1(x) = {} & \sum_g \sum_t \text{CGT}_{gt} * \text{HGT}_{gt} \\ & + \sum_g \sum_r \text{CGR}_{gr} * \text{HGR}_{gr} \\ & + \sum_g \sum_d \text{CGD}_{gd} * \text{HGD}_{gd} \\ & + \sum_t \sum_r \text{CTR}_{tr} * \text{HTR}_{tr} \\ & + \sum_r \sum_d \text{CRD}_{rd} * \text{HRD}_{rd} \\ & + \sum_t \sum_d \text{CT}_{td} * \text{HTD}_{td} \\ & + \sum_r \text{OCR}_r * \text{RW}_r + \sum_t \text{OCT}_t * \text{TW}_t \\ & + \sum_d \text{OCD}_d * \text{DW}_d + \sum_t \text{ICT}_t * Y_t \\ & + \sum_r \text{ICR}_r * X_r + \sum_d \text{ICD}_d * Z_d \\ & - \sum_r \text{IN}_r * \left(\text{RW}_r - \sum_d (\text{HRD}_{rd}) \right). \end{aligned} \tag{2}$$

In this objective, transportation cost is calculated by multiplying unit transportation cost and amount of transported waste between two nodes. The operation cost is the cost of processing hazardous waste in treatment, recycling, and disposal centers that can be obtained by unit operation cost product to quantity of hazardous waste in each center. Initial investment cost is capital cost if a new facility is established for a node. Also, total cost saving or income from the recycled waste is estimated in the cost function.

Risk Objective

The transportation risk is considered for all routes between generation nodes, recycling centers, treatment centers, and disposal centers. To quantify risk, the exposure risk method is used. In this approach, the total number of people along a route during a hazardous waste transportation are considered. This objective considers environmental aspect to transport hazardous wastes and waste residues:

$$\begin{aligned}\text{Minimize}\, f_2(x) = {} & \sum_g \sum_r \text{PGR}_{gr} * \text{HGR}_{gr} \\ & + \sum_g \sum_t \text{PGT}_{gt} * \text{HGT}_{gt} \\ & + \sum_g \sum_d \text{PGD}_{gd} * \text{HGD}_{gd} \\ & + \sum_r \sum_d \text{PRD}_{rd} * \text{HRD}_{rd} \\ & + \sum_t \sum_d \text{PTD}_{td} * \text{HTD}_{td} \\ & + \sum_t \sum_r \text{PTR}_{tr} * \text{HTR}_{tr}. \end{aligned} \tag{3}$$

Constraints

The model is under several reasonable constraints.

Mass Balance Constraint

There are flow balance constraints for each node of graph:

$$\text{GEN}_g = \sum_r \text{HGR}_{gr} + \sum_t \text{HGT}_{gt} + \sum_d \text{HGD}_{gd}, \quad \forall g \in G, \tag{4}$$

$$\sum_r \text{HGR}_{gr} = f_g * \text{GEN}_g, \quad \forall g \in G, \tag{5}$$

$$\text{RW}_r = \sum_g \text{HGR}_{gr} + \sum_t \text{HTR}_{tr}, \quad \forall r \in R, \tag{6}$$

$$\text{TW}_t = \sum_g \text{HGT}_{gt}, \quad \forall t \in T, \tag{7}$$

$$\sum_r \text{HTR}_{tr} = b_t * \text{TW}_t, \quad \forall t \in T, \tag{8}$$

$$\sum_d \text{HGD}_{gd} = k_g * \text{GEN}_g, \quad \forall g \in G, \tag{9}$$

$$\sum_d \text{HRD}_{td} = a_t * \text{TW}_t - \sum_r \text{HTR}_{tr}, \quad \forall t \in T, \tag{10}$$

$$\sum_d \text{HRD}_{rd} = (1 - c_r) * \text{RW}_r, \quad \forall r \in R, \tag{11}$$

$$\text{DW}_d = \sum_r \text{HRD}_{rd} + \sum_t \text{HTD}_{td} + \sum_g \text{HGD}_{gd}, \quad \forall d \in D. \tag{12}$$

Constraint (4) shows flow balance from generation nodes to other facilities. In this constraint, the amount of hazardous waste at generation nodes is equal to transported waste to recycling centers, treatment centers, and disposal centers. Constraint (5) says total amount of transported hazardous waste to the recycling centers has to be equal to the percentage of generated waste. Constraint (9) is also about transported hazardous waste from generation nodes to disposal centers. Constraints (6), (7), and (12) are amount of hazardous wastes that are recycled, treated and disposed in their centers, respectively. Constraint (8) denotes percentage of recyclable hazardous waste after treating. Constraints (10) and (11) present flow of hazardous waste between treatment centers and recycling centers to disposal centers, respectively.

Capacity Constraints

The capacity of facilities is limited to some factors such as equipment and manpower. Hence, a group of constraints are needed to display the maximum capacity of treatment, recycling, and disposal centers for a unit time. Constraints (13), (14), and (15) show this limitation. Also, to establish a new center minimum amount of hazardous waste or waste residue is required. Constraints (16), (17), and (18) are utilized as minimum capacity constraints for treatment, recycling and disposal centers, respectively:

$$RW_r \leq CAR_r^u X_r, \quad \forall r \in R, \tag{13}$$

$$TW_t \leq CAT_t^u Y_t, \quad \forall t \in T, \tag{14}$$

$$DW_d \leq CAD_d^u Z_d, \quad \forall d \in D, \tag{15}$$

$$RW_r \geq CAR_r^l X_r, \quad \forall r \in R, \tag{16}$$

$$TW_t \geq CAT_t^l Y_t, \quad \forall t \in T, \tag{17}$$

$$DW_d \geq CAD_d^l Z_d, \quad \forall d \in D. \tag{18}$$

Variables Constraints

The problem is a mixed integer programming model; therefore, binary and nonnegative variables are utilized to formulate that. Constrain (19) is used for no negativity variables, and constraint (20) is considered for binary variables:

$$\begin{aligned} &HGR_{gr} \geq 0, \quad \forall g \in G, \forall r \in R,\\ &HGT_{gt} \geq 0, \quad \forall g \in G, \forall t \in T,\\ &HGD_{gd} \geq 0, \quad \forall g \in G, \forall d \in D,\\ &HTR_{tr} \geq 0, \quad \forall t \in T, \forall r \in R,\\ &HRD_{rd} \geq 0, \quad \forall r \in R, \forall d \in D,\\ &HTD_{td} \geq 0, \quad \forall t \in T, \forall d \in D, \end{aligned} \tag{19}$$

$$\begin{aligned} &TW_t \geq 0, \quad \forall t \in T,\\ &RW_r \geq 0, \quad \forall t \in T,\\ &DW_d \geq 0, \quad \forall d \in D,\\ &Y_t \in \{0,1\}, \quad \forall t \in T,\\ &X_r \in \{0,1\}, \quad \forall r \in R,\\ &Z_d \in \{0,1\}, \quad \forall d \in D. \end{aligned} \tag{20}$$

If g is the number of generation nodes, t is the number of potential treatment nodes, r is the number of potential recycling nodes, and d is the number of potential disposal nodes, then the proposed model has$(r+t+d)$ binary variables and $(gr + gt + gd + tr + rd + td + r + t + d)$ real decision variables. The number of total constraints of the proposed model without variable constraints is $(3g + 4r + 5t + 3d)$.

PROPOSED SOLUTION TO MULTIOBJECTIVE PROBLEM

Here, a multiobjective model with two conflict objectives was proposed. Multiobjective optimization [26] is to optimize two or more conflicting objectives with regard to a set of group constraints simultaneously. If the optimization of one objective leads to the automatic optimization of the other, it is not a multiobjective optimization. There is a group of solution methods for multiobjective problems called classical multiobjective optimization methods. To solve the model, the weighted sum method (WSM) is utilized. This method is a classical method that is used widely to solve multiobjective models. WSM combines two objectives to a single objective by multiplying each objective with a decision maker supplied weight. This method scalarizes the set of objectives with a convex combination. $f_1(x)$ and $f_2(x)$ have different units; after the objectives are normalized, a combined objective function can be created by summing the weighted normalized objectives and the multiobjective model is converted to a single objective model as follows [26]:

$$\text{Min} f(x) = \frac{w_1 * f_1(x)}{f_1^*(x)} + \frac{w_2 * f_2(x)}{f_2^*(x)}, \tag{21}$$

$$w_1 + w_2 = 1. \tag{22}$$

Here w_1 and w_2 are the weights of $_{f1}(x)$ and $_{f2}(x)$ objective function, respectively. First each objective is solved separately and optimal objectives are named $f_1^*(x)$ and $f_2^*(x)$. Thereafter $f_1(x)/f_1^*(x)$ and $f_2(x)/f_2^*(x)$ are weighted by $w1$ and $w2$; then two objectives are summed together. The single objective model is solved by considering different weights and constraints to have Pareto frontier of solutions. Amounts of obtained variables are inserted to the objectives to compute value of each objective function.

To have Pareto frontier of solutions the model is solved with different weights for each objective. There are two theorems in regard to classical method and also (21).

Theorem 1. The solution to the problem represented by (21) is Pareto-optimal if the weight is positive for all objectives.

Theorem 2. If x^* is a Pareto-optimal solution of a convex multiobjective optimization problem, then there is a nonzero positive weight vector w such that x^* is a solution to the problem in (21) (Figure 3)

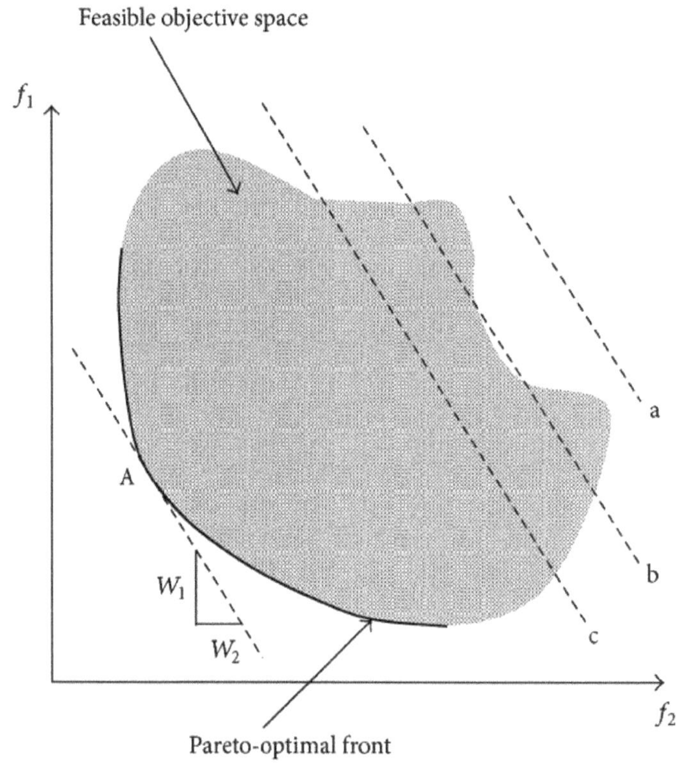

Figure 3: The weighted sum method on convex Pareto-optimal front.

ANALYSIS AND APPLICATION FOR PROVINCE OF IRAN

Analysis of model requires a real data; therefore, application of model is done in a province of Iran called Markazi. This province has 12 counties with a population of 1325159 and 29400?km^2 area. There are fundamental industries in this province such as refinery, petrochemical industries, and aluminum manufacturers. There are 15 industrial areas in Markazi province with a total area of 14?km^2. Amount of industrial hazardous waste in all counties is about 160000?ton per year. It is estimated to be around 700 tons per day considering 255 workdays in Iran. Out of 12 counties 7 are chosen for this model due to producing industrial hazardous waste. The information about the amount of hazardous waste in each county is not attainable. These amounts have been assumed proportional to industrial zone area in each county. For instance, the whole of industrial zone area in Arak county is 3?km^2; hence the amount of produced industrial hazardous waste is 150?ton per day. The same method has been utilized for other industrial zones in different counties.

To select proper site for treatment, disposal, and recycling centers a geographical information system tool is applied. For GIS analysis Arc view (3.2) software has been used. Some input information is needed to carry out this software analysis. In this paper, common and significant criteria have been considered to select suitable area. These criteria are geology, topography, climate, land slope, distance from urban and rural area, protected area, infrastructure and distance to highway. Thereafter, analyzing with software, final site selection has been achieved. In a general manner all proper areas have been put in 2 groups: first is excellent area and second group is good area. These areas are illustrated in Figure 4. Based on amount of produced hazardous waste and capacity of facilities 5, 5, and 3 of selected areas are considered for treatment, disposal and recycling centers respectively. We tried to choose these 13 areas from excellent areas and also closer to big industrial zones. Selected areas are illustrated in Figure 5. To calculate transportation cost, fuel cost and amount of fuel that a truck consumes are considered. Average of the fuel consumption for a truck is about 0.5 liter per km in Iran, and price of fuel is \$0.3 per liter. Hence transportation cost is \$0.15 per km, but in reality fuel cost is only 50 percent of trucks cost in transportation; therefore, it is around \$0.3 per km. This cost is different for hazardous waste due to using specific equipment and trucks. Therefore, based on [2, 19] the unit transportation cost for CTR_{tr}, CDR_{rd}, and CTD_{td} is 70% of hazardous waste. Capacity of these centers with regard to other centers in Iran is 100 tons, 250 tons, and 250 tons for recycling centers, treatment centers, and disposal centers, respectively. The investment cost to open a new center for hazardous waste based on previous price in Iran is around \$10 million for treatment centers, \$5 million for recycling center, and \$7 million for disposal centers. Operation costs in various centers are considered to be \$200, \$250, and \$50 per ton for treatment, recycling, and disposal centers, respectively. Income from recycled waste is about \$700. The population exposure bandwidth is considered to be 1000?m along the road here due the fact that ReVelle et al. [16] believed that different bandwidth does not have any effect on their solution. For this aim Arc GIS (3.2) has been used to estimate population in 1000?m bandwidth. In generation nodes 15% of generated hazardous waste is assumed recyclable and also 10% of generated hazardous waste is directly transported to disposal centers. Rate of mass reduction in incineration is about 80% based on [2, 19]. In treatment centers 10% of treated waste is considered recyclable and is transported to recycling centers. Also in recycling centers 95% of hazardous waste is recycled and transported to the market or other factories.

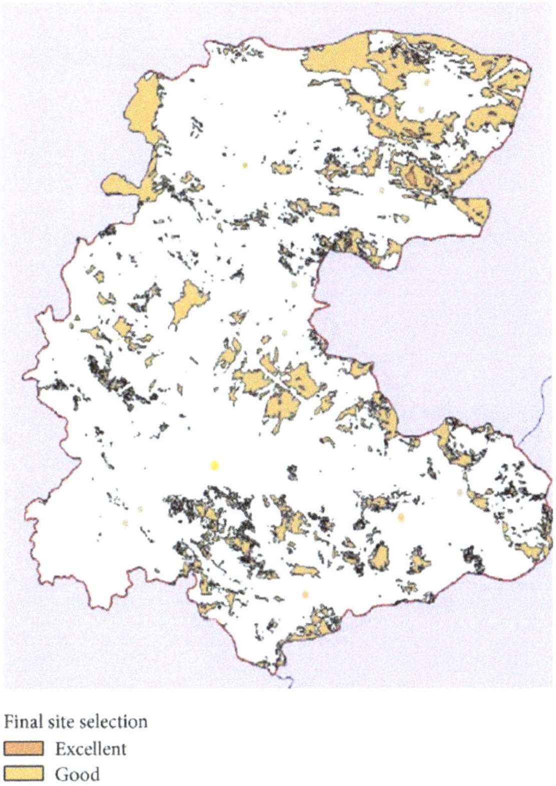

Figure 4: Final site selection by GIS.

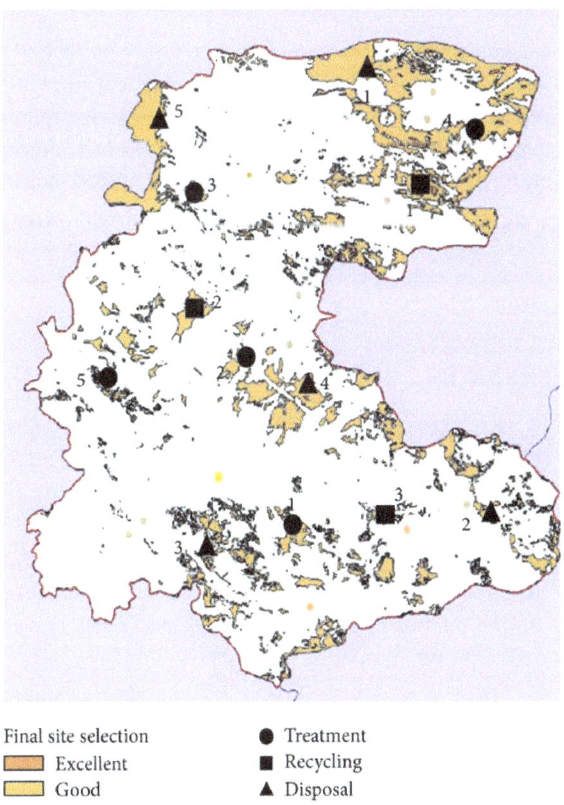

Figure 5: Appropriate site for waste management facilities.

Considering input data and problem framework, the model has been solved via GAMS software with CPLEX solver version 12.5 using a computer with Intel core i5-2.5?GHZ, 4?GB memory.

To solve model different weights are utilized. Thus, with changing w_1 and w_2, also considering $w_1 + w_2 = 1$, groups of weight vectors are produced to achieve efficient solution of the model from the Pareto set. In this paper 11 weight vectors have been used. In Table 1 weight vector groups, amount of each objective function based on its special weight, and generation time to solve model using GAMS software are illustrated. In Table 2, location of treatment centers, disposal centers and recycling centers with regard to 11 weight groups are presented.

Table 1: Amount of two objective functions with regard to their weight vectors.

Solution number	1	2	3	4
w_1	0	0.1	0.2	0.3
w_2	1	0.9	0.8	0.7
$f_1(x)$	$6.839707 * 10^7$	$6.839707 * 10^7$	$6.8397707 * 10^7$	$6.442543 * 10^7$
$f_2(x)$	763254	763254	763254	849007
CPU time	0.047	0.094	0.039	0.047
Solution number	5	6	7	8
w_1	0.4	0.5	0.6	0.7
w_2	0.6	0.5	0.4	0.3
$f_1(x)$	$6.442543 * 10^7$	$6.342844 * 10^7$	$6.342791 * 10^7$	$6.342791 * 10^7$
$f_2(x)$	849007	864802	878519	878519
CPU time	0.078	0.078	0.047	0.031
Solution number	9	10	11	
w_1	0.8	0.9	1	
w_2	0.2	0.1	0	
$f_1(x)$	$6.23202 * 10^7$	$6161363 * 10^7$	$6.136445 * 10^7$	
$f_2(x)$	895567	899487	977830	
CPU time	0.036	0.093	0.062	

Table 2: Location of different facilities considering various weight vectors.

Weighs w_1	w_2	Treatment centers	Disposal centers	Recycling centers
0	1	1, 3, 4	1, 2, 3, 5	1, 3
0.1	0.9	1, 3, 4	1, 2, 3, 5	1, 3
0.2	0.8	1, 3, 4	1, 2, 3, 5	1, 3
0.3	0.7	1, 3, 4	1, 2, 5	1, 3
0.4	0.6	1, 3, 4	1, 2, 5	1, 3
0.5	0.5	2, 3, 4	1, 2, 5	1, 3
0.6	0.4	2, 3, 4	1, 2, 5	2, 3
0.7	0.3	2, 3, 4	1, 2, 5	2, 3
0.8	0.2	1, 3, 4	1, 3, 5	2, 3
0.9	0.1	1, 3, 4	1, 3, 5	1, 2
1	0	1, 2, 3	1, 3, 4	1, 3

The result in Table 1 illustrates a tradeoff between cost and risk objectives. The conflict between two objectives appeared by varying amount of w_1 and w_2 between 0 and 1 increments of 0.1. It is clear that the cost value increases in order to decrease the risk value in Table 1. The same result can be seen in Table 2. The location of treatment, recycling and disposal centers is changing by using different weight factors

A sample of solved model with $w_1 = 0.5$ and $w_2 = 0.5$ is displayed in Figure 6; in this figure relation between chosen facilities is presented.

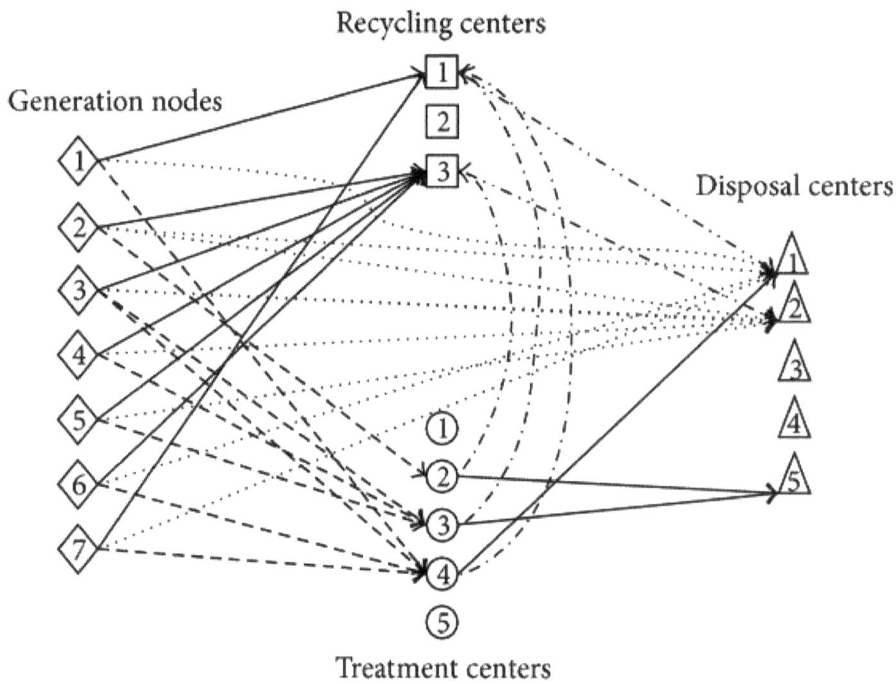

Figure 6: A sample of efficient solution for the model with equal weights.

CONCLUSION

This assay has developed a multiobjective model to locating-routing hazardous waste. This model is based on literature and studied models with some difference. First, considering a part of generated hazardous waste as nonrecyclable and nontreatable waste and routing it to disposal center directly. Second, to have more comprehensive model, operation cost for each facility and saving cost from selling recycled waste were considered. The framework of this problem is location of treatment centers, recycling centers and disposal centers in suitable sites and routing industrial hazardous waste between them. Two conflict objectives have been considered. First, total cost minimization consists of transportation cost, operation cost, initial investment cost, and also cost saving parameter form earned income of selling recycled waste. Second objective is routing risk minimization using risk exposure method. To combine multiobjective model and the single objective model, the weighted sum method that is a kind of scalarization method was utilized. Application of model has been tested in Markazi province in Iran considering some assumptions for unavailable input data. The result of the solved example shows conflict between transportation risk and total cost objectives. For example, it is possible to decrease the cost objective with allocating higher weight (priority), while risk objective with lesser weight has a marginal increase. The same result can be taken in location of facilities. Their locations are changing when the weight vectors are varying. In this study, GAMS software with applying CPLEX solver has been used to solve problem.

The finding from this study suggests some future research. First, amount of hazardous waste depends on the number of industry and population and they vary in course of time. Hence, developing a model with long term planning for hazardous waste is necessary. Second, to solve multiobjective model in the field of locating-routing hazardous waste, scalarization method is common and popular. For such problem with conflicting criteria using metaheuristic method to encounter these problems with conflict objectives can be good as mentioned in Samanlioglu [2] and Alumur and Kara [19]. This study developed a new mathematical model based on a proposed framework; therefore, a data-driver framework [27] can be applied for future research. The last suggestion for future research is about amount of produced hazardous waste; in previous research amount of hazardous waste was considered deterministic, whereas amount of produced waste is uncertain in reality; hence fuzzy set theory can be useful in this way.

REFERENCES

1. A. K. Nema and S. Gupta, "Optimization of regional hazardous waste management systems: an improved formulation," Waste Management, vol. 19, no. 7, pp. 441–451, 1999.
2. F. Samanlioglu, "A multi-objective mathematical model for the industrial hazardous waste location-routing problem," European Journal of Operational Research, vol. 226, no. 2, pp. 332–340, 2013.
3. S. J. Ratick and A. L. White, "A risk-sharing model for locating noxious facilities," Environment & Planning B, vol. 15, no. 2, pp. 165–179, 1988.
4. E. Erkut and S. Neuman, "A multiobjective model for locating undesirable facilities," Annals of Operations Research, vol. 40, no. 1, pp. 209–227, 1992.
5. M. M. Wyman and M. Kuby, "Proactive optimization of toxic waste transportation, location and technology," Location Science, vol. 3, no. 3, pp. 167–185, 1995.

6. J. Rakas, D. Teodorović, and T. Kim, "Multi-objective modeling for determining location of undesirable facilities," Transportation Research Part D, vol. 9, no. 2, pp. 125–138, 2004.
7. L. Alçada-Almeida, J. Coutinho-Rodrigues, and J. Current, "A multiobjective modeling approach to locating incinerators," Socio-Economic Planning Sciences, vol. 43, no. 2, pp. 111–120, 2009.
8. H. A. Eiselt, "Locating landfills-Optimization vs. reality," European Journal of Operational Research, vol. 179, no. 3, pp. 1040–1049, 2007.
9. J. Krarup, D. Pisinger, and F. Plastria, "Discrete location problems with push-pull objectives," Discrete Applied Mathematics, vol. 123, no. 1–3, pp. 363–378, 2002.
10. J. J. Saameño Rodríguez, C. Guerrero García, J. Muñoz Pérez, and E. Mérida Casermeiro, "A general model for the undesirable single facility location problem," Operations Research Letters, vol. 34, no. 4, pp. 427–436, 2006.
11. E. Erkut and V. Verter, "Modeling of transport risk for hazardous materials," Operations Research, vol. 46, no. 5, pp. 625–642, 1998.
12. J. Zhang, J. Hodgson, and E. Erkut, "Using GIS to assess the risks of hazardous materials transport in networks," European Journal of Operational Research, vol. 121, no. 2, pp. 316–329, 2000.
13. E. Erkut and A. Ingolfsson, "Transport risk models for hazardous materials: revisited," Operations Research Letters, vol. 33, no. 1, pp. 81–89, 2005.
14. S. Bonvicini and G. Spadoni, "A hazmat multi-commodity routing model satisfying risk criteria: a case study," Journal of Loss Prevention in the Process Industries, vol. 21, no. 4, pp. 345–358, 2008.
15. V. Kazantzi, N. Kazantzis, and V. C. Gerogiannis, "Risk informed optimization of a hazardous material multi-periodic transportation model," Journal of Loss Prevention in the Process Industries, vol. 24, no. 6, pp. 767–773, 2011.
16. C. ReVelle, J. Cohon, and D. Shobrys, "Simultaneous siting and routing in the disposal of hazardous wastes," Transportation Science, vol. 25, no. 2, pp. 138–145, 1991.
17. T. L. Jacobs and J. M. Warmerdam, "Simultaneous routing and siting for hazardous-waste operations,"Journal of Urban Planning & Development, vol. 120, no. 3, pp. 115–131, 1994.
18. I. Giannikos, "A multiobjective programming model for locating treatment sites and routing hazardous wastes," European Journal of Operational Research, vol. 104, no. 2, pp. 333–342, 1998.
19. S. Alumur and B. Y. Kara, "A new model for the hazardous waste location-routing problem," Computers and Operations Research, vol. 34, no. 5, pp. 1406–1423, 2007.
20. J. Zhao, "Model and algorithm for hazardous waste location-routing problem," in Proceedings of the International Conference of Logistics Engineering and Management (ICLEM ‹10), pp. 2843–2849, 2010.
21. Y. Xie, W. Lu, W. Wang, et al., "A multimodal location and routing model for hazardous materials transportation," Journal of Hazardous Materials, vol. 227, pp. 135–141, 2012.
22. G. Wang, L. Qin, G. Li, and L. Chen, "Landfill site selection using spatial information technologies and AHP: a case study in Beijing, China," Journal of Environmental Management, vol. 90, no. 8, pp. 2414–2421, 2009.
23. P. V. Gorsevski, K. R. Donevska, C. D. Mitrovski, and J. P. Frizado, "Integrating multi-criteria evaluation techniques with geographic information systems for landfill site selection: a case study using ordered weighted average," Waste Management, vol. 32, no. 2, pp. 287–296, 2012.
24. I. Mahamid and S. Thawaba, "Multi criteria and landfill site selection using GIS: a case study from Palestine," Open Environmental Engineering Journal, vol. 3, pp. 33–41, 2010.
25. G. Nagy and S. Salhi, "Location routing: issues, models and methods," European Journal of Operational Research, vol. 177, no. 2, pp. 649–672, 2007.
26. K. Deb, "Multi-objective optimization," in Multi-Objective Optimization Using Evolutionary Algorithms, pp. 13–46, 2001.
27. S. Yin, S. X. Ding, A. Haghani, et al., "A comparison study of basic data-driven fault diagnosis and process monitoring methods on the benchmark Tennessee Eastman process," Journal of Process Control, vol. 22, no. 9, pp. 1567–1581, 2012.

CHAPTER 7

OPTICAL NONDESTRUCTIVE VISUAL AND NEAR-INFRARED METHODS FOR FOOD QUALITY AND SAFETY

Jarmo T. Alander, Vladimir Bochko, Birgitta Martinkauppi, Sirinnapa Saranwong, and Timo Mantere

INTRODUCTION

Consumers are the driving force in the food market. They have become more health conscious, demanding and willing to pay for the "good quality." The consumers' trust for food industry has been diminished due to food scandals thus making it important to improve the safety monitoring.

The "quality" and "safety" have different meanings and aspects which depend on the food class, target market, criteria, and application. In some sense, the safety is part of the quality but here we separate these two because we like to emphasize the difference between how a human perceive the food and how we can evaluate threats of health. Some unsafe food cannot be detected by manual inspection. The food quality and safety evaluation has become more important and the need for more comprehensive assessment for all food batches is adequate.

Even though a lot of research and development work has been done on food safety and quality, more needs to be done to find economic ways of monitoring food safety. In this paper, our purpose is to go through the subject according to Food and Agriculture Organizations of the United Nations (FAO) categories (http://www.fao.org/) and publications related to quality and safety. These two factors can be evaluated in many different methods and techniques.

The traditional methods are sensory evaluation, chemical analysis, and microbiological analysis. A more recent approach in technical sense is optical techniques, especially the multispectral methods in visual and near-infrared wavelengths. The optical methods can be used for nondestructive, fast, real-time, online monitoring of all samples (Huang et al. [1]). Table 1 gives a general comparison of these four approaches.

Table 1: Comparison of method properties

Method	Subjective/objective	Sample/all
Sensory	Subjective	All
Chemical	Objective	Sample
Microbiological	Objective	Sample
NIR	Objective	All

The sensory evaluation is the oldest method and still used every day. The evaluation is done by both consumers and professional food observers. It is relatively fast but hardly suitable for a large amount of samples due to many reasons like observer's fatiguing. The reliability varies with food item group and by person. The results are typically very subjective, depending on many non-food-related factors, and difficult to evaluate quantitatively.

Citation: Jarmo T. Alander, Vladimir Bochko, Birgitta Martinkauppi, Sirinnapa Saranwong, and Timo Mantere, "Optical Nondestructive Visual and Near-Infrared Methods for Food Quality and Safety," International Journal of Spectroscopy, vol. 2013, Article ID 341402, 36 pages, 2013. doi:10.1155/2013/341402

The chemical analysis is an objective technique which often provides reliable results if the monitored batch is represented well by the samples measured. It has some limitations—the processes may not work correctly due to some substances, the need of accurate and correct calibrations, the need of complex multiphased processing, and the need of experts for the analysis. It might not be suitable if the distribution of the substance content is nonuniform. It cannot provide a point measurement of the sample, only an average over the whole sample. Since the sample is destroyed, chemical analysis cannot be used for monitoring all food units. In addition, the price of the analysis per measurement may be quite high and it produces toxic waste; thus, the number of samples for monitoring must be limited due to economic and ecological reasons. It is generally not very well suited for fast, online product management.

The microbiological analysis is usually applied to detect biological organism like fungi or bacteria but not so much for identifying other content. It gives often objective and reliable results provided that, for example, the culture do grow, the growth is detectable, or the sample is not contaminated by some other sources. The analysis is typically slow and more tedious and expensive than the previous two methods. The microbiological analysis is suitable only for low speed, offline monitoring using a representative sample set of the batch. The main problem is that it allows defective products to be manufactured even though the cheapest solution might be to immediately stop processing and clean up the manufacturing unit.

To overcome the disadvantages and limitations of the previous methods, new economic, fast, and environment friendly techniques are sought after. As a result, optical multispectral measurements and imaging have recently become more and more popular in food measurements because they lend themselves easily to online monitoring of all samples even without touching the samples. The price of the measurement instruments has also decreased so they provide a viable option for the traditional methods. However, analysis of spectral measurements is often not easy and requires expertise. The mathematical and statistical models created might not be general and need to be adjusted to new conditions and products. Nowadays, a typical optical system employs ultraviolet (fluorescence) and visible wavelengths (VIS), near-infrared (NIR) or infrared (IR) areas or a combinations of them. The near-infrared spectrum is especially interesting since it is related to overtones and combinations of such chemical bonds as C–H, O–H, and N–H which have influence on many properties of food.

History

Food and agriculture have been subjects for near-infrared (NIR) applications from the beginning of NIR measurement history. The earliest work which involved nondestructive moisture content determination of grain and seeds is published in [2] of the United State Department of Agriculture (USDA). Later, the success of NIR spectroscopy is marked by the establishment of wheat protein analysis throughout Canada by Phil Williams of the Canadian Grain Commission in 1975 [3]. In this area of application, at first, NIR was restricted to quality evaluation of low-moisture products such as cereal grains including wheat, corn, and soybean due to interference of water absorption in the long-wavelength region, especially in the combination area of 1900–2500 nm. In the latter stages of NIR applications, the development of silicon sensors allowed acquisition in the short-wavelength region of 700–1100 nm, leading to the analysis of thicker and higher-moisture samples such as fruit, fish, meat, and intact kernels. The use of the short-wavelength region allows nondestructive measurement of relatively thick samples since the electromagnetic wave in this region has better penetration because the absorption of water in this third overtone band becomes weak.

According to Osborne [4], the major limitation of using NIR spectroscopy for food analysis is its dependence on less precise reference methods. Since the chemical information of sample is analyzed from the changes in spectra, in most food samples NIR spectroscopy needs calibration against a reference method with the ingredient of interest.

Other optical methods have been studied for food quality lately by many researchers, for example, Farkas and Dalmadi [5], Kaffka [6], and Huang et al. [7]. The optical methods could alleviate these problems although they are not so accurate. Another advantage of optics is that it can be imaging and thus characterizing the sample better. A popular spectral area for food research has been near-infrared (NIR) in which many publications can be found like Huang et al. [7], Pallav et al. [8], Ozaki et al. [9], and Osborne [4].

Berrueta et al. [10] have recently made a review of supervised pattern recognition methods in food analysis. Reid et al. [11] have made a review of different food authentication methods including NIRS. Cen and He [12] give a review of NIR reflectance spectroscopy in determination of food quality.

Du and Sun [13] give a review of machine learning techniques such as neural networks and evolutionary optimization in computer vision for food quality.

Optical NIR Measurements for Food

NIR Measurement Modes

There are four standard measurement modes for the acquisition of NIR spectra from a sample: transmission, reflection, transflection, and interaction (interactance). Illustrations of these are shown in Figure 1. The selection of measurement mode depends on many factors like sample type or installment location. Generally, the design and implementation of measurement system and the selection of measurement mode require at least some expertise.

In the transmission measurement mode (Figure 1(a)), incident light propagates through the sample and is measured as it exits the sample at a point directly opposite to the light source. Occasionally, in the case of transmission for solid samples, the detector is placed at an angle of 90 degrees to the source.

For the reflection mode (Figure 1(b)), the light source and the detector are placed on the same side of sample. To avoid surface reflection, detectors are usually placed at an angle of 45 degrees to the sample plane. In some cases, an integrating sphere is used to concentrate light reflected from the sample before it reaches the detector by capturing and integrating light from various directions.

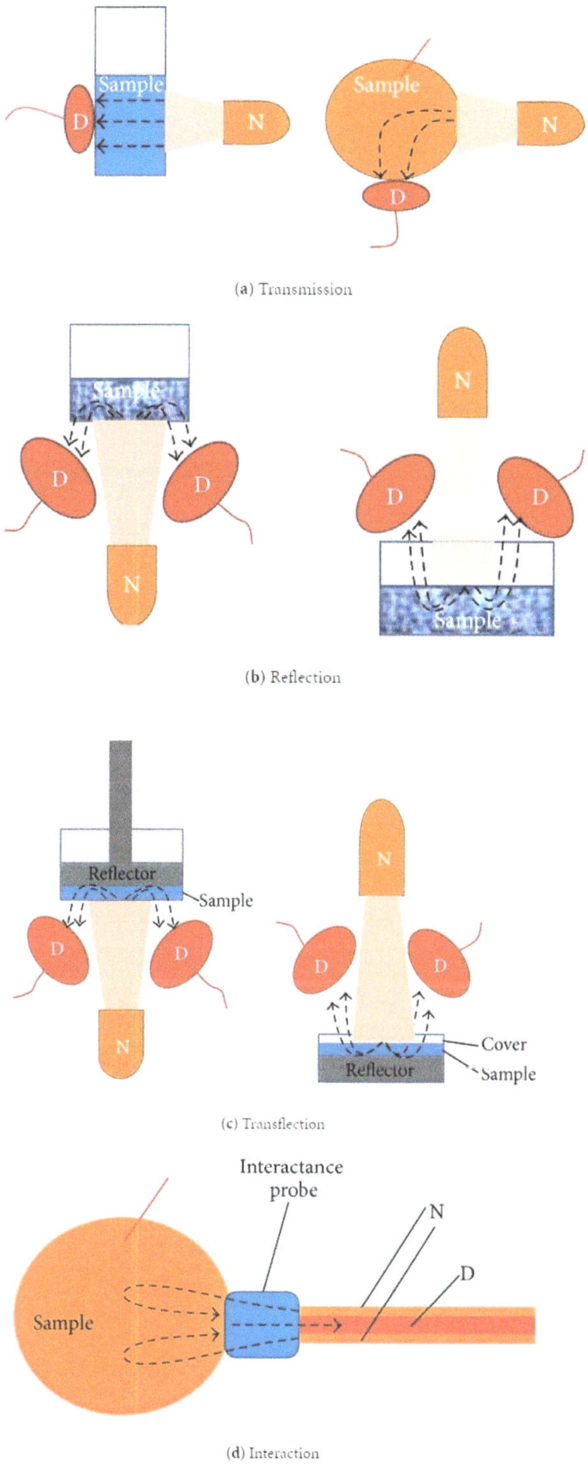

Figure 1: Sample presentation for near-infrared (NIR) spectroscopy. N: NIR light source, D: detector.

Transflection (Figure 1(c)) is the combination of transmission and reflection modes. The sample is positioned as if for reflection mode measurement, but a reflector is situated opposite to the light source on the far side of the sample, reflecting light that is transmitted through the sample back through the sample again to the detector. The samples, generally in liquid form, are placed in a holder of 0.1–0.3 mm width along the direction of light propagation. Therefore, the physical path length of transflection mode is double the sample thickness. Transflection is often used to reduce path length as the thinnest cuvette (standard sample holder for transmission measurements) is generally 1.0 mm thick. In some cases, transflection is employed for NIR instruments without transmission capability.

The fourth configuration, the interactance mode (Figure 1(d)), is also a combination of transmission and reflection but for solid samples. In this mode, the sample is illuminated by a fiber-optic cable(s) in direct contact with the sample surface or the sample holder. Light propagates inside the sample and interacts with the interior through reflection, transmission, and/or absorption. Fibre bundles that are placed on the same side as the source subsequently return the light that propagates inside the sample to the detector. The direct contact between the fibre bundles (for both the source and the detector) and the sample eliminates the effect of surface reflection and maximizes the penetration depth.

Types of Food Samples

From the product viewpoint, NIR applications in food and agriculture could be classified into three groups by the state of the sample: (i) liquid samples, (ii) ground and relatively small solid samples, and (iii) relatively large samples that require nondestructive or noninvasive measurement.

A benefit in working with samples from the first group (liquid samples) is the handling of optical path length. In the case of a clear liquid such as soy sauce, filtrated fruit juice, or vegetable oil, samples can be contained in a quartz cuvette, a disposable glass vial, or a transflectance sample holder, all of which would provide a relatively high consistency of optical path length due to the low scattering characteristics of the samples. The measurements used are transmission or transflection. The sample thickness can be adjusted based on the absorptivity of major composition(s) and the constituent of interest.

Figure 2 shows NIR spectra in the long-wavelength region of water and oil acquired in various physical path lengths. According to the figure, if the sample analyzed is water-rich and the constituent of interest has absorption bands overlapped with water absorption, the appropriate physical path length would be 0.2 mm for the combination region measurement (1800–2100 nm), 1.0 mm for the first overtone region (1400–1500 nm), and 10 mm for the second overtone region (900–1000 nm). However, as aforementioned, the decision for optical path length is also dependent on the molar absorptivity, that is, the ability of a particular substance to absorb light at a given wavelength, and the concentration of that constituent of interest. In many cases, the constituent of interest has absorption bands on the shoulder of water bands such as in the region of 2100–2300 nm. In such cases, sample cells with longer physical path length, say 1 mm, could be more appropriate than the 0.2 mm sample cell due to lower absorptivity of substances in this region.

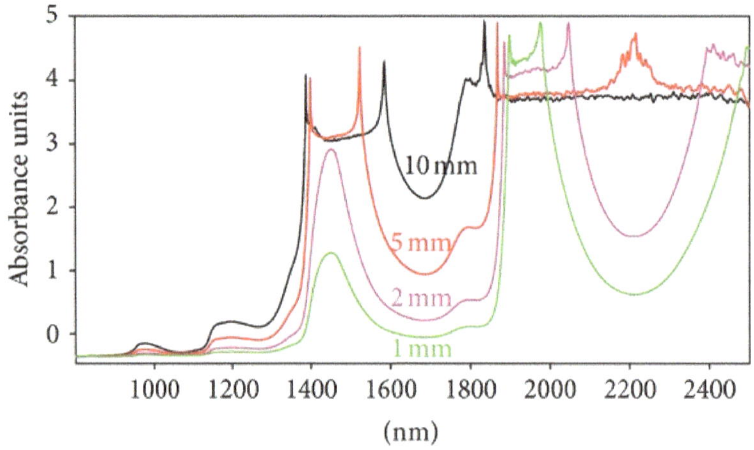

(a) Water measured at 1, 2, 5, and 10 mm path length

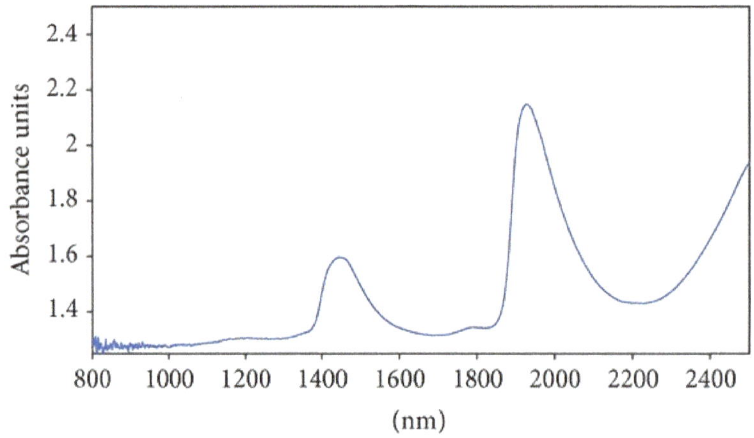

(b) Water measured at 0.2 mm path length

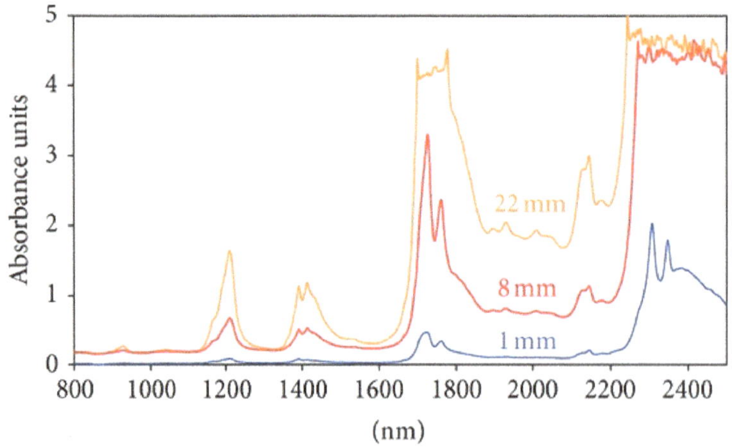

(c) Vegetable oil measured at 1, 8, and 22 mm path length

Figure 2: NIR spectra of water and oil measured at different path lengths. (a) Water spectra when the path length is 1, 2, 5, or 10 mm. (b) Water spectra when the path length is 0.2 mm. (c) Vegetable oil spectra when the path length is 1, 8, or 22 mm. The shape and level of measured spectra depend on the path length.

On the other hand, if the sample is oil-based, sample cells with significant longer path length could be used due to the lower absorptivity of oil compared to water. In the case where the constituent of interest has very low absorptivity and/or concentration, for example, acid value or peroxide value in fresh vegetable oil, increasing physical path length to obtain adequate information is necessary.

However, one should keep in mind that the analysis of clear samples typically differs from that of the analysis of scattering samples. When colloids or suspensions are analyzed, the scattering characteristics of the samples are stronger, and thus the optical path length, that is, the actual distance that NIR wave travels, is much longer than that of clear samples. Due to this, the path length would have to be reduced. Figure 3 shows the NIR spectra of homogenized milk (colloid) and vegetable-mix juice (suspension) using a 4 mm path length. In this case, milk has a stronger absorption due to its composition and the scattering characteristics of the constituent small fat globules. Vegetable-mix juice has less absorbance compared to milk but stronger than that of water at 5 mm path length (for water, see Figure 2). This is due to the scattering characteristics of the constituent suspended particles.

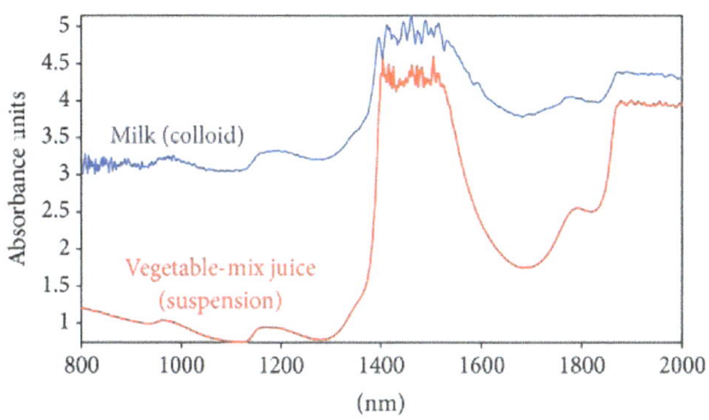

Figure 3: NIR of milk (colloid) and vegetable-mix juice (suspension) measured with 4 mm physical path length. Note: the preamplifier gain used for milk was stronger than that of vegetable-mix juice due to very weak transmitted light.

Some Important Considerations for Measurements

Another point to be aware of for liquid samples with water as a major constituent is the importance of temperature control. It is well documented [14–17] that water absorption bands in the NIR region shift in both peak position and amplitude according to sample temperature, as the number of hydrogen bonds in the water molecules is temperature dependent. Increasing temperature results in a higher percentage of water molecules with fewer hydrogen bonds due to the higher kinetic energy. Therefore, in many cases, temperature control is necessary to perform precise NIR measurements. Since many oil-based samples are in solid or opaque form at room temperature, temperature control is required not mainly to maintain spectral consistency but to increase sample transparency.

For ground or small-particle samples, NIR spectra are usually acquired in reflection mode due to strong scattering characteristics. Such samples are generally placed in sample holders with quartz or glass windows. Quartz is more expensive but has lower OH absorption compared to glass resulting in better precision in some constituents, especially those with OH functional groups. However, in the case where the absorption of the constituent of interest occurs at different wavelength region than the OH absorption in glass and there is enough concentration of those constituents, it is possible to use glass-window sample holder or glass Petri dish. For many food manufacturers, the use of glass or quartz is prohibited and plastic such as a polystyrene dish is used as a sample holder. In that case, the absorption band of the constituent would fall within the absorption bands for polystyrene, especially in the combination region beyond 2000 nm. If the absorption bands of the constituent can be precisely measured in the shorter-wavelength region by arranging an adequate path length and detector, the use of plastic dish would be applicable. For samples with a relatively coarse particle size, there is the possibility that light could pass through the space between samples and be lost. One way to reduce this problem is to place a reflector on the back side of the sample cell. During the measurement, spinning the sample cell will increase the acquisition area, thus reducing error due to sample nonuniformity especially in relatively coarse or large-particle samples. Examples of samples belonging to this group include ground and intact cereal grains, cereal meals such as soy bean meals or corn gluten meals, crushed processed foods/snacks, and powder milk or drink.

Finally, we consider samples that require nondestructive measurement. Samples in this group include meat, fish, intact fruit, and intact nut kernels. For the most part, these samples are sold intact and the constituent of interest is not evenly distributed between the surface and the sample core. To obtain information from deep inside the sample, two modes of measurement are used for this group: the interaction mode and the transmission mode. As mentioned earlier, most of the applications in this group use silicon detectors that allow spectral acquisition in the short-wavelength region of 700–1100 nm. The measurement parameters could be moisture and sugar contents for intact fruit and fat, protein, and moisture content in meat and kernels, as well as detection of various internal disorders or pest in agricultural products.

To employ NIR spectroscopy for routine quality control, it is not enough to develop calibration equations. A system developer needs to consider various factors affecting calibration results including instrument status (source strength, detector responses, and wavelength selection mechanism), atmospheric effects (temperature and humidity), and physical conditions (dust, vibration, and stray light) as well as transferability of the equations developed between various instrument in the same period of time and the transferability to the newer version of instrument and software upon the end of instrument life span.

Chemometrics

Optical measurement results connected with chemical analysis and statistics are called chemometrics. Chemometrics has become an established technique for handling chemical data. Good reviews of chemometrics are provided by Geladi and Grahn [18], and Workman and Springsteen [19], and Koljonen et al. [20], which present a review of genetic algorithm optimization in chemometry. The next text is a short introduction for the subject, and for a deeper insight, the previous reviews are recommended.

The first step of chemometrics is to collect the data. This includes sample selection, measurements, and chemical analysis. The measurement geometry, measurement conditions, number of samples, and ingredients need to be very carefully designed and determined. The samples need to be representative of the food batches because interpolation or generalization of the model most probably does not work. The samples are first subjected to optical measurement and after that chemical analysis. One should also notice that the chemical analysis might sometimes provide unreliable results. Sometimes even the samples may change during time. The data should be divided into a training and test sets.

After the measurements, the equations to transform optical data to chemical data should be set up. Since many functions and equations are available, it might be useful to test a couple of them to select the best results. After the selection, the equations need to be solved and coefficients to be determined with the training set. A typical approach for modeling is regression. The training of the model can be understood as a "calibration." The optical methods are calibrated to produce the amount of a given substance or substances.

The training data should be subjected to detection of residuals, outliers, and nonlinearity. The outliers should be excluded and nonlinearity corrected if present. Otherwise the results might be distorted. Typically the number of wavelengths from the optical measurements needs to be reduced due to their high amount and interdependence. Many techniques are available for wavelength reduction (e.g., splines or Savitzky-Golay filtering).

Maybe the most often used regression models are multiple linear regression (MLR) or "ordinary" least squares (LS), ridge regression (RR), principal component regression (PCR), and partial least squares (PLS). The general equation for the regression is

$$y = b_0 + b_1x_1 + b_2x_2 + \cdots + b_kx_k + f$$
$$= b_0 + \sum_{i=1}^{k} b_ix_i + f = b_0 + f + BX, \quad (1)$$

where y is the response (from chemical analysis), x_k are the predictors (from optical measurements, k pieces), b_k are the regression coefficient to be determined, b_0 is the offset and a constant factor, and f is the residual. If X and y are meancentered, then $b_0 = 0$.

The solution for regression coefficient for the LS is

$$b = \left(X^TX\right)^{-1}X^Ty, \quad (2)$$

where T means transpose of the matrix. The LS may not work since the inverse might not exist or may be unstable. It is also notoriously sensitive to noise. The results should be also diagnosed using the following method: coefficient of determination and root mean squared error of calibration.

A better solution for problems related to inversion can be RR:

$$b = \left(X^TX + rI\right)^{-1}X^Ty, \quad (3)$$

where I is an identity matrix and r is a ridge constant. The ridge constant should be as small as possible. However, there is no established method for the selection of the ridge constant or diagnostic of results.

PCR also provides an easy inversion of matrices. It requires applying the principal component analysis (PCA) to reduce dimensions of X and to obtain the eigenvectors. The eigenvectors are orthogonal. The PCA also reduces the noise present in X. The equation for regression can be formulated as

$$y = Vd + f, \quad d = \left(V^TV\right)^{-1}V^Ty, \quad (4)$$

where V is the principal components and dcontains coefficients. The number of components needs to be determined by testing and checking.

Assume that y matrix consists of several measurements of y. PLS decomposes both y and x matrices. It also removes noise in both of these matrices. Its formulae are

$$y = UQ^T + F,$$
$$X = TP^T + E, \quad (5)$$

where T and U are projections (principal components) of X and y on eigenvectors and P and Q are eigenvectors of X and y. The projections are obtained by decomposing the corresponding matrix. Error terms are marked as E and F.

Both PLS and PCR calculate one component at a time using regression between t and u.Then a residual is calculated and the calculation of the next component is based on this residual

The obtained results for equations need to be evaluated by using the test set. The predicted values are calculated using test values from the optical measurement set and parameters obtained from the training phase. The difference between actual and predicted values in the test set can be evaluated using the root mean squared error of prediction and the coefficient of determination.

The visualization gives a good view of how the model is working. This could be done, for example, by plotting root mean squared error of calibration, root mean squared error of prediction, calculated loading plot, residual versus sample, and residual versus predicted test value.

Food Quality and Safety

Safety Factors

We consider here food safety to factors which may cause danger to health either directly or indirectly.

(1) Foreign Bodies. A foreign body can be anything unwanted in a food product. It can be from the raw material for the product or external material added indeliberately; for example, the raw material can contain bones which are shattered to sharp bone fraction pieces during the manufacturing process. The product may be littered with machine parts due machine blistering or residues from other sources like pieces of plastic.

(2) Incorrect, Contaminated, or Phony Ingredients in Product. Incorrect food ingredients refer here to any material which should not be in the food. They can be residues from previous production like peanut remnants or accidently added wrong material. The food can be

contaminated from external sources like manure or too much pesticide. The phony ingredients are deliberately added to generate more income from consumers or manufacturers. The phony items can be rip-offs like cheap wine sold as a more expensive version or even very dangerous ones like melanin milk.

(3) Spoilage. The spoilage is considered in this paper as a decomposition of the food. It is often a byproduct of the activity of harmful bioagents like fungi or bacteria. Other causes are wrong handling, like overheating, freezing or wrong harvesting time, wrong environment, like too hot environment or environment that has gases like ethene which mature fruits too fast, or age.

(4) Bioagents: Fungi, Yeast, and Molds; Microbes, Bacteria, and Viruses. Fungi, yeast, or molds can be used in the production of certain food, but usually they are serious quality problems. This paper concentrates on articles on harmful effects of the bioorganisms like food poisoning or infections. The detection of smaller bioorganisms like bacteria or viruses is usually very hard due to their small size.

Quality Factors

Food quality consists of several factors. The validity or importance of a factor depends on the food group and on the evaluator. For example, the fat content is irrelevant for apples but very relevant for meat. The factors for this paper are appearance and taste.

(1) Appearance. The appearance of food is related to the visual experience of the product. It is affected by the perceived colors and textures as well as personal preferences and opinions. The environment in which the food is presented has its own effect. Since the color perceived depends on the prevailing illumination, for example, there are already available special lamps to emphasize the color of meat. Good texture can be achieved with proper processing.

(2) Taste. Taste is composed of several ingredients and is affected greatly by smell and also appearance. In this paper, we limit the taste to the actual food ingredients or state-like amount of sugar or acidity.

Smell. The smell can be used to separate bad food from good food. It is also essential to make food more delicious.

Juiciness. One attribute especially for fruit is juiciness. It is related to the water content and distribution of fruits.

Noise. Consumers often expect certain noise when eating food like cereals.

Freshness. Freshness is important especially for fish and other sea products.

Authentication. Some food products have been falsified. An example of this kind of products is wine. Spectral methods can be used to separate authentic wine from nonauthentic wine. The NIR spectra is used for food authentication by Sun [21].

Essential Ingredients. Food can contain many elements which have positive effects on health. These elements include, for example, flavonoids and vitamins in plants flavonoids and vitamins or omega-6 fatty acid in fish. It is generally thought that the higher the amount obtained from food the better the effect on health. Ingredients can be correlated using chemometrics.

Unwanted Ingredients. Food contains also ingredients which consumers want to avoid for real or imaginary reasons. These ingredients can be preservatives which can in fact make the food safer. Other harmful ingredients include excessive amount of fat or salt.

Unwanted Objects. Although food can be perfectly safe and edible, it can still contain unwanted objects. This kind of objects can range from machine parts to stones and insect parts.

In their book, Workman and Weyer [22] explain theoretical backgrounds on how different qualities can be analyzed and found from NIR spectra; these qualities include alkanes and cycloalkanes, methylene groups, alkenes and alkynes, aromatic compounds, hydroxyls, water, carbonyls, amines and amides, carbohydrates, amino acids, peptides, and proteins.

High Quality. Affordable price consistently.

Chemometric Models. None known.

Optimization

Optimization with NIR spectroscopy is mostly connected with the selection of optimal wavelengths that correlate best with the object of interest. Usually we want to use as few wavelengths as possible, since the less the data needs to be analyzed the faster it is. However, we need to be sure that the selected wavelengths are adequate for reliable results.

A special type of genetic algorithms is called genetic programming (GP). It uses genetic algorithms to find a function or a program to solve a given problem. GP has been demonstrated to find solutions to problems that have been found by the most skilled engineers as patents and other inventions [23]. Another view of GP is to say that it is symbolic regression, that is, a method to find a function that best satisfies the given constraints. Yet another usage of GP is to find rules to explain data. GP is an extremely flexible method to solve computational intelligence problems. The main drawback of GP is its slow processing which prevents its usage in most online applications. However, in the systems design or similar phase, GP can be a nice tool to find good solutions automatically after the proper fitness function is implemented.

While spectroscopy produces huge amounts of raw data, there is a strong need to automatically process this data to compress it and to find the features that are relevant or correlate to those that are looked for. Norris and Williams [24] used wavelength selection so that they selected 12 wavelengths that produced the closest prediction with the required constituent in wheat measurements, particle size, and protein content. Goodacre and Kell [25] have demonstrated spectroscopical imaging GP applications related directly or indirectly to food quality. More specifically the application was to find the algebraic formula of the wavelengths that give the best discrimination between the bacterial samples examined. More generally the authors claim that GP is one potential tool to decrypt metabolomic data for biomedical applications.

With several materials, the VIS/NIR reflectance works best if the wavelengths used are optimized between 400 and 1800 nm [26]. This was also true with wholegrain measurement, where they found wavelengths between 1100 and 1400 nm giving the highest correlations with most of the constituents of interest with wholegrain measurements. However, the protein content was best determined from the 1700 to 2500 nm wavelength band.

Illustrative Example of Tomato Classification

The spectral methods lend themselves very well to fruit quality inspection. In this example, the transmittance spectra of tomatoes were measured using a spectrophotometer (Ocean optics). The spectra were further handled: they were smoothed to reduce noise and then downsampled by splines. An example of tomato spectra is shown in Figure 4. The transmittance is measured from 400 nm to 1000 nm. The spectra differ in both visible and NIR areas. The difference is caused by the ripening process.

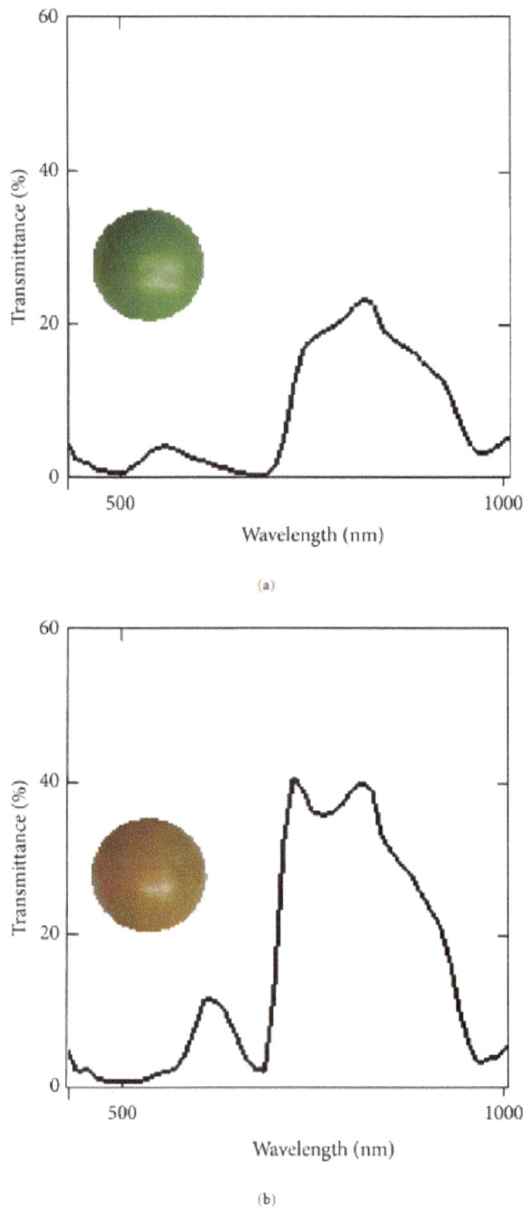

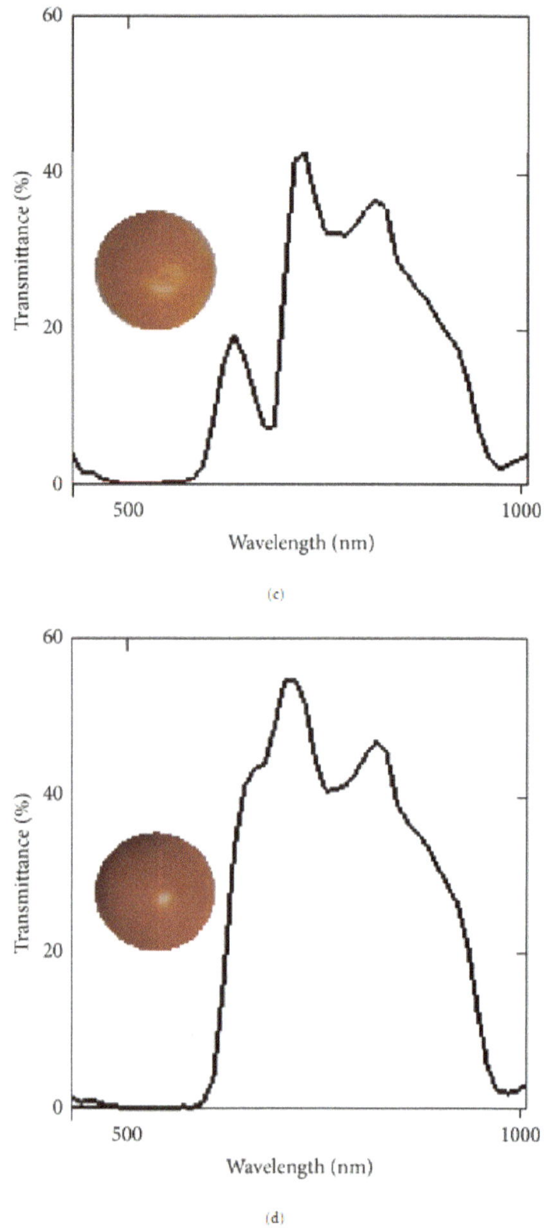

Figure 4: Transmittance spectra for different tomatoes of the same cultivar from 400 nm to 1000 nm. (a) A raw tomato, (b) a partially raw tomato, (c) an acceptably ripe tomato, and (d) a ripe tomato.

The spectral data itself contains a lot of information, part of which is in some cases irrelevant or redundant. The dimensionality of data can therefore be reduced. Here the locally linear embedding was used for this purpose in a nonlinear manner. The visualization of results can be seen in Figure 5, in which the feature space is spanned by the first two embedded components.

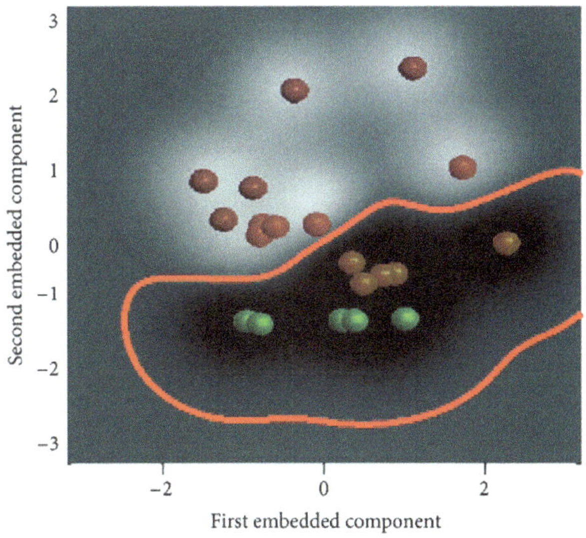

Figure 5: The feature space is spanned by the first two components. The red curve indicates the boundary between good and bad tomatoes. The good tomatoes are outside the red curve.

The kernel Fisher discriminant (KFD) was used to classify tomatoes into good and bad ones. As shown in Figure 5, the tomatoes are classified 100% correctly into these two groups. The red-colored tomatoes are ripe while the yellow and green are raw. The boundary between the groups is visualized with red curve (Figure 5).

REVIEW OF HYPERSPECTRAL/MULTISPECTRAL METHODS FOR DIFFERENT FOOD GROUPS

This paper follows the FAO classification in grouping food which divides the food and animal feed into five classes:(i) class A—primary food commodities of plant origin,(ii)class B—primary food commodities of animal origin,(iii)class C—primary animal feed commodities,(iv)class D—processed foods of plant origin,(v)class E—processed foods of animal origin.

Each of these classes has subclasses. Class A consists of five subclasses: fruits, vegetables, grasses, nuts and seeds, and herbs and spices. Class B has the following classes: mammalian products, poultry products, aquatic animal products, amphibians and reptilians, and invertebrate animals. Class C has only one class, primary feed commodities of plant origin. Subclasses, which belong to class D, are secondary food commodities of plant origin, derived products of plant origin, manufactured foods (single ingredient) of plant origin, and manufactured foods (multi-ingredient) of plant origin. Class E is made of the following subclasses: secondary food commodities of animal origin, derived edible products of animal origin, manufactured food (single ingredient) of animal origin, and manufactured food (multi-ingredient) of animal origin.

More detailed reviews of the selected papers for each of the classes can be found at the Appendices A (for class A), B (for class B), C (for class D), and D (for class E). The papers in class C are not considered in this paper.

Class A—Primary Food Commodities of Plant Origin

Class A has five groups: fruits, vegetables, grasses, nuts and seeds, and herbs and spices.

The fruit group consists of 6 subgroups: citrus, pome and stone fruits, berries and other small fruits, and assorted tropical and subtropical fruits with edible or inedible peels. Optical and spectroscopic methods [27] and near-infrared spectroscopy [28] present a review of promising methods for fruit quality.

Vegetables include bulb vegetables, cabbage (Brassica (cole or cabbage) vegetables, Head cabbages, Flowerhead cabbages), fruiting vegetables either cucurbits or other than cucurbits, leafy vegetables (including Brassica leafy vegetables), legume vegetables, pulses, root and tuber vegetables. Mushrooms and fungi belong also to this group. Reviews of vegetable quality have been made by Butz et al. [27] and Nicolaï et al. [28].

The grasses have two groups: cereal grains and grasses used for sugar or syrup production. The nuts and seeds' group consists of tree nuts, oilseed, and seed for beverages, and sweets. The herbs and spices' group consists of herbs and spices as the name indicates.

Table 2 provides a short description for each of the subgroups of class A. We list parameters which have successfully been evaluated for the group and methods applied.

Table 2: Different food groups, their parameters and methods applied for class A.

Food group/appendix	Examples of parameters evaluated and one of related research articles	Methods applied

Fruit/A.1	Quality	
	Taste (Mehinagic et al. [46])	
	Content	
	(i) Sugar (Walsh et al. [47])	MLR, PCR, PLS, PCAGenetic algorithms for wavelength and calibration model selection
	(ii) Soluble solid (Carlini et al. [48])	
	(iii) acid content (León et al. [51])	
	(iv) water/moisture (Blakey et al. [50])	
	(v) titratable acidity and ph (Ying and Liu [56])	
	(vi) vitamin C and polyphenol (Pissard et al. [57])	
	Translucency (Terdwongworakul et al. [53])	
	Internal quality (Magwaza et al. [54])	
	Maturity (Tanaka and Kojima [58])	
	Defects (Blasco et al. [262])	
	Safety	
	Internal insect detection (Xing et al. [60])	
	Fecal and other contaminations (Kim et al. [61] and Mehl et al. [263])	
	Storage (Veraverbeke et al. [63])	
Vegetables/A.2	Quality	Texture metrics, wavelength selection, PLS, MLR, PCA
	Appearance (Flores Rojas [65] and Gowen et al. [84])	
	Content	
	(i) Cyst nematode (Fernández Pierna et al. [64])	
	(ii) Sugar (Tsuta et al. [67])	
	(iii) Soluble solid content, lycopene, -caro-tene,	
	and carotenoid (Pedro and Ferreira [68])	
	(Iv) Protein (Liu et al. [69])	
	(v) Phytic acid (Delwiche et al. [73])	
	(vi) Dry matter (Van Dijk et al. [70])	
	(vii) Protein (Pazdernik et al. [75])	
	(viii) Freshness (Esquerre et al. [81])	
	(ix) Moisture (Lamb and Hurburgh Jr. [78])	
	Structural characteristics (Siska and Hurburgh Jr. [83])	
	Damage (Esquerre et al. [81])	
	Safety	
	Incorrect ingredients (Roggo et al. [72])	
	Bioagents (Pearson and Wicklow [85])	

Grasses/A.3	Quality	PLS, MLR, PCA
	Content	
	(i) Moisture, gluten, alveograph parameters, and protein (Miralbés [90])	
	(ii) Scab, vomitoxin, and ergosterol (Dowell et al. [89])	
	(iii) Rutin and D-chiro-inositol (Yang and Ren [91])	
	(iv) Carotenoids (Brenna and Berardo [92])	
	(v) Phenolics and antioxidant capacity (Zhang et al. [94])	
	(vi) Neutral acid detergent fibers and acid detergent lignin (Stubbs et al. [95])	
	(vii) Nitrogen and protein (Wiley et al. [96])	
	structural characteristics	
	(i) Softness/hardness (Manley et al. [98])	
	(ii) Density (Siska and Hurburgh Jr. [83])	
	Taste (Kawamura et al. [99])	
	Safety	
	many impurities like plastics or faeces (Fernández Pierna et al. [64])	
Nuts and seeds/A.4	Quality	PLS
	Content	
	(i) Moisture, fat, stearic acid, and oleic acids (Davrieux et al. [100])	
	(ii) Sinigrin, gluconapin, and glucosinolate (Font et al. [101])	
	(iii) Lipids, water, protein, and chlorophyll (Font et al. [102])	
	(iv) Oil and fatty acid content (Kim et al. [103])	
	(v) Starch (Hacisalihoglu et al. [104])	
	(vi) Amino acids (Wang et al. [105])	
	Authentication (Downey et al. [107])	
Herbs and spices/A.5	Quality:	PLS
	Content	
	(i) Kava pyrones (Gautz et al. [110])	
	(ii) Moisture, volatile content, and coloring strength (Zalacain et al. [111]	
	(iii) Curcuminoids (Tanaka et al. [112])	
	(iv) Ginsenosides (Ren and Chen [113])	
	(v) Lignans (Kwan et al. [114])	
	Origin (Zalacain et al. [111])	

Class B—Primary Food Commodities of Animal Origin

Class B consists of the following groups: mammalian products, poultry products, aquatic animals, amphibians and reptilians, and invertebrate animals. Examples of publications can be found at Table 3.

Table 3: Class B groups

Food group/appendix	Examples of parameters evaluated and one of related research articles	Methods applied
Mammalian products/B.1	*Quality* Taste (Byrne et al. [120]) Color (Van den Oord and Wesdorp [124]) Diet (Prieto et al. [130]) Content (i) pH (Andrés et al. [128]) (ii) Moisture (Chan et al. [135]) (iii) Fat (Sierra et al. [136]) (iv) Iron (Hong and Yasumoto [139]) (v) Protein and dry content (Prieto et al. [140]) Maturity (Prieto et al. [154]) Shear force (Yancey et al. [155]) Tenderness (Shackelford et al. [126]) Texture (Byrne et al. [[120]) Sensory characteristics (Prieto et al. [159]) Structural properties (Shackelford et al. [126]) Drip loss (Geesink et al. [160]) Authentication (Dian et al. [168] and Arvanitoyannis and Houwelingen-Koukaliaroglou [264]) Freshness (Downey and Beauchêne [173]) *Safety* Spoilage (Horváth et al. [116]) Contamination (Windham et al. [118]) Bioagents (Ellis et al. [117]) Adulteration of milk (Sato et al. [181])	Color, geometrical features, texture Multivariate data analysis, ratios, MLR, PCA, PLS
Poultry products/B.2	*Quality* Content (fat, protein, iron) Authenticity/origin (Hong and Yasumoto [139]) Storage and handling (Liu et al. [192]) *Safety* Safety in general (Chen et al. [182]) Fecal contamination (Park et al. [188]) Spoilage (Lin et al. [190])	Wavelength selection PLS Correlation
Aquatic animal products/B.3	*Quality* Temperature Appearance (olsen [195]) Identification/authentication (Cozzolino et al. [197]) Freshness (Uddin et al. [198]) Content (i) Water, fat, and protein (Gjerde and Martens [204]) (ii) Free fatty acids (Zhang and Lee [203]) (iii) Salt and moisture (Huang et al. [7]) (iv) Crude lipid (Mathias et al. [209]) Defects like bruises (Lin et al. [211]) *Safety* Incorrect ingredients (Font et al. [212]) Spoilage (Lin et al. [202])	Wavelength PCA, linear discriminant analysis PLS
Invertebrate animals and spices/B.5	*Quality* Authenticity (Gayo and Hale [213]) Content (Brown [215])	PLC, PCA

Mammalian products consist of nonmarine meat, mammalian fats, edible offals, and milks. It seems that NIR technologies are more acceptable, for example, to the AOAC (Association of Official Agricultural Chemists).

The poultry can be studied by other optical techniques than just reflectance or transmittance. Autofluorescence has been suggested for food by Andersen et al. [29] and for poultry by Wold and Mielnik [30].

FAO includes in aquatic animals fish (freshwater, marine and diadromous, i.e., migrating between fresh and saltwater), fish roe and milt, fish offals, marine mammals, and crustaceans. Seafood and its products are an important source of nutrition [31]. Their special challenge is the changes in properties from the moment of catch or harvest due to metabolic processes, microbiological processes, and microbiological fauna change. This change is described as a function of spoilage and freshness parameters.

The invertebrate animals' group consists of molluscs (including cephalopods) and other invertebrate animals.

Class D—Processed Foods of Plant Origin

Class D has four groups: secondary food commodities of plant origin, derived products of plant origin, manufactured foods (single ingredient) of plant origin, and manufactured foods (multi-ingredient) of plant origin.

Secondary food commodities of plant origin are dried fruits, vegetables and herbs, secondary food commodities of plant origin, and miscellaneous secondary food commodities of plant origin.

Derived products of plants for human consumption include cereal grain milling fractions, teas, vegetable oils: crude, edible, or refined, origin including miscellaneous derived edible products of plant origin, and fruit juices.

Manufactured foods (single ingredient) of the plant origin group naturally consist of fruits, vegetables, or miscellaneous manufactured single-ingredient foods from plants. Manufactured foods (multi-ingredient) of the plant origin group are all manufactured, multi-ingredient cereal products. Table 4 shows examples of published works in this area.

Table 4: Class D food groups, their parameters, and methods applied.

Food group/appendix	Examples of parameters evaluated and one of related research articles	Methods applied
Secondary food commodities of plant origin/D.1	Content (i) Starch, sugar, nitrogen, and ash (Lebot et al. [216]) (ii) Hollow heart (Dacal-Nieto et al. [217]) (iii) Bioactive components (Jourbert et al. [218]) (iv) Glucan and protein (Bellato et al. [219]) Adulteration (Laasonen et al. [224])	Wavelength selection PLS PCA
Derived products of plant origin/D.2	Content (i) Dry extract and sugars (Li et al. [225]) (ii) Acids (Chen et al. [226]) (iii) Alcohol (González-Sáiz et al. [227]) (iv) Soluble solids and pH (Liu and He [229]) (v) Oils (Ng et al. [230])	Multivariate calibration PLS, MLR
Manufactured foods (single ingredient) of plant origin aquatic animal products/D.3	Content (i) Carbohydrate (Chen and Lei [232]) (ii) Soluble solids and pH (Liu et al. [238]) (iii) Methanol (Dambergs et al. [233]) (iv) Acids (Fan et al. [236]) Protein-lipid-carbohydrate interactions (Bruun [237]) Authentication (Cozzolino et al. [239]) Age	Wavelength selection PLS, LS, PCA
manufactured foods (multi-ingredient) of plant origin/D.4	Content Fat (Vines et al. [243]) Moisture (Kays et al. [245]) Impurity (Fernández Pierna et al. [64]) Total fiber amount (Kays et al. [246]) Structural properties (de Alencar Figueiredo et al. [253]) Handling and fermentation (Bruun et al. [254])	PLS Wavelength selection

Class E—Processed Foods of Animal Origin

Class E has the following subgroups: secondary food commodities of animal origin, derived edible products of animal origin, manufactured food (single-ingredient) of animal origin, and manufactured food (multi-ingredient) of animal origin.

Secondary food commodities of animal origin consist of dried meat and fish products and secondary milk products. We include here also products using dried meat and fish. An example of this is ham.

Derived edible products of the animal origin group include processed crustaceans and animals, milk, and derived fats. Manufactured food (single ingredient) of the animal origin group is the manufactured milk products using only single ingredient.

Manufactured milk products using multi-ingredient belong to manufactured food (single ingredient) of the animal origin group. Here we consider dairy products like cheese or yogurts. Some publications has been listed in Table 5.

Table 5: Different food groups, their parameters, and methods applied for class E.

Food group/appendix	Examples of parameters evaluated and one of related research articles	Methods applied
Secondary food commodities of animal origin/E.1	Appearance (García-Rey et al. [257])	Wavelength
Manufactured food (multi-ingredient) of animal origin/E.5	Appearance (Lucas et al. [258]) Smell and taste (Çelik [259]) Content (Lucas et al. [258], Arvanitoyannis and Van Houwelingen-Koukaliaroglou [264]) pH (Navrátil et al. [260]) Origin (Cozzi et al. [261]) Authenticity (Woodcock et al. [265])	PLS, wavelength discriminant analysis

Other Foods Not Included in FAO Criteria

Some foods are not included in FAO classification although they are commonly used. For example, honey is not included. There is no clear category for processed foods like hamburgers or sausages. Some of the research done for these foods is introduced here.

Authentication

There have been cases in which the ingredients of manufactured food have been replaced by a cheaper version. For example, beef hamburger can be adulterated with mutton, pork, skim milk powder, or wheat flour. Ding and Xu [32] used near-infrared for detecting and evaluating the adulteration level in the adulterated hamburgers with 5–25% phony ingredients. The accuracy increases as a function of adulteration rate since there were more ingredients to be detected. The highest detection rate was 92.7%.

ElMasry et al. [33] have applied hyperspectral imaging (Imspector N17E, Specim, Spectral Imaging Ltd., Oulu, Finland) in the region 900–1700 nm for predicting colour, pH, and tenderness of fresh beef. With more improvement in terms of speed and processing, their approach has potential for nondestructive quality measurements.

Liao et al. [34] have used VIS/NIR reflectance spectroscopy for the online assessment of the pH value in fresh pork. Their result show that VIS/NIR spectroscopy can be used for the online prediction of pH and that wavelength selection can provide a cost-effective calibration model. For wavelength selection, they compared variable importance in projection (VIP), uninformative varible elimination (UVE), genetic algorithms, and success project algorithm (SPA).

Collell et al. [35] have used NIR reflectance spectroscopy to predict moisture, water activity, and salt content on the surface of dry-cured ham during the process. Their results show that the proposed method could be used in the dry-curing process of ham. However, the optical properties of the meat surface seem to strongly affect the measured spectra.

Authenticity and origin are also important issues for honey especially when it is claimed to be collected by bees from certain flowers. One example is avocado honey. Dvash et al. [36] applied near-infrared reflectance spectroscopy to obtain the concentration of specific sugar (perseitol) for this honey.

Downey and Kelly [37] studied sulfited strawberry and raspberry purées which have been adulterated with cooked apples (10–75% w/w). They applied VIS and NIR transreflectance measurement on 0.1 mm thick samples. The best wavelength bands were 400–1098 nm with 75% correct classification rate for strawberry and 750–1098 nm with 95% rate for raspberry. PLS regression was used for the prediction of apple content with 400–1880 nm range for raspberry and with 1100–1880 nm range for strawberry. They concluded that apple adulteration can be detected if its content exceeds 25% for raspberry and 20% for strawberry purées.

Fügel et al. [38] made a review about authenticity control of fruit purées, fruit preparations, and jams.

Content

Qiu et al. [39] applied near-infrared transmittance to quantify the chemical content of honey. They found that 1 mm optical path length was the best for this purpose. Modified partial least squares were applied for the determination of content except for moisture. Moisture was determined using just partial least squares. The models were successful for fructose, glucose, sucrose, and maltose content but poor and unreliable for free acid, lactone, and hydroxymethylfurfural.

NIR spectroscopy is applicable also for wholemeals. Fontaine et al. [40] used near-infrared for detecting essential amino acids for different meals. They concluded that the NIRS calibration model was good for evaluating the content of protein-rich feed ingredient for amino acids methionine, cystine, lysine, and threonine, tryptophan with 85–98% variation explained and was better than crude protein regression. The protein and moisture contents were also evaluated.

Kim et al. [41] used NIR reflectance (400–2498 nm) and partial least-square modeling to detect the total dietary fiber in meals. Meals have different handling alternatives: homogenization, homogenization and drying, and homogenization, drying, and defatting. The reference method for total dietary fiber evaluation was AOAC 991.43. The predictions were good. The main advantage over the NIR-based method was faster evaluation. The traditional method takes 4 days.

Zeng et al. [42] developed a NIR transmission system for determining the amount of starch in liquid gravies with low concentration. The best wavelengths were 605, 623, 919, 1017, and 1031 nm for the regression model. The starch content makes possibility to evaluate the viscosity of liquid gravies.

Authenticity

Sausages can be made with different fats. In some cases, sausages can be identified based on their fat substitutes [43]. The quality of them has been studied, for example, by Gaitán-Jurado et al. [44] and Ortiz-Somovilla et al. [45].

CONCLUSIONS

Nondestructive methods for food quality measurement are warmly welcomed by the food industry. Optical engineering provides a large variety of measurement techniques, some of which, like optical and near-infrared spectroscopy and imaging, have especially high potential for various food-quality-related measurements. And indeed, during the last few decades, it has seen an increasing number of applications, not only in food quality laboratories, but also in factories, and even in farm fields and orchards, where the food quality arises. In this paper, we have shortly described the most prominent application examples in this rapidly growing area of food quality techniques. Unfortunately there are already so many important works in this area that we were able to consider only a small fraction of the total literature existing, which means that too many important contributions had to be omitted from this paper. However, the authors hope that even in its current imperfect state this paper could help those who are trying to find solutions or model approaches to their food quality problems from optics, imaging, and spectroscopy.

APPENDICES

A.

Class A, Type 01: Fruits

Taste

Taste is an elemental parameter for fruits. Mehinagic et al. [46] demonstrated promising results in combining taste and spectra for apples.

Content and Factors Related to Content

A popular ingredient sought after for a fruit is sugar, for example, Walsh et al. [47]. Another important content is soluble solid content. Carlini et al. [48] evaluated the soluble solid content using VIS/NIR spectra. They modeled spectra using partial least-square regression.

The time-of-flight technique with NIR was applied to detect sugar and acid contents of apples [49]. The modeling used multiple linear regression, principal component regression, and partial least squares. The prediction precision was high and especially superior for acid content.

NIR has been used to predict water content of several different fruits. The water content of avocados was used by Blakey et al. [50] to predict ripening of the fruit. The water content was predicted using PLS (partial least squares) model with 14 factors in the region of 1100–2000 nm.

León et al. [51] studied the intact olive VIS/NIR spectra (400–1700 nm) for evaluation of oil, moisture, oleic acid, and linoleic acid contents. They compared spectra from different female parent trees and harvest year. The partial least-square calibration models were not transferable across year.

Zhang et al. [52] have recently used partial least-square (PLS) calibration model for the measurement of soluble solids content (SSC) of fresh jujube (Ziziphus zizyphus L.). The device used was FieldSpec3 (Analytical Spectral Device, USA). The correlation found was over 0.88 indicating the feasibility of the method for SSC measurement.

Terdwongworakul et al. [53] have used NIR transmittance spectroscopy for the quality assessment of mangosteen fruits. The goal was to develop a method for nondestructive detection of translucency in mangosteen fruit. The water peak absorbance at 955 nm seems to be an important factor for discriminating translucent and normal fruits.

Magwaza et al. [54] give a recent review of NIRS in external and internal quality analysis of citrus fruits.

Shi et al. [55] have used the genetic algorithm (GA) for wavelength selection for the determination of SSC in Fuji apples. They used NIRS in the range from 1065 to 1625 nm and were able to reduce the number of wavelengths from 88 to 17 with their GA approach.

Ying and Liu [56] have used FT-NIR spectroscopy for the internal quality assessment of pears. For wavelength selection, they also used genetic algorithms (GAs). The quality factors tested were sugar content (SC), titratable acidity (TA), and pH. Despite the complexity of the spectral data, the GA found calibration models that significantly outperformed those based on full-spectrum analyses. Pissard et al. [57] suggested the use of NIR spectroscopy and multivariate calibration to determine the levels of vitamin C, total polyphenol, or sugar content. They stated that the prediction performance was very good.

Maturity

Maturity or fruit ripeness affects the quality and taste of the fruit. Tanaka and Kojima [58] studied the growth stages and maturity of Japanese pear fruit using the sugar concentration as an indicator. Pear juice from different maturity stages produced different spectra, the absorption bands of which were related to carbohydrate. Multiple linear regression models were applied to determine the content of sucrose, fructose, sorbitol, and glucose sugars with 3–7 wavelength variables. Sirisomboon et al. [59] used NIR with principal component analysis (PCA) to classify tomatoes. The model provided a prediction for maturity.

Unwanted Objects

Insects cause damage to fruits in many ways. Some of them can grow inside the fruit like in tart cherry. Xing et al. [60] have used genetic algorithms to find the optimal wavelengths for detection of internal insect (plum curculio (Coleoptera: Curculionidae)) infestation that in tart cherry. Based on the GA wavelength selection on the reflectance spectra (580–980 nm), three to four wavelength regions were selected. The results were checked using partial least-squared discriminant analysis (PLSDA) and they showed that the wavelengths selected resulted in classification efficiency similar to that of the full spectral range. Different GA runs resulted in somewhat different results, which are typical for the wavelength selection problem.

Contamination

Fruit skin can be contaminated by fecal material which can have pathogens like E. coli. Kim et al. [61] applied multispectral imaging 450–851 nm to detect fecal with thin and thick layer on apples with different green and red colorations. This was somehow successful for thick patches. Apple coloration caused problems with thin layer. Kim et al. [62] studied also fluorescence spectral information. They identified four optimal bands (450, 530, 685, and 735 nm) for fecal detection and suggested the use of a simple ratio.

Storage

Veraverbeke et al. [63] utilized NIR transmission spectra to solutions of single wax components and extracted apple wax. The wax was collected from apples with different parameters: apple cultivar (Jonagold, Jonagored, and Elshof), picking data (early, commercial, and late), storage time (0, 4, or 8 months in controlled atmosphere), and shelf life periods (0, 1, and 2 weeks). They applied canonical discriminant analysis on the first derivative of spectra. The proposed NIR technique could identify cultivar and storage duration but not very well the picking date or shelf life period. It seems that storage time and cultivars are related to differences in the number of aliphatic chains, like alkanes and esters, and the presence of α-farnesene.

Class A, Type 002: Vegetables

Fernández Pierna et al. [64] have used NIR hyperspectral

imaging and chemometrics for the detection of cyst nematode (Heterodera schachtii) infected sugar beets. They used a pushbroom system (SWIR XEVA CL 2.5 320 TE4 camera from XENICS using an Imspector N25E spectrograph with a cooled, temperature-stabilised mercury-cadmium-telluride (MCT) detector from Specim Ltd., Oulu, Finland) combined with HyperPro and HyperSee software (BurgerMetrics SIA, Riga, Latvia). The samples were moved with a conveyor belt at 2 mm/s speed. The result of the study was that NIR imaging has potential for, for example, plant breeders to select cyst-nematode-resistant traits.

Appearance

The quality of a vegetable is sometimes evaluated by its texture. One of these kinds of vegetables is Asparagus, the texture of which has been evaluated by Flores Rojas [65]. The textural quality parameters considered in their research were maximum shear force and cutting energy. The evaluation was based on results from a scanning monochromator of 400–2500 nm and a combination of diode array and scanning monochromator of 350–2500 nm.

Content

Sugar concentration is important for many vegetables because it affects their sensory evaluation and storage ability. Sugar content of melon was evaluated and visualized using NIR absorption spectra by the autor in [66]. His paper indicated a strong inverse correlation for a melon between sugar content and specific wavelength (676 nm) near to the chlorophyll absorption band. The visualization was implemented by taking an image through a filter with band around 676 nm and converting the pixel values to another unit (absorbance) to assign them with a color. However, there was still a bias caused by melon color. Later, they developed another visualization method using deeper NIR region to prevent bias from melon color [67]. They noticed that the two 2nd derivative absorbances in 874 nm and 902 correlated strongly to the sugar content. For this visualization, two band-pass filters were employed.

Pedro and Ferreira [68] applied NIR and multivariate calibration to evaluate soluble solid, lycopene, β-carotene, and carotenoid contents of tomato. The PLS-1 was used for calibration. Prediction accuracy values were over 0.99.

Protein content of edible fungi (Auricularia auricular) was studied by Liu et al. [69] They applied partial least-square, multiple linear regression, and least square support vector machine on data and compared performances of different preprocessing including Savitzky-Golay smoothing, standard normal variate, multiplicative scatter correction, first derivative, second derivative, and direct orthogonal signal correction. Wavelength selection was based on successive projections algorithm. Liu et al. [69] concluded that the best full spectrum partial least-square model was obtained using "raw" data whereas optimal successive projection algorithm with multiple linear regression, successive projection algorithm with partial least squares, and successive projection algorithm with least-square support vector machine models were achieved with spectra preprocessed multiplicative scatter correction. The optimal model was produced by successive projection algorithm with least-square support vector machine combination.

For cooked potatoes, dry matter content is important. Van Dijk et al. [70] studied three potato cultivars, cv. Nicola (a firm cooking potato), cv. Irene (a mealy cooking potato), and cv. Bintje (intermediate between firm and mealy cooking potato). They related NIR spectra to dry mater content and sensory perceived texture of steam cooked potatoes. The dry mater content affected most on the texture while the effect of chosen cultivar is minimal. Storage did not have any effect on the relationships.

Dacal-Nieto et al. [71] have used genetic algorithm (GA) for feature selection in potato classification done by the help of computer vision.

Roggo et al. [72] applied NIR for sucrose content with very small sucrose ester of fatty acid (SEF, fresh beet) and got satisfactory results for brix, marc, sugar in molasses, and juice purity but lower accuracy results for glucose, nitrogen, sodium, and potassium.

Soybean is an important vegetable in many countries. Soybeans have phytic acid which has nutritional and environmental importance. To evaluate the levels of the phytic acid, Delwiche et al. [73] measured its complement, inorganic phosphorus, using diffuse reflectance of near-infrared (1100–2500 nm) for ground meal and transmittance (600–1900 nm) for single beans. The results were not very good with partial least-square regression but give an indication. Kovalenko et al. [74] developed calibration equations for estimating amino acid composition of whole soybeans from near-infrared spectra. They also concluded that cysteine and tryptophan could not be estimated from spectra while leucine and lysine have a high correlation.

Pazdernik et al. [75] applied near-infrared reflectance spectroscopy to evaluate crude glycinin fraction of soybean protein. They find that the fraction was positively correlated to crude protein concentration and negatively to oil concentration. The high crude protein and fraction content was related to higher growing temperatures while the lower growing temperature produced higher oil concentration. Pazdernik et al. [76] studied also amino and fatty acid compositions. Hollung et al. [77] applied NIR reflectance on soybeans to find samples with a low content of oligosaccharides and nonstarch polysaccharides. They were able to predict that total content of nonstarch polysaccharides could be predicted using partial least-square regression.

The moisture content of whole soybean in the range of 5 to 22% can be determined with great accuracy using near-infrared transmittance (800 to 1100 nm) and difference in absorbance, stepwise multiple linear regression, or partial least-square modeling [78].

The classification based on soybean position can be useful, too. Bennett et al. [79] analyzed field-grown soybean seeds with different nodal positions with NIR. They found that seed protein content varied as a function of place: those from plant basal region have higher oil content and lower protein content while the opposite was true for the apex. Later in [80], they studied the seed quality traits (protein, oil) utilizing NIR to determine percentage of oil and protein. They noticed that early-planted crop has increased the amount of oil (3–8%) while there was not much change in protein content.

Freshness

Quality of mushroom can decrease if too long time elapsed from harvest. Esquerre et al. [81] used VIS and NIR absorption spectra to detect bruises and evaluate the freshness. The partial least-square model was employed and it was mostly influenced by changes in water matrix and enzymatic browning.

Authentication

Li et al. [82] have recently tested classification methods for the authentication of pure camellia oil NIR spectra. Sample authenticities were confirmed by gas chromatography. The classification methods were unsupervised principal component analysis (PCA), hierarchical cluster analysis (HCA), supervised classification techniques based on discriminant analysis (DA), and radical basis function neural network (RBFNN). The classification rate of all methods reached 98.3% when smoothing, first derivative, and autoscaling were used. The good performance showed that NIR spectroscopy could be successfully used as a rapid, simple, and nondestructive method to discriminate pure camellia oil.

Structural Characteristics

The physical and chemical properties of soybean and corn are correlated with size and density. However, the change on size and density has no effect in accuracy of determination of soybean protein and oil content. The effect is very small for determining protein, oil, and starch contents for corn [83].

Damage

Vegetables, mushrooms, and fungi can suffer physical damages for too rough handling. VIS/NIR absorption seems to be useful for quantifying bruise damages for mushrooms [81, 84].

Incorrect Ingredients

Roggo et al. [72] suggested the use of NIR spectroscopy for decreasing pollution of heavy metals, lead acetate, caused by the legal methods, polarimetric measurement for sucrose content, and wet chemical analysis for quality control.

Bioagents

The detection of fungi from food assumes those the spectra of fungi differ from that of the food itself. This assumption has been shown to be valid, for example, by Pearson and Wicklow [85] who detected the fungi in corn kernels, Gordon et al. [86] for food grain, and Berardo et al. [87] for maize.

Jin et al. [88] have used GAs to find the optimal wavelengths when trying to discriminate aflatoxin-producingAspergillus flavus (AF13) fungi and non-toxin-producing Aspergillus strains (AF2038, AF283, and AF38). Their hyperspectral image was taken of several strains cultured 7 days on potato dextrose agar using both broad-spectrum UV light source and tungsten halogen bulbs. GA was used to find the most important principal components (PCs) among the first 40 PCs. The selected PCs were given to a support vector machine (SVM) for training for classification between toxigenic and atoxigenic fungi species. The discrimination accuracy varied between the strains and the light source used ranging from 70% to 95% pixel-wise.

Berardo et al. [87] applied NIR for the detection and identification of mycotoxigenic fungi and their toxic metabolites. They analyzed the samples for fungi infection, ergosterol, and fumonisin B_1 content. They concluded that NIR is suitable for monitoring mold contamination and separating the contaminated lots from clean lots.

Class A, Type 003: Grasses

Content

Cereal grains have many important ingredients which are searched in many studies. Also moisture is an important parameter. Dowell et al. [89] have applied NIRS for scab, vomitoxin, and ergosterol determination in single-wheat kernels. The results obtained show that the NIRS method was able to find all scab-damaged kernels that were identified by official inspectors.

Miralbés [90] measured NIR transmittance 850–1048 nm of whole wheat from different countries and evaluated their moisture, protein, wet gluten, dry gluten, and alveograph parameters (W, P, and P/L) using partial least squares. The alveograph parameters required division of wheat samples into two groups according to their deformation energy values.

Tartary buckwheat has beneficial ingredients, rutin and D-chiro-inositol, against hemorrhagic diseases and insulin-resistant diseases. Yang and Ren [91] applied NIR reflectance to evaluate their content. They developed a partial least-square method and used the results from HPLC as a reference. A good chemometric correlation existed.

Brenna and Berardo [92] studied maize whose antioxidant compounds like carotenoids are useful against reactive oxygen substances. They combined NIR and HPLC results for determining carotenoid concentration of grinded maize using modified partial least squares.

Rice is another very important food plant. Kawamura et al. [93] used NIR transmittance and PLS for successful evaluation of moisture and protein contents of undried rough rice. However, they found that their method was not applicable to determine amylase content. Zhang et al. [94] suggested use of NIR to predict total phenolics, flavonoid content and antioxidant capacity of dehulled grain of brown rice. The phenolic content, and antioxidant capacity were successfully determined using mPLS and PLS while this failed for flavonoids. Stubbs et al. [95] measured residue from cultivars of spring wheat (Triticum aestivum L.), winter wheat, and spring barley (Hordeum vulgare) with NIR and made models to estimate neutral detergent fiber (NDF), acid detergent fiber (ADF), acid detergent lignin (ADL), carbon (C), sulfur (S), nitrogen (N), and C : N. The models were successful for neutral acid detergent fibers and acceptable for acid detergent lignin, for others less successful.

Wiley et al. [96] studied barley mutants. They evaluated grain nitrogen and protein composition. The grains were measured with NIR reflectance spectra, which were resolutionally enhanced by derivative spectroscopy. The advantage of derivative spectroscopy is to allow wavelengths to be identified with clear differences in contribution from the associated chemical bonds. They compare the 4th derivative of wholemeal flour with high and low N containing grains.

Structural Characteristics

Structural characteristics like hardness have effect on quality. For maize, the hardness was evaluated using NIR spectra in [97]. Manley et al. [98] used hyperspectral images of 1000 to 2498 nm wavelength range to classify yellow maize kernels. They found clusters for different endosperms (vitreous, floury, and 3rd type) which present varying ratios of soft and hard kernels. Cereal composition reveals the properties of inner structure. Siska and Hurburgh Jr. [83] applied NIR transmittance and PLS to obtain corn density which is related to starch and protein.

Taste

Taste from cereal food is affected by the grains themselves. The study of Kawamura et al. [99] indicates that VIS/NIR reflectance could be useful for rough classification of wholegrain milled rice using multiple linear regression (MLR) analysis into three groups: poor, better, and best taste.

Safety

Fernández Pierna et al. [64] applied near-infrared hyperspectral images for separating impurities from cereal grains. The cereal grains were wheat, spelt, and barley while the contaminants were a variety of different matters (straw, broken grains, grains from other crops, weed seeds, insects, plastic, and stones, pieces of wood and paintings, and animal feces) classified into different categories. For each category, a discriminant model was developed. The method was based on PCA and support vector machine.

Class A, Type 004: Nuts and Seeds

Content

Nuts of shea tree can be used as a source of nut butter. The moisture, fat, stearic acid, and oleic acid can be determined using NIRS [100].

Front et al. [101] studied the sinigrin, gluconapin, 4-hydroxyglucobrassicin, and total glucosinolate of Indian mustard seed. They modeled the NIR spectra with modified partial least-square regression for chemometrics. The coefficients of determination were over 0.8 for all other except for 4-hydroxyglucobrassicin which was only 0.33. Font et al. [102] measured intact and ground seed of Indian mustard, Ethiopian mustard, and rapeseed and applied modified partial least-square regression to combine near-infrared spectra and acid detergent fiber values. The sample type did not have any significance on the evaluation. According to them, the most important factors for modeling were C–H and O–H groups of lipids, water, protein, and chlorophyll.

Kwan et al. [103] determined the oil and fatty acid contents of intact perilla seed using NIR reflectance. They concluded that mPLS produced good predictions for oil, oleic acid, and linolenic acid contents whereas the predictions for palmitic, stearic, and linoleic acid were unsuitable for screening purposes.

The content of intact common bean was subjected to NIR reflectance in order to determine seed weight, protein, and starch for 91 genotypes [104]. PLS model was able to predict protein content with high accuracy (SEP) = 1.6% and reasonable accuracy for weight and starch. Seed color had a correlation to spectra; black seed has very different spectra. Color did not affect model accuracy.

Wang et al. [105] used near-infrared and PLS to study proteins and amino acids in peanuts. They found the results acceptable. Yang and Sun [106] demonstrated that the hardness of licorice seeds can be done with NIR.

Authentication

Downey et al. [107] applied near- and mid-infrared spectra for authentication of lyophilized coffee samples of Arabica and Robusta cultivars. They used factorial discriminant analysis and partial least-square models.

Downey et al. [108] applied VIS and NIR transflectance spectra to detect adulteration of pure extra virgin olive oils. The adulteration rates for the olive oil were 1% (w/w) and 5% (w/w) sunflower oil. The best classification was obtained using 1st derivative of 400–2498 nm spectra. The adulterant content was predicted with 0.8% standard error accuracy.

Woodcock et al. [109] studied European olive oil samples to specify their geographical origins with near-infrared transflectance spectroscopy (1100–2498 nm). The samples were collected from harvests of 2005, 2006, and 2007. The detection rate for Ligurian ones was 92.8% and non-Ligurian ones 81.5% when using the first-derivative data pretreatment and discriminant partial least-square regression.

Class A, Type 005: Herbs and Spices

Content

The herbs and spices contain many characteristic ingredients, the amount of which is often a concern. Often the goal is to find an alternative method for HPLC. The herbs and spices can come in many forms like leaves, roots, or powders. Chemometrics can be applied to all these forms.

Another example is kava which is an herb with anxiolytic properties caused by active compounds called kavalactones or kava pyrones. As a natural product, the amounts of active compounds vary between specimens. The NIR reflectance and partial least squares can be used to detect these compounds reliablly [110]. The results from the developed method are reproducible and have repeatability comparable to HPLC methods.

NIR reflectance can be used to determine several factors of the chemical composition of saffron according to Zalacain et al. [111]: moisture and volatile content, coloring strength, $E^{1\%}_{1\,cm}$ (250 nm), and $E^{1\%}_{1\,cm}$ (330 nm), established on the ISO 3632 Technical Specification Normative and used to certify saffron quality in the international market.They also analysed five main crocetin glycosides (compounds causing saffron color). The best correlations obtained for trans-crocetin di (β-d-gentibiosyl) ester was $R^2 = 0.93$, for trans-crocetin (β-d-glucosyl)-(β-d-gentibiosyl) $R^2 = 0.94$, and for picrocrocin (saffron bitterness) $R^2 = 0.92$

Tanaka et al. [112] quantified curcuminoids (total curcuminoid content, curcumin, demethoxycurcumin, and bisdemethoxycurcumin) of turmeric. They applied 2nd derivative preprocessing and standard normal variate to NIR data and after that partial least-square regression. Characteristic absorption of curcuminoids was detected in approximately 1700 nm and 2300 nm–2320 nm. The results obtained have a high correlation with HPLC.

Many herbs and spices are often milled to powder-like Ginseng. Chemical methods like HPLC are expensive and sometimes they can be replaced by NIR/VIS spectroscopy 400–2500 nm. The authors in [113] studied it for American Ginseng root powder to detect ginsenosides. They found that it is applicable for major ginsenosides, Rb_1, Re, and m-Rb and total ginsenosides.

Sesame has some antioxidative-activity-related ingredients like lignans (including lipid-soluble lignans sesamin and sesamolin) and water-soluble lignin glycosides (sesaminol triglucoside and sesaminol diglucoside). Kim et al. [114] developed a NIR reflectance method for evaluating the lignin content of intact sesame seeds. The modeling was done using mPLS regression using results from high-performance liquid chromatography analysis as reference values. The results were good ingredients except for sesaminol diglucoside owning to its small amount.

Xin et al. [115] suggested a method to separate raw and processed herbs. They measured the herbs in the NIR area and used support vector machine, random forests, and linear discriminant analysis in analysing and classification. The obtained results were excellent.

Origin

Sometimes the origin is important for consumers. Zalacain et al. [111] suggested the use of near-infrared to determine the origin of saffron (Iran, Greece, and Spain).

B.

Class B, Type 06: Mammalian Products

Spoilage

Meat spoils very easily and thus produces a health risk not to mention unpleasantness. So excluding certain delicacies, the spoilage should be avoided or at least the spoilt meat should be removed before going to consumer. The spoilage for different meats has been studied for pork meat [116] and beef [117]. These papers considered also bacteria and microbiological factors.

Contamination

The fecal material is sometimes found on animal carcasses. Windham et al. [118] suggested the use of four optimal wavelengths: 434, 517, 565, and 628 nm. These wavelengths were selected using multivariate data analysis and VIS/NIR reflectance data. Different spectral ratios were tested to find optimal classification. The ratios 565 nm/517 nm provide 100% detection rate in carcasses of broilers fed a corn or soybean meal diet. Another ratio, 574 nm/588 nm was shown to be optimal for a larger variety of diets.

Taste (including Juiciness)

One of the earliest references to taste of meat was already published in 1969 [119]. Byrne et al. [120] used NIR reflectance (750–1098 nm) to examine the taste of heifer cow. The samples from the longissimus dorsi (2.5 cm thick) were taken at 2, 7, and 14 days post mortem. They were vacuum-packaged in plastic bags and stored at −20°C.

The results from taste panel and chemical processes for lamb meat were evaluated against NIR reflectance by Andrés et al. [121]. They evaluated a set of 232 muscle samples from Texel and Scottish Blackface lambs. They concluded that absorbance is correlated to juiciness, flavour, abnormal flavor, and overall liking of lamb meat, but the correlation does not provide very clear classification. It was applicable for separating extreme samples.

Appearance

The appearance consists of color, geometrical features, and texture of meat. Texture of meat here is the general surface appearance.

Color of meat depends on water content and concentrations of myoglobin and also the relative proportions of myoglobin's different forms, oxymyoglobin, myoglobin, and ferric myoglobin [122, 123].

Meat color has been studied for decades and several methods have been suggested. For example, Van den Oord and Wesdorp [124] suggested the use of reflectance differences at 630 and 580 nm. Later, beef color prediction has been studied by Leroy et al. [125] and Shackelford et al. [126]. Shackelford et al. [126] showed that the best color prediction with NIR was obtained in reflectance mode. Leroy et al. [125] studied also beef, longissimus thoracis, and obtained similar results. For pork, color is an especially important parameter for customers which make their buy/not buy partially based on it. Cozzolino et al. [127] applied VIS/NIR for prediction.

Andrés et al. [128] applied VIS/NIR (400–2500 nm) for determining $L^*a^*b^*$ colour parameters. The results for luminosity at 0 and 60 min had a good predictability.

Diet

The diet affects the composition of beef: their subcutaneous fat and its fatty acid composition [129]. Prieto et al. [130] report that NIR can be used to separate whether or not the beef cows has been fed with flaxseed.

pH

pH is affected by light scattering [131–133] study produced over 0.8 values for coefficient of correlation when using first derivative and multiplicative scatter correction for calibration and validation sets. The obtained precision is good; only 4.27% of predictions had difference more than 0.2 pH. Andrés et al. [128] predicted pH of beef with NIR with good accuracy.

Content

Important factors for meat are lean-to-fat ratio, meat percentage, intramuscular fat, marbling, protein, water contents and muscle area. Certain ingredient, or their amount can be used as a guarantee for quality and used as a mark of quality.

Moisture of meat has been evaluated by Liao et al. [133] with coefficient of correlation near 0.8 for validation set using the first derivative and multiplicative scatter correction. Tornberg et al. [134] with no processing showed very poor predictability (0.21) for both VIS (400–800 nm) and NIR (800–2500). Chan et al. [135] used first derivate and preprocessing of spectra (400–1700 nm) and obtained results over 0.7 for the validation set.

Liao et al. [133] showed over 0.75 values for the calibration, and the validation sets for protein content. Values around 0.7 were obtained by Chan et al. [135].

The intramuscular fat is important quality factor which has been studied for beef [130, 136, 137] and pork [133,137]. Sierra et al. [136] studied the composition and content of intramuscular fat has been studied by using NIR transmittance 850 to 1050 nm. They evaluated the fatty acid profile for samples of ground beef from thelongissimus thoracis of yearling bulls. The study indicates that it is possible with relatively good accuracy to evaluate saturated, branched, and monounsaturated fatty acids. NIR transmittance also provides information about conjugated linoleic acids. Prieto et al. [130] used NIR reflectance 400 to 2498 nm for determining concentration of polyunsaturated fatty acids and their biohydrogenation products in beef cows which have been fed on special diet (flaxseed).

Studies of intramuscular fat for pork [132–135, 137, 138] have dissimilar results. The best results were from Prevolnik et al. [137] where the coefficient of determination value was over 0.9. Very poor results were obtained by Tornberg et al. [134] when no preprocessing was used with VIS (400–800 nm) whereas NIR (800–2500 nm) provided almost twice as good results. The first derivative of spectra with preprocessing provided further improvement [135]. One reason for difference could be different wavelength areas: Tornberg et al. [134] used visible range 400–800 nm, Barlocco et al. [138] used NIR (800–2500 nm), Savenije et al. [132] used VIS/NIR (400–2500 nm), and Liao et al. [133] used VIS/NIR (350–1100 nm). The homogeneity of samples varies [133,137, 138]; for example, Liao et al. [133] and Savenije et al. [132] used intact pork samples. Minced pork samples were used, for example, in [137, 138]. Liao et al. [133] studied in addition to intramuscular fat protein and water content for pork in 350–1100 nm.

The intramuscular fat and water for lamb meat were predicted at a higher accuracy and they seem to correlate to the sensory parameters [121].

Iron-related substances can also be evaluated from meat. Hong and Yasumoto [139] used NIR diffuse reflectance from 1100 to 2500 nm to study heme and nonheme iron in raw meats: beef, pork, chicken, rabbit, duck, horse, lamb, and mutton. The second derivate was taken from data and after that multiple regression equations were applied. The result indicates that the multiple correlation coefficients are over 0.9 for both hemes and total iron. Hong and Yasumoto [139] obtained the best results with multiple regression equation at wavelengths where NH groups in protein molecules and C–C and CH in the porphyrin ring of heme have absorption.

The same NIR wavelength range (1100–2500 nm) has been tested for evaluating crude protein (CP), myoglobin, collagen, ether extract (EE), gross energy (GE), dry matter (DM), and ash content of oxen meat [140]. Prieto et al. [140] found that myoglobin, collagen, and ash contents could be evaluated with poor predictability. They speculate that this was due to lack of correlation between these parameters and ether extract content. However, ether extract evaluation was successful and important since this parameter is highly related to the organoleptic characteristics (like juiciness, flavor, and texture) in oxen meat samples. Also other parameters like crude protein, dry matter, and gross energy, which have a high correlation with ether extract content, were predicted with an acceptable degree of accuracy.

Pérez-Marín et al. [141] applied NIR for all stages of pigs: the live animal, carcass in slaughterhouse, subcutaneous fat sample with skin, skin-free subcutaneous fat sample, and transverse section. They develop prediction model for the four main fatty acids in Iberian pig fat, obtaining for palmitic acid SECV values of 1.24% for in vivo analysis and 0.82% for carcass analysis, for stearic acid 0.67% and 0.94%, for oleic acid 1.42% and 1.48%, and for linoleic 0.36% and 0.55%. The model proved the feasibility of NIR because it accounted for between 60% and 74% of the variation of spectra of live animals and between 31% and 87% of variation of carcasses. For Iberian pigs, also intact fat or adipose tissues are studied [142–144].

Lamb muscles appear to have special absorption bands at 424 nm and 550 nm. These bands are related to the Soret band and the oxymyoglobin absorption band [145, 146]. The absorption bands (at 540 nm and at 580 nm) correlate with both myoglobin and oxymyoglobin [145, 147, 148], but myoglobin has also another absorption band at 555 nm [145–150] with the absorption band at 574 nm with oxyhaemoglobin absorption.

According to Murray [151], an absorption band at 762 nm is caused by the OH third overtone or according to Swatland [145] and Delpy and Cope [149], an absorption band produced by the oxidation of the myoglobin (deoxymyoglobin). Pigments and other characteristics (intramuscular fat, fattyacids, and moisture) cause most of the spectral information used for the VIS region (400–700 nm) and the NIR region around 1400–2300 nm.

Similar results are reported by others like Downey et al. [152]; Lomiwes et al. [153] predicted glycogen concentration of prerigor muscle.

Maturity

The age of animal can affect the meat quality since there can be differences in intramuscular fat and water contents. Prieto et al. [154] applied successfully NIR reflectance 1100–2500 nm to separate adult, over 4 years old castrated steers and young, under 14 months old cattle. The samples were taken from longissimus thoracismuscle.

Shear Force

Shear force value indicates how much force is needed to cut the meat and it is related to tenderness. In fact, Yancey et al. [155] studied the connection between shear force (Warner-Bratzler shear force (WBSF), and Meullenet-Owens razor shear (MORS)) and tenderness (consumer evaluation, also with overall impression) for beef ribeye rolls taken from Select, low Choice, top (upper 2/3) Choice, and Prime quality grade carcasses with visible and near-infrared spectroscopy reflectance. Steaks were cut 2.54 cm thick and put to aging for 14 or 28 days. The longer aging of samples was related to increased tenderness. The 2nd derivatives of VIS/NIR measurements were found to be capable for predicting tenderness and overall impression than shear force measures.

Many studies for shear force have been made, for example, Byrne et al. [120] 1998, Park et al. [156] Rodbøtten [157] and Venel et al. [158]. Byrne et al. [120] evaluated Warner-Bratzler shear force using NIR 750–1098 nm for Heifer cows. They obtained over 0.7 value of correlation coefficient for both groups, for samples taken at 2, 7 and 14 days post mortem, vacuum-packaged in plastic bags and stored at −20°C. However, the results seem to vary. Prieto et al. [159] studied beef and cooked pork and suggest significant variation as a function of position. Shackelford et al. [126] applied shear force detection online for beef. The NIR transmission measurement provided better results than reflectance measurements [125]. The ability was limited longissimus thoracis at 2 or 8 days post mortem. Venel et al. [158] suggest using additional features (age, grade, pH, and sex) to improve results.

The results for pork meat have different conclusions. Study of Geesink et al. [160] indicates poor predictability. Liao et al. [133] obtained relatively good prediction for the calibration set but over 2.5 times poorer for the validation set. It has been discussed that the problem is the significant variations at different positions of the samples.

Tenderness

Tenderness of a beef describes the resistance to bite or shear meat and is related to the hardness. It should be noted that the characteristics can correlate to each other, like tenderness of meat that has a positive correlation to flavor and juiciness [161].

Generally the customers prefer beef with low effort needed and thus tenderness is an important factor related to quality. NIR models have predictions rating fair to good for the tenderness of postrigor longissimus muscle in many publications [120, 148, 156, 162], and [157]. Common measurement ranges were 1100–2500 nm and 750–1100 nm. It seems that absorption increases when the meat becomes tougher. Tenderness predictions for prerigor longissimus muscle were not so good [134, 157] which may indicate that the change from before rigor to after rigor and parameters related to it might be critical for tenderness. The applicability of techniques is different: extracted part of a muscle sample [156, 162–164] or a very small area (4 cm^2) [120]. Shackelford et al. [165] have extended the technique for exposed longissimus cross-section of ribbed beef. Later, an online detection of tenderness has been tested by Shackelford et al. [126].

Texture of Meat

Byrne et al.'s [120] results showed correlation coefficient values of 0.5 and 0.7 for heifer meat texture by NIR. Texture of lamb meat has been evaluated by Andres et al. [121], but although absorbance does correlate to texture, it was not suitable for classification.

Acceptability

The acceptability of heifer meat seems to be very poorly predicted by NIR according to Byrne et al. [120]. The correlation coefficient values were under 0.5.

Sensory Characteristics other than Appearance

It would be advantageous to classify the beef based on its quality directly in the abattoir. Prieto et al. [159] applied VIS/NIR fiber optics measurements (350 to 1800 nm) to very fresh beef muscle to correlate it with lab color, cooking loss, instrumental texture (Volodkevitch, 10-day aged meat; slice shear force, 3 and 14-day aged meat), and sensory characteristics. However, the results indicated that only the sensory characteristics correlation was useful.

Structural Properties

Structural properties are related to functional quality. They can be water holding capacity, isometric tension, muscle fiber shortening, pH, and cooking loss.

Shackelford et al. [126] evaluated NIR for water-holding capacity and cooking loss for beef. The average reflectance showed higher absorbance for samples with highest cooking loss with meat 2 days post mortem. The difference between spectra diminished when the meat was 8 days post mortem. The transmission spectra improved the results. The drip loss prediction was not very high. It was best for reflectance at day 8 with a correlation coefficient of 0.54. Tornberg et al. [134] studied the drip loss for porcine meat using wavelength 802–2500 nm. They obtained a correlation coefficient of 0.64.

The study of Prieto et al. [159] suggest that VIS/NIR may not be very useful for determining cooking loss. Drip loss and shear force for pork were studied by Geesink et al. [160]. The data (reflectance spectra (1000–2500 nm), water-holding capacity, shear force, ultimate pH, and colour ($L*$, $a*$, $b*$-value) of 96 pork longissimus) were obtained two days post mortem and subjected to stepwise multiple linear regression (SMLR) and partial least-square regression (PLSR). The drip loss was predicted successfully while shear force was not.

Other Subjects

Shackelford et al. [165] studied optimal measurement for NIR and VIS. They recommended for online VIS/NIR studies of longissimus quality traits of ribbed beef carcasses: a high-intensity probe (with a 50 mm diameter hole in the limiter plate), data acquirement was done with averaging 20 spectra per observation, measurement placelongissimus from one side per carcass, and at a standardized bloom time for spectra taking. Offline measurement application to lamb and pork and smaller muscles may require a reduction in the size of the opening in the limiter plate.

Information about sample characteristics, process, and slaughter conditions can be found, for example, in Cozzolino et al. [166], Cozzolino et al. [146], Cozzolino et al. [167], Mitsumoto et al. [148] 1991, Swatland [145], and Cozzolino and Murray [150].

Authentication

Authenticity of meet has several viewpoints. One viewpoint is the origin, another adulteration, and the last one ecological point. The common factor is that consumers should get what they intend to buy, even when the food bought is otherwise acceptable. Some people prefer organic food for their perceived superiority. The composition of meat, flavor, color, and fatty acid composition [168] of cattle and sheep depend heavily on diet.

Dian et al. [168] used VIS 400–700 nm and VIS/NIR 400–2500 nm with laboratory monochromator for separating pasture-fed and concentrate-fed lamb carcasses. They evaluate for this purpose three methods: (1) composition based on carotenoid absorption and perirenal fat reflectance at 450–510 nm, (2) composition based on perirenal fat reflectance at 400–700 nm, and (3) full spectrum of 400–2500 nm. The measured target was especially the perirenal fat because it is always present and can be removed without affecting carcass quality, and it provides better results than subcutaneous caudal fat [169]. The methods assessed the pasture-fed lambs by classification rates of 89.1%, 90.8%, and 97.5% and for concentrate-fed lambs by 98.6%, 98.6%, and 97.8%, respectively.

Cozzolino and Murray [170] applied VIS/NIR 400–2500 nm spectra for classifying the meat from beef, lamb, pork, and chicken. The achieved accuracy was over 80% with PCA and PLS.

Ding and Xu [171] obtained ideal results while separating beef and kangaroo samples. All kangaroo samples were identified with 100% accuracy.

Drip Loss/Water Holding Capacity

The water-holding capacity is a major pig meat technological property. Prevolnik et al. [172] used artificial neural network to predict protein drip loss from NIR with error of 2.5–2.6%.

Meat Freshness

The meat freshness can be thought to be related to its butchery date and possible freezing/thawing. Downey and Beauchêne [173] successfully applied VIS/NIR for separating the fresh meat from frozen/thawed beef using factorial discriminant procedure. Another study to separate frozen and unfrozen beef was done by Thyoldt and Isaksson [174].

Spoilage

Spoilage of beef has been studied by Ellis et al. [117]; Horváth et al. [116] applied NIR spectroscopy to pork meet.

Bioagents

Hobson et al. [175] suggested the use of spectral methods for detecting the metabolic state of microorganisms.

Milks

General Quality. Milk quality measured using NIR has been suggested by Kawasaki et al. [176]. Al-Qadiri et al. [177] studied the quality with visible (starting at 600 nm) and short infrared NI Karoui and De Baerdemaeker [178] give a review of analytical methods including NIRS coupled with chemometric tools for the quality determination and identity of dairy products.

Content. Laporte and Paquin [179] used partial and overall PLS calibration of near-infrared transmission data to evaluate its fat, crude protein, true protein, and casein content. The results indicated that NIR provides reliable results for the evaluation purposes and overall PLS calibration can improve the results. Šašić and Ozaki [180] subjected raw milk to SW-NIR (800–1100 nm) measurements. They predicted the fat, protein, and lactose contents using PLS regression and band assignment. The correlation of spectra and milk protein was sufficiently high. Fat content prediction needed pretreatment. Prediction of lactose content was limited.

Adulteration. Milks are also subject to adulteration. Some cases of the adulteration can be subtle like adding foreign fat. Sato et al. [181] used NIR analysis to detect these fats.

Class B, Type 07: Poultry Products

Chicken meat products need to be controlled because of their safety risk and quality. These issues have been considered in many papers such as Chen et al. [182], Chen and Marks [183], Chen and Marks [184], Chen et al. [185], McElhinney et al. [186], and Rannou and Downey et al. [187].

Park et al. [188] have recently constructed a prototype multispectral imaging system for online poultry carcass fecal contamination detection. They used a line-scan imaging system using 517, 565, 600, and 802 nm. The longest wavelength was used for removing artifacts caused by cuticle.

Incorrect Ingredients

The poultry meat can be contaminated by many things. One of the most dangerous things is the manure which contains many disease-causing microorganisms.

Content

The poultry meat is evaluated for ash, dry matter (DM), crude protein (CP), and fat content. The meat can be in different formats-like fresh or dried. Freeze dried ostrich meat tenderloin (M. ambiens), big drum (M. iliofibularis), and fan fillet (M. gastrocnemius) were evaluated by Viljoen et al. (2005) using NIR 1100–2500 nm and PLS 0. The results were accurate for crude protein and fat, but the prediction of ash and dry matter was not very successful.

Iron-related substances can also be evaluated from meat. Hong and Yasumoto [139] studied heme and nonheme iron in raw meats (e.g., chicken and duck) using NIR diffuse reflectance. They applied multiple regression equations, the second derivative data (1100 to 2500 nm) with multiple correlation coefficients over 0.9 for both hemes and total iron. The best MREs were obtained with the NIR data recorded at wavelengths where NH groups in protein molecules and C–C and CH in the porphyrin ring of heme give absorption.

Authenticity/Origin

The origin of chicken meat affects its value. Ding et al. [189] applied VIS/NIR (750–1100 nm) to separate broilers from Chinese local chickens using pieces of chicken meat cut.

Spoilage

Lin et al. [190] used chicken as a material for investigating spoilage. Poultry can be spoilt with too much heating. It seems that the intensity of wavelength bands around 445 nm and 560 nm decreases with cooking time when using correlation analysis [191]. After this decrease, the intensity begins to vary at 475, 520, and 585 nm.

Storage and Handling

It has been indicated that thawing may swallow the reabsorption of certain fluids like soluble proteins, vitamins, and salts by the meat. However, there is a risk of meat discoloration and lipid oxidation if the time for thawing is too long. Liu et al. [192] studied the effect of chilled and frozen storages (temperature, time) and shear force on chicken breasts using generalized 2D correlation analysis of VIS and NIR spectra. They noticed that the temperature of frozen storage changed intensities at 445 and 560 nm. The NIR spectra decreased. Tender and tough muscles caused difference in NIR spectral intensity. Liu et al. [193] examined the thawing behavior of frozen chicken meat with 2D VIS/NIR correlation.

Class B, Type 08: Aquatic Animal Products

Visible spectroscopy may sometimes be more useful than NIRS as found by Stormo et al. [194] who used visible spectroscopy for assessing endpoint temperature of heat-treated surimi. Their method shows potential for use in the food processing industry.

Appearance

Blood in fish is sometimes an unwanted characteristic. Olsen [195] and Olsen et al. [196] suggested that it might be possible to develop a method using VIS/NIR to access residual blood in intact cod muscle.

Identification and Authentication

Fishmeal batches can be made from different fish species. Cozzolino et al. [197] used NIR reflectance (1100 nm–2500 nm) and chemometrics to identify and authenticate the patches. They concluded that dummy partial least-square regression achieves 80–82% correct classification. Linear discriminant analysis based on principal component analysis scored achieved over 80% correct rate.

Freshness

Fish quality is largely determined by freshness. The freshness parameter can be linked to the time of catch or harvest or freezing. It can refer to instrumental analysis or sensory evaluation.

Uddin et al. [198] suggest using of VIS/NIR spectroscopy to determine whether the fish were frozen-thawed or fresh. Bøknaes et al. [199] used NIR spectroscopy and sensory quality evaluation to rate the freshness of thawed and chilled cod fillets packed in modified atmosphere. They studied the estimation of storage temperature, storage period, and chill storage period for fish. The storage time for cod and salmon was also investigated with VIS/NIR spectra by Nilsen et al. [200]. Several techniques like visible spectroscopy, texture, and image analysis can be combined to improve results [201]. The spoilage of rainbow trout has been studied by Lin et al. [202]. Zhang and Lee [203] used near-infrared for determining free fatty acids content of fish oil and mackerel. Since the change in free fatty acids is related to hypoxanthine, they suggested it to be used as an indicator of freshness. They found that the partial least-square model produced better results than multiple linear regression.

Content

Earliest fish study seems to be by Gjerde and Martens [204], they demonstrated that NIR can be used to predict water, fat, and protein contents in rainbow trout. They used reflectance spectroscopy to minced and freeze-dried samples. Since then, a lot of fish studies have been published and some at-line and online food quality controls. However, the technique is not yet a standard everyday instrumental tool. This might be due to high instrumentation costs and the need of models between spectroscopic reading and quality parameters in order to analyze them [31].

The fat content of fish has and still is under research. It has been applied on different species and different forms of fish: frozen and thawed glazed bigeye tuna [205], albacore [206], frozen skipjack [207, 208], raw and frozen horse mackerel [206], frozen and thawed mackerel. Fish are sometimes ranked by their fat using visual inspection which seems to be an inaccurate method [205].

Mathias et al. [209] suggested the use of NIR for protein content analysis of freeze-dried freshwater fish. The standard deviation between results from NIR (mean lipid content) and chemical analysis was 3.7%. Mathias et al. [209] suggested also the use of NIR for lipid content analysis of freeze-dried freshwater fish. The standard deviation between results from NIR (mean lipid content) and chemical analysis was 10%.

Application of short-wavelength near-infrared spectra (600–1100 nm) can be used to determine the salt and moisture contents of salmon roe [7].

Lee et al. [210] applied diffusive reflectance of short-wavelength near-infrared in the range of 700–1000 for predicting the amount of crude lipid of intact, whole rainbow trout. The best correlation between measured and reference values was obtained from measurements at a midposition between dorsal and adipose fins above the lateral line. The modeling was done using multivariate calibration models, multiple linear regression, and partial least squares.

Defects

Lin et al. [211] used VIS and SW-NIR (600–1100 nm) to find bruises in intact, whole Pacific pink salmon. The measurements were done in diffuse reflectance mode and without removing skin or scales. The spectral results were compared against digital images of filleted fish with partial least-square model. The correct classification rate for nonbruised fish spectra was 84%, while for bruised it was 81%.

Incorrect Ingredients

Inorganic arsenic is an unwanted and unhealthy ingredient. Font et al. [212] subjected freeze-dried samples of red crayfish to inorganic arsenic content analysis. They applied a modified partial least-square regression for modeling with acid-digestion solvent-extracted hydride generation atomic absorption spectrometry measurements. The data was standardized by applying standard normal variate and detrending algorithms and preprocessed with 2nd derivative.

Bioagents

Fish can contain parasites which range from plain ugly appearance to dangerous for human health. Some food-borne illnesses caused by microorganisms can be prevented by proper handling like cooking.

Spoilage

Fish is spoiled easily and the spoiled fish is usually unsuitable for human consumption. Lin et al. [202] studied the spoilage for rainbow trout. They used 600–1100 nm region which may be applicable to detect various fundamental molecular vibrations of several functional groups like CH, O–H, N–H, and C=O. The absorption, reflectance, and transmittance of biochemical components help to determine the chemical composition and thus spoilage.

Class B, Type 009: Amphibians and Reptilians

The amphibians and reptilians group includes frogs, lizards, snakes, and turtles. This area seems to be lacking research papers.

Class B, Type 010: Invertebrate Animals

The invertebrate animals group consists of molluscs (including cephalopods) and other invertebrate animals.

Authenticity

Some edible invertebrate animals are more valuable and sought after than others. The processing of seafood causes appearance changes which make it very hard for a consumer to separate between authentic and adulterated products. This increases the risk of adulteration. Gayo and Hale [213] applied VIS and NIR spectra to find adulteration of crab meat. The crab meats (Atlantic blue and blue swimmer) were mixed with surimi-based adulterant in 10% increase. The adulterant increase caused decreasing water absorbance. They concluded that both PLC and principal component analysis can detect successfully adulteration. In another study, by Gayo et al. [214], VIS/NIR was applied to separate the Atlantic blue crabmeat adulterated with blue swimmer crabmeat (10% increment). Adulteration of meat changed the water absorption bands which could be used for adulteration detection.

Content

Brown [215] applied VIS and NIR (350– 2500 nm) for homogenized oyster meat and analyzed the data quantitatively and qualitatively. The results for moisture (r^2 = 0.92, SEP = 0.53% wet weight), protein (r^2 = 0.97, SEP = 0.18% wet weight), fat (r^2 = 0.97, SEP = 0.11% wet weight), and glycogen (r^2 = 0.94, SEP = 0.24% weight) indicate that VIS/NIR is useful for oyster evaluations.

C.

Class D, Type 12: Secondary Food Commodities of Plant Origin

Content

Lebot et al. [216] applied NIR spectra (350–2500 nm) for flours from tropical root and tuber crops (cassava, sweet potato, taro, and yam). They evaluated starch, total sugars, cellulose, total nitrogen, and ash (total minerals) contents of the flours. The prediction accuracy achieved over 85% confidence for starch, sugar, and nitrogen, while for ash (minerals) it was 71%. The method was unsuitable for cellulose. The advantage of NIR is its cheapness.

Dacal-Nieto et al. [217] have recently used NIR multispectral imaging to detect the hollow heart of potato tubers. The camera used is sensitive from 900 nm to 1700 nm and used Specim Imspector N17E (Specim Oy, Oulu, Finland) spectrograph. Several wavelength selection methods were used, out of which the genetic algorithm was marginally the best, that is, gave the best classification accuracy in the classification step.

Green honeybush contains mangiferin and hesperidin which are bioactive components. Jourbert et al. [218] predicted content of these ingredients in dried plant material using diffuse NIR reflectance with results suitable for screening purposes.

Bellato et al. [219] studied wholegrains and flour of oats. The purpose was to determine β-glucan and protein content. They applied NIR reflectance for flour and grains and NIR transmittance for flours. The model for chemometric relationship was a partial robust M-regression, a version of partial least squares. The models using reflectance values stood out best.

For example, Schimleck et al. [220] evaluated successfully air-dried leaves from Melaleuca cajuputi tree for their 1.8-cineole (Eucalyptol) content and oil concentrations using NIR.

Evaluating roots provides an attractive possibility to select the individual containing the highest amount of ingredients or classify the roots in different groups according to some preset content classes. It is also possible to compare different root processing methods. Joubert et al. [221] studied Devil's claw root for evaluating iridoids (harpagoside and 8-ρ-coumaroyl harpagide (8ρCHG)) and moisture content from different drying conditions. The NIR study indicates that the harpagoside retention was lower in sun-dried samples than tunnel dried or freeze-dried samples. They concluded that the best retention was obtained from tunnel-drying at 50°C. NIRS was very suitable for predicting moisture and suitable for semiquantitative classification using the iridoids.

Leaves of green tea were subjected to NIR reflectance measurement and PLS modeling to predict polyphenol and alkaloid compounds (Schulz et al. [222]). The prediction of parameters of green tea was very successful for gallic acid, epicatechin, epigallocatechin, epicatechin gallate, epigallocatechin gallate, caffeine, and theobromine, but a lower accuracy was obtained for total polyphenol content probably due to the colorimetric method (lack of specificity).

The dry matter of leaves can be evaluated very precisely (Schulz et al. [222]) for green tea R^2 = 0.94; SD/SECV = 4.12. Furthermore, it is possible to discriminate tea leaves of different age by principal component analysis on the basis of the received NIR spectra.

Hernández-Hierro et al. [223] examined visible-NIR spectroscopy for freeze-dried broccoli. They concluded that it is potentially usable for screening of almost all individual glucosinolates and total glucosinolates.

Adulteration

Laasonen et al. [224] used PLS and pretreated NIR reflectance spectra to identify fast Echinacea purpurea dried and milled root from samples adulterated/substituted by other species (Echinacea angustifolia, Echinacea pallida, or Parthenium integrifolium). Minimum adulteration rate for detection was 10%.

Class D, Type 13: Derived Products of Plant Origin

Content

Juices contain many important ingredients whose concentration needs to be evaluated. Li et al. [225] measured transmission spectra of orange juice dry extract in the range of 1100–2500 nm. They used multivariate calibration to evaluate the amount of sugars (glucose, fructose, and sucrose) and acids (citric and malic acids). The reference values were obtained using enzymatic assays. The best model was obtained using partial least-square regression. Fruit juice from raw Japanese apricot which is also known as Japanese plum was also subjected to evaluation of citric and malic acid contents. Chen et al. [226] used NIR spectra of 1100–1850 nm and partial least-square regression for producing acceptable prediction results for the evaluation.

Sometimes the concentration of juice is important and one needs to get a high yield. For example, González-Sáiz et al. [227] applied successfully NIR and multivariate calibration for monitoring alcoholic fermentation of onion juice. The goal was the valorization of worthless onions.

Cen et al. [228] used VIS/NIR (325–1075 nm) for determining the soluble solids contents and pH in orange juice. The data was preprocessed with Wavelet packet transform, standard normal variate transformation (SNV), and Savitzky-Golay first-derivative transformation. Partial least-square (PLS) regression was applied and it produced best results with SNV.

We include in this juice group also cola beverages. For a cola beverage, important parameters are, for example, pH. Liu and He [229] applied chemometrics to VIS/NIR spectra to get the soluble solid content and pH of cola beverage with excellent prediction precision. Least squares with support vector machine provided better results than partial least squares.

Different oils have also ingredients which need to be evaluated. Ng et al. [230] used NIR (1100–2500 nm) for the evaluation of degradation products (including total polar material and free fatty acids) of frying oils. According to their results, partial least-square regression is better in 700–1100 nm area than forward stepwise multiple linear regression, while in the whole range, they are equal. The best correlations were over 0.98.

Class D, Type 014: Manufactured Foods (Single Ingredient) of Plant Origin

Content

Cozzolino et al. [231] did a review about near-infrared spectroscopy applied to grapes and wine.

Chen and Lei [232] applied VIS and NIR spectra to obtain the amount of carbohydrate in soy milk powder. Their goal was to find a high speed method for this. They searched optimal wavelength band in spectra using simulated annealing and wavelet packet transform tree. After this, different partial least-square models were tested using 20 variables.

Dambergs et al. [233] determined the methanol content of wine-fortifying spirit. They developed partial least-square model for NIR spectra with results from gas chromatography used as a reference.

Sáiz-Abajo et al. [234] used NIR to classify the samples into red/white wine vinegar and alcohol vinegar samples. They used orthogonal signal correction (OSC) method to remove information related to tartaric acid. Sáiz-Abajo et al. [235] studied 14 parameters which are important for the quality of wine vinegar.

Fan et al. [236] have recently used NIRS with several wavelength selection methods to predict the total acid of vinegar. Their main conclusion was that wavelength selection is essential for the creation of better regression models for acid prediction.

Bruun [237] has studied protein-lipid-carbohydrate interactions and their importance to the food quality.

Feng Liu et al. [238] have recently used VIS/NIR spectroscopy and least-square support vector machine (LS-SVM) for the soluble solid content (SSC) and pH determination of fruit vinegar. The results indicate that the method could be used in the online monitoring of vinegar fermentation.

Authenticity

Cozzoline et al. [239] used VIS and NIR spectra to classify Chardonnay and Riesling white wines according to their Australian varietal origins. Discriminant partial least-square (DPLS) regression was the most successful model tested; achieving 100% correct labeling for Riesling. Later, Liu et al. [240] classified Tempranillo wines from Australia and Spain according to their geographical origins. They accompanied both VIS and NIR data (400–2500 nm) for multivariate analysis PCA, DPLS, PLS-DA, and LDA from which PLS-DA was the best with high correct classification rate.

Buratti et al. [241] have tested NIR and IR spectroscopy combined with electronic nose and tongue for the offline monitoring of wine fermentation. The results are promising for the development of an online fermentation monitoring method based on these techniques.

Age

Yu et al. [242] applied LS-support vector machine (SUM) to NIR spectra to determine age and oenological parameters of rice wine. The inputs to LS-SVM applied were scores of ten principal components (PCs) and radial basis functions. LS-SVM was a little better than PLS for alcohol content, titratable acidity, and pH.

Class D, Type 015: Manufactured Foods (Multi-Ingredient) of Plant Origin

Content

Information about fat is often searched by consumer, for

example, in cereals. However, the information is typically an estimate because the methods used like calorimetry destroy samples. Vines et al. [243] predicted the total amount of fat using near-infrared and modified partial least-square regression with good results when compared against reference values from the AOAC 996.01 method. Windham et al. [244] tested the effect of moisture to the calibration models and found that it has a significant effect. However, they developed a new partial least-square regression model to be robust enough to provide good prediction in a wide range of residual moistures. Kays et al. [245] used NIR reflectance (1100 to 2498 nm) to determine the total dietary fiber in cereal and grain products. AOAC (991.43) enzymatic-gravimetric method produced the reference samples.

Fernández Pierna et al. [64] have recently used NIR hyperspectral imaging and chemometrics for the detection of impurities in cereals. They used a whiskbroom system MatrixNIR Chemical Imaging instrument (Malvern instruments, Analytical Imaging, Columbia, MD, USA)) with an InGaAs FPA along with two liquid crystal tuneable filters (LCTF) for wavelength selection. The results show potential for NIR imaging in impurity detection. However, the whiskbroom system might not be as good in this task as a pushbroom system that they used in another case study with sugar beets.

Total Fiber Amount

Kays et al. [246] evaluated the total fiber amount of cereals using NIR spectra. The samples with high sugar and crystalline sugar were selected because a unique spectrum of crystalline sugar affects NIR spectra. The developed partial least-square model predicted the content well. Kays et al. [247] extended the NIR model to cereals with high fat or sugar and fat. The extension was necessarily because high fat or sugar may differ substantially from spectral properties.

Kays and Barton II [248] used NIR reflectance (1104–2494 nm) to determine the chemometric model for predicting gross energy and available energy of cereals. The reference values were obtained from bomb calorimetry. In another study, Kays and Barton II [249] determined accurately soluble and insoluble dietary fiber using NIR reflectance spectra (NIRSystems 6500 monochromator) and reference values from the AOAC Method 991.43. The modeling was done using modified PLS models.

Another aspect is how much energy comes from fat of ground cereals. Kays and Barton II [250] applied successfully near-infrared in the range of 1104–2494 nm to determine this using a version of PLS regression model. The sample reference was obtained using calorimetry. Archibald and Kays [251] analyzed whole intact breakfast cereals, snack foods, wholegrains, and milled products. They found that total dietary fiber could be evaluated from NIR reflectance spectra and moisture has not any effect on the results. Transmission and reflectance 700–1100 nm NIR spectra of cookies were measured by Ozanich et al. [252] to determine moisture and oil contents. Moisture content has highest correlation (and absorbance) with 965 nm wavelength (a combination band of H_2O) while oil content with 930 nm (a 3rd overtone CH stretch of methylene) in the normalized spectra.

Structural Properties

de Alencar Figueiredo et al. [253] studied a wide range of sorghum cores with high genetic diversity. They used NIR reflectance to determine amylase, protein, and lipid contents, endosperm texture, and hardness.

Handling/Fermentation

Bruun et al. [254] used NIR for analyzing gluten powder with different handling (heat, moisture). The NIR data was processed with second-derivative transformation and extended multiplicative signal correction. The moisture caused shifts and intensity changes in the protein bands. Bruun et al. [254] assumed this to be due to changes in water binding and microenvironments of the amino acid side chains. Heat caused an increase at 2209 nm and decrease at 2167–2182 nm which could be due to an α-helix to β-sheet transformation. Next, in [255] different salts were added to the hydrated gluten. The same preprocessing was used and PCA was applied on this data. The results showed a correlation between spectra and salt type and concentration.

Sinelli et al. [256] used NIR (750–1100 nm) and MIR (4000–600 cm^{-1}) with a fiberoptic surface interactance for dough proofing. Dough proofing is a phase during which the dough starts to ferment after the mixture of ingredients, it affects the quality of bread, and it was also noticed that NIR spectra changed.

D.

Class E, Type 016: Secondary Food Commodities of Animal Origin

Secondary food commodities of animal origin consist of dried meat and fish products and secondary milk products. We include here also products using dried meat and fish. An example of this is ham.

Appearance

García-Rey et al. [257] studied appearance (texture and color) of sliced samples from deboned and cross-sectioned dry-cured in VIS/NIR 400–2200 nm with a fiber-optic probe. The classification accuracy for pastiness was 88.5% and 79.7% when compared against sensory evaluation. They classified into two classes: defect class and no-defect class.

Class E, Type 017: Derived Edible Products of Animal Origin

Derived edible products of the animal origin group include processed crustaceans and animals, milk, and derived fats.

Class E, Type 018: Manufactured Food (Single Ingredient) of Animal Origin

Manufactured food (single ingredient) of the animal origin group is the manufactured milk products using only single ingredient.

Class E, Type 019: Manufactured Food (Multi-Ingredient) of Animal Origin

Manufactured milk products using multi-ingredient belong to manufactured food (single ingredient) of the animal origin group. Here we consider dairy products like cheese or yogurts.

Appearance

Lucas et al. [258] applied VIS and NIR for estimating color parameters (brightness, redness, and yellowness) with good results.

Smell

NIR seems to be suitable. Çelik [259] used NIR to evaluate yogurts' aroma and its fermentation [260].

Taste

Taste depends on the goodness of the fermentation process. NIR can be combined with electric nose as a method for online fermentation monitoring and visualizing the state of the fermentation [260].

Content

The titratable acidity of the cultures for yogurt and sour milk can be predicted by forming a model using PLS with NIRS signals [260].

Lucas et al. [258] applied VIS/NIR (400–2500 nm) for 4 cheese varieties. They estimated the content of cheese for the following items: dry matter, fat, pH, retinol, α-tocopherol, β-carotene, xanthophylls, sodium chloride, calcium, potassium, magnesium, zinc, and total antioxidant capacity. The data was handled with modified partial least squares. The coefficient of determination and residual predictive deviation indicated that the spectra provided a good estimate for dry matter, fat, β-carotene, sodium chloride, calcium, and zinc in fresh cheeses while it was poor for the rest (pH, retinol, α-tocopherol, xanthophylls, potassium, magnesium, and total antioxidant capacity).

pH

pH of the dairy product is important. Navrátil et al. [260] predicted its value using NIR.

Origin

Cozzi et al. [261] developed a NIR- (1100–2500 nm) based method using discriminant analysis to discriminate Asiago d'Allevo cheese from different production chains (alpine farms, mountain, and lowland factories). Alpine products were discriminable from factory products due to their different fatty acids profile and color. The results were confirmed by chemical data.

ACKNOWLEDGMENTS

The authors want to acknowledge Tekes (the Finnish Funding Agency for Technology and Innovation) and industrial partners Riitan Herkku, Oy Ari Green Plant Ltd., Jepuan Peruna Oy, Etnovia Oy, and Specim, Spectral Imaging Ltd., for funding this research in a project called TarhaNIRSo (FruitForce) in Sapuska Research Program of Tekes. They also like to acknowledge Professor Paul Geladi for his many advices on NIRS applications.

REFERENCES

1. H. Huang, H. Yu, H. Xu, and Y. Ying, "Near infrared spectroscopy for on/in-line monitoring of quality in foods and beverages: a review," Journal of Food Engineering, vol. 87, no. 3, pp. 303–313, 2008.
2. K. H. Norris and J. R. Hart, "Direct spectrophotometric determination of moisture content of grain and seeds," in Proceedings of the International Symposium on Humidity and Moisture, 1965.
3. P. Williams, "Grains and seeds," in Near-Infrared Spectroscopy in Food Science and Technology, Y. Ozaki, W. F. McClure, and A. A. Christy, Eds., chapter 7.1, pp. 165–217, John Wiley & Sons, 2007.
4. B. G. Osborne, "Near-infrared spectroscopy in food analysis," in Encyclopedia of Analytical Chemistry, John Wiley & Sons, 2006.
5. J. Farkas and I. Dalmadi, "Near infrared and fluorescence spectroscopic methods and electronic nose technology for monitoring foods," Progress in Agricultural Engineering Sciences, vol. 5, no. 1, pp. 1–29, 2009.
6. K. Kaffka, "How the NIR technology came to and spread in Europe for quality assessment and control in the food industry," Acta Alimentaria, vol. 37, no. 2, pp. 141–145, 2008.
7. Y. Huang, T. M. Rogers, M. A. Wenz et al., "Detection of sodium chloride in cured salmon roe by SW-NIR spectroscopy," Journal of Agricultural and Food Chemistry, vol. 49, no. 9, pp. 4161–4167, 2001.
8. P. Pallav, G. G. Diamond, D. A. Hutchins, R. J. Green, and T. H. Gan, "A near-infrared (NIR) technique for imaging food materials," Journal of Food Science, vol. 74, no. 1, pp. E23–E33, 2009.
9. Y. Ozaki, W. F. McClure, and A. A. Christy, Eds., Near-Infrared Spectroscopy in Food Science and Technology, Wiley-Interscience, 2006.
10. L. A. Berrueta, R. M. Alonso-Salces, and K. Héberger, "Supervised pattern recognition in food analysis,"Journal of Chromatography A, vol. 1158, no. 1-2, pp. 196–214, 2007.
11. L. M. Reid, C. P. O›Donnell, and G. Downey, "Recent technological advances for the determination of food authenticity," Trends in Food Science and Technology, vol. 17, no. 7, pp. 344–353, 2006.
12. H. Cen and Y. He, "Theory and application of near infrared reflectance spectroscopy in determination of food quality," Trends in Food Science and Technology, vol. 18, no. 2, pp. 72–83, 2007.
13. C.-J. Du and D.-W. Sun, "Learning techniques used in computer vision for food quality evaluation: a review," Journal of Food Engineering, vol. 72, no. 1, pp. 39–55, 2006.
14. M. Iwamoto, J. Uozumi, and K. Nishinari, "Preliminary investigation of the state of water in foods by near

infrared spectroscopy," in NIR/NIT Conference, Budapest, Hungary, 1987.

15. M. Iwamoto, S. Kawano, and H. Abe, "Water in foods. Analysis of hydrogen bonding related to water in foods," NIR News, vol. 6, no. 3, pp. 10–12, 1995.
16. H. Maeda, Y. Ozaki, M. Tanaka, N. Hayashi, and T. Kojima, "Near infrared spectroscopy and chemometrics studies of temperature-dependent spectral variations of water: relationship between spectral changes and hydrogen bonds," Journal of Near Infrared Spectroscopy, vol. 3, pp. 191–201, 1995.
17. H. Büning-Pfaue, "Analysis of water in food by near infrared spectroscopy," Food Chemistry, vol. 82, no. 1, pp. 107–115, 2003.
18. P. Geladi and H. Grahn, Multivariate Image Analysis, John Wiley & Sons, 1996.
19. J. Workman Jr. and A. Springsteen, Applied Spectroscopy: A Compact Reference for Practitioners, Academic Press, 1998.
20. J. Koljonen, T. E. M. Nordling, and J. T. Alander, "A review of genetic algorithms in near infrared spectroscopy and chemometrics: past and future," Journal of Near Infrared Spectroscopy, vol. 16, no. 3, pp. 189–197, 2008.
21. D.-W. Sun, Modern Techniques for Food Authentication, Elsevier/Academic Press, 2008.
22. J. Workman Jr. and L. Weyer, Practical Guide to Interpretive Near-Infrared Spectroscopy, CRC Press, Boca Raton, Fla, USA, 2008.
23. J. R. Koza, "Genetic programming as a means for programming computers by natural selection," Statistics and Computing, vol. 4, no. 2, pp. 87–112, 1994.
24. K. Norris and P. Williams, "Optimization of mathematical treatment of raw near-infrared signal in the measurement of protein in hard red spring wheat. I. Influence of particle size," Cereal Chemistry, vol. 61, pp. 158–165, 1983.
25. R. Goodacre and D. B. Kell, "Evolutionary computing for the characterisation and qualification of microbial systems from hyperspectral data," in FT-IR Meeting in Berlin, 2000.
26. P. Williams and D. Sobering, "Comparison of commercial near infrared transmittance and reflectance instruments for analysis of whole grains and seeds," Journal Near Infrared Spectroscopy, vol. 1, no. 1, pp. 25–32, 1993.
27. P. Butz, C. Hofmann, and B. Tauscher, "Recent developments in noninvasive techniques for fresh fruit and vegetable internal quality analysis," Journal of Food Science, vol. 70, no. 9, pp. R131–R141, 2005.
28. B. M. Nicolaï, K. Beullens, E. Bobelyn et al., "Nondestructive measurement of fruit and vegetable quality by means of NIR spectroscopy: a review," Postharvest Biology and Technology, vol. 46, no. 2, pp. 99–118, 2007.
29. C. M. Andersen, J. P. Wold, and S. B. Engelsen, "Autofluorescence spectroscopy in food analysis," inHandbook of Food Analysis Instruments, chapter 15, pp. 347–364, CRC Press, 2009.
30. J. P. Wold and M. Mielnik, "Nondestructive assessment of lipid oxidation in minced poultry meat by autofluorescence spectroscopy," Journal of Food Science, vol. 65, no. 1, pp. 87–95, 2000.
31. L. Nollet and F. Toldra, Handbook of Seafood and Seafood Products Analysis, CRC Press, Boca Raton, Fla, USA, 2010.
32. H. B. Ding and R. J. Xu, "Near-infrared spectroscopic technique for detection of beef hamburger adulteration," Journal of Agricultural and Food Chemistry, vol. 48, no. 6, pp. 2193–2198, 2000.
33. G. ElMasry, D.-W. Sun, and P. Allen, "Near-infrared hyperspectral imaging for predicting colour, pH and tenderness of fresh beef," Journal of Food Engineering, vol. 110, no. 1, pp. 127–140, 2012.
34. Y. Liao, Y. Fan, and F. Cheng, "On-line prediction of pH values in fresh pork using visible/near-infrared spectroscopy with wavelet de-noising and variable selection methods," Journal of Food Engineering, vol. 109, no. 4, pp. 668–675, 2012.
35. C. Collell, P. Gou, J. Arnau, and J. Comaposada, "Non-destructive estimation of moisture, water activity and NaCl at ham surface during resting and drying using NIR spectroscopy," Food Chemistry, vol. 129, no. 2, pp. 601–607, 2011.
36. L. Dvash, O. Afik, S. Shafir et al., "Determination by near-infrared spectroscopy of perseitol used as a marker for the botanical origin of avocado (Persea americana Mill.) honey," Journal of Agricultural and Food Chemistry, vol. 50, no. 19, pp. 5283–5287, 2002.
37. G. Downey and J. D. Kelly, "Detection and quantification of apple adulteration in diluted and sulfited strawberry and raspberry purées using visible and near-infrared spectroscopy," Journal of Agricultural and Food Chemistry, vol. 52, no. 2, pp. 204–209, 2004.
38. R. Fügel, R. Carle, and A. Schieber, "Quality and authenticity control of fruit purées, fruit preparations and jams-a review," Trends in Food Science & Technology, vol. 16, no. 10, pp. 433–441, 2005.
39. P. Y. Qiu, H. B. Ding, Y. K. Tang, and R. J. Xu, "Determination of chemical composition of commercial honey by near-infrared spectroscopy," Journal of Agricultural and Food Chemistry, vol. 47, no. 7, pp. 2760–2765, 1999.
40. J. Fontaine, J. Hörr, and B. Schirmer, "Near-infrared reflectance spectroscopy enables the fast and accurate prediction of the essential amino acid contents in soy, rapeseed meal, sunflower meal, peas, fishmeal, meat meal products, and poultry meal," Journal of Agricultural and Food Chemistry, vol. 49, no. 1, pp. 57–66, 2001.
41. Y. Kim, M. Singh, and S. E. Kays, "Near-infrared spectroscopy for measurement of total dietary fiber in homogenized meals," Journal of Agricultural and Food Chemistry, vol. 54, no. 2, pp. 292–298, 2006.
42. W. Zeng, H. Zhang, and T. Lee, "Direct determination of the starch content in gravy by near-infrared spectroscopy," Journal of Agricultural and Food Chemistry, vol. 44, no. 6, pp. 1460–1463, 1996.
43. M. R. Ellekjaer, T. Naes, T. Isaksson, and R. Solheim, "Identification of sausages with fat-substitutes using near

infrared spectroscopy," in Near Infrared Spectroscopy: Bridging the Gap Between Data Analysis and NIR Applications, pp. 320–326, Ellis Horwood, London, UK, 1992.

44. A. J. Gaitán-Jurado, V. Ortiz-Somovilla, F. España-España, J. Pérez-Aparicio, and E. J. De Pedro-Sanz, "Quantitative analysis of pork dry-cured sausages to quality control by NIR spectroscopy," Meat Science, vol. 78, no. 4, pp. 391–399, 2008.

45. V. Ortiz-Somovilla, F. España-España, A. J. Gaitán-Jurado, J. Pérez-Aparicio, and E. J. De Pedro-Sanz, "Proximate analysis of homogenized and minced mass of pork sausages by NIRS," Food Chemistry, vol. 101, no. 3, pp. 1031–1040, 2006.

46. E. Mehinagic, G. Royer, D. Bertrand, R. Symoneaux, F. Laurens, and F. Jourjon, "Relationship between sensory analysis, penetrometry and visible-NIR spectroscopy of apples belonging to different cultivars,"Food Quality and Preference, vol. 14, no. 5-6, pp. 473–484, 2003.

47. Walsh, J. Guthrie, and J. Burney, "Application of commercially available, low-cost, miniaturised NIR spectrometers to the assessment of the sugar content of intact fruit," Australian Journal of Plant Physiology, vol. 27, no. 12, pp. 1175–1186, 2000.

48. P. Carlini, R. Massantini, and F. Mencarelli, "Vis-NIR measurement of soluble solids in cherry and apricot by PLS regression and wavelength selection," Journal of Agricultural and Food Chemistry, vol. 48, no. 11, pp. 5236–5242, 2000.

49. S. Tsuchikawa and T. Hamada, "Application of time-of-flight near infrared spectroscopy for detecting sugar and acid contents in apples," Journal of Agricultural and Food Chemistry, vol. 52, no. 9, pp. 2434–2439, 2004.

50. R. J. Blakey, J. P. Bower, and I. Bertling, "Influence of water and ABA supply on the ripening pattern of avocado (Persea americana Mill.) fruit and the prediction of water content using Near Infrared Spectroscopy," Postharvest Biology and Technology, vol. 53, no. 1-2, pp. 72–76, 2009.

51. L. León, A. Garrido-Varo, and G. Downey, "Parent and harvest year effects on near-infrared reflectance spectroscopic analysis of olive (Olea europaea L.) fruit traits," Journal of Agricultural and Food Chemistry, vol. 52, no. 16, pp. 4957–4962, 2004.

52. S. Zhang, Y. Zhao, W. Guo et al., "Robust model of fresh jujube soluble solids content with near-infrared (NIR) spectroscopy," African Journal of Biotechnology, vol. 11, no. 32, pp. 8133–8140, 2012.

53. A. Terdwongworakul, N. Nakawajana, S. Teerachaichayut, and A. Janhiran, "Determination of translucent content in mangosteen by means of near infrared transmittance," Journal of Food Engineering, vol. 109, no. 1, pp. 114–119, 2012.

54. L. S. Magwaza, U. L. Opara, H. Nieuwoudt, P. J. Cronje, W. Saeys, and B. Nicolaï, "NIR spectroscopy applications for internal and external quality analysis of citrus fruit-a review," Food and Bioprocess Technology, vol. 5, no. 2, pp. 425–444, 2012.

55. B. Shi, B. Ji, D. Zhu, Z. Tu, and Z. Qing, "Study on genetic algorithms-based NIR wavelength selection for determination of soluble solids content in fuji apples," Journal of Food Quality, vol. 31, no. 2, pp. 232–249, 2008.

56. Y. Ying and Y. Liu, "Nondestructive measurement of internal quality in pear using genetic algorithms and FT-NIR spectroscopy," Journal of Food Engineering, vol. 84, no. 2, pp. 206–213, 2008.

57. A. Pissard, J. A. Fernández Pierna, V. Baeten et al., "Non-destructive measurement of vitamin C, total polyphenol and sugar content in apples using near-infrared spectroscopy," Journal of the Science of Food and Agriculture, vol. 93, no. 2, pp. 238–244, 2013.

58. M. Tanaka and T. Kojima, "Near-infrared monitoring of the growth period of Japanese pear fruit based on constituent sugar concentrations," Journal of Agricultural and Food Chemistry, vol. 44, no. 8, pp. 2272–2277, 1996.

59. P. Sirisomboon, M. Tanaka, T. Kojima, and P. Williams, "Nondestructive estimation of maturity and textural properties on tomato "Momotaro" by nearinfrared spectroscopy," Journal of Food Engineering, vol. 112, no. 3, pp. 218–226, 2012.

60. J. Xing, D. Guyer, D. Ariana, and R. Lu, "Determining optimal wavebands using genetic algorithm for detection of internal insect infestation in tart cherry," Sensing and Instrumentation for Food Quality and Safety, vol. 2, no. 3, pp. 161–167, 2008.

61. M. Kim, A. Lefcourt, K. Chao, Y. Chen, I. Kim, and D. Chan, "Multispectral detection of fecal contamination on apples based on hyperspectral imagery: part I application of visible and near-infrared reflectance imaging," Transactions of the ASABE, vol. 45, no. 6, pp. 2027–2037, 2002.

62. M. Kim, A. Lefcourt, Y. Chen, I. Kim, D. Chan, and K. Chao, "Multispectral detection of fecal contamination on apples based on hyperspectral imagery: part II application of hyperspectral fluorescence imaging," Transactions of the American Society of Agricultural Engineers, vol. 45, no. 6, pp. 2039–2047, 2002.

63. E. A. Veraverbeke, J. Lammertyn, B. M. Nicolaï, and J. Irudayaraj, "Spectroscopic evaluation of the surface quality of apple," Journal of Agricultural and Food Chemistry, vol. 53, no. 4, pp. 1046–1051, 2005.

64. J. A. Fernández Pierna, P. Vermeulen, O. Amand, A. Tossens, P. Dardenne, and V. Baeten, "NIR hyperspectral imaging spectroscopy and chemometrics for the detection of undesirable substances in food and feed," Chemometrics and Intelligent Laboratory Systems, vol. 117, pp. 233–239, 2012.

65. K. U. Flores Rojas, Determinación no destructiva de parámetros de calidad de frutas y hortalizas mediante espectroscopía de reflectancia en el infrarrojo cercano [doctoral thesis], Universidad de Córdoba, Córdoba, Spain, 2009.

66. J. Sugiyama, "Visualization of sugar content in the flesh of a melon by near-infrared imaging," Journal of Agricultural and Food Chemistry, vol. 47, no. 7, pp. 2715–2718, 1999.

67. M. Tsuta, J. Sugiyama, and Y. Sagara, "Near-infrared

imaging spectroscopy based on sugar absorption band for melons," Journal of Agricultural and Food Chemistry, vol. 50, no. 1, pp. 48–52, 2002.

68. A. M. K. Pedro and M. M. C. Ferreira, "Nondestructive determination of solids and carotenoids in tomato products by near-infrared spectroscopy and multivariate calibration," Analytical Chemistry, vol. 77, no. 8, pp. 2505–2511, 2005.

69. F. Liu, Y. He, and G. Sun, "Determination of protein content of Auricularia auricula using near infrared spectroscopy combined with linear and nonlinear calibrations," Journal of Agricultural and Food Chemistry, vol. 57, no. 11, pp. 4520–4527, 2009.

70. C. Van Dijk, M. Fischer, J. Holm, J. Beekhuizen, T. Tolle-Smits, and C. Boeriu, "Texture of cooked potatoes (Solanum tuberosum). 1. Relationships between dry matter content, sensory-perceived texture, and near-infrared spectroscopy," Journal of Agricultural and Food Chemistry, vol. 50, no. 18, pp. 5082–5088, 2002.

71. A. Dacal-Nieto, E. Vázquez-Fernández, A. Formella, F. Martin, S. Torres-Guijarro, and H. González-Jorge, "A genetic algorithm approach for feature selection in potatoes classification by computer vision," in Proceedings of the 35th Annual Conference of the IEEE Industrial Electronics Society (IECON ‹09), pp. 1955–1960, November 2009.

72. Y. Roggo, L. Duponchel, and J. Huvenne, "Quality evaluation of sugar beet (Beta vulgaris) by near-infrared spectroscopy," Journal of Agricultural and Food Chemistry, vol. 52, no. 5, pp. 1055–1061, 2004. ·

73. S. R. Delwiche, L. O. Pordesimo, A. M. Scaboo, and V. R. Pantalone, "Measurement of inorganic phosphorus in soybeans with near-infrared spectroscopy," Journal of Agricultural and Food Chemistry, vol. 54, no. 19, pp. 6951–6956, 2006.

74. I. V. Kovalenko, G. R. Rippke, and C. R. Hurburgh Jr., "Determination of amino acid composition of soybeans (Glycine max) by near-infrared spectroscopy," Journal of Agricultural and Food Chemistry, vol. 54, no. 10, pp. 3485–3491, 2006.

75. D. L. Pazdernik, S. J. Plehn, J. L. Halgerson, and J. H. Orf, "Effect of temperature and genotype on the crude glycinin fraction (11s) of soybean and its analysis by near-infrared reflectance spectroscopy (near-IRS)," Journal of Agricultural and Food Chemistry, vol. 44, no. 8, pp. 2278–2281, 1996.

76. D. L. Pazdernik, A. S. Killam, and J. H. Orf, "Analysis of amino and fatty acid composition in soybean seed, using near infrared reflectance spectroscopy," Agronomy Journal, vol. 89, no. 4, pp. 679–685, 1997. ·

77. K. Hollung, M. Øverland, M. Hrustić et al., "Evaluation of nonstarch polysaccharides and oligosaccharide content of different soybean varieties (Glycine max) by near-infrared spectroscopy and proteomics," Journal of Agricultural and Food Chemistry, vol. 53, no. 23, pp. 9112–9121, 2005.

78. D. T. Lamb and C. R. Hurburgh Jr., "Moisture determination in single soybean seeds by near-infrared transmittance," Transactions of the American Society of Agricultural Engineers, vol. 34, no. 5, pp. 2123–2129, 1991.

79. J. O. Bennett, A. H. Krishnan, W. J. Wiebold, and H. B. Krishnan, "Positional effect on protein and oil content and composition of soybeans," Journal of Agricultural and Food Chemistry, vol. 51, no. 23, pp. 6882–6886, 2003.

80. J. O. Bennett, O. Yu, L. G. Heatherly, and H. B. Krishnan, "Accumulation of genistein and daidzein, soybean isoflavones implicated in promoting human health, is significantly elevated by irrigation,"Journal of Agricultural and Food Chemistry, vol. 52, no. 25, pp. 7574–7579, 2004.

81. C. Esquerre, A. A. Gowen, C. P. O›donnell, and G. Downey, "Initial studies on the quantitation of bruise damage and freshness in mushrooms using visible-near-infrared spectroscopy," Journal of Agricultural and Food Chemistry, vol. 57, no. 5, pp. 1903–1907, 2009. ·

82. S. Li, X. Zhu, J. Zhang, G. Li, D. Su, and Y. Shan, "Authentication of pure camellia oil by using near infrared spectroscopy and pattern recognition techniques," Journal of Food Science, vol. 77, no. 4, pp. C374–C380, 2012.

83. J. Siska and C. R. Hurburgh Jr., "Corn density measurement by near-infrared transmittance,"Transactions of the American Society of Agricultural Engineers, vol. 38, no. 6, pp. 1821–1824, 1995.

84. A. A. Gowen, C. P. O›Donnell, M. Taghizadeh, P. J. Cullen, J. M. Frias, and G. Downey, "Hyperspectral imaging combined with principal component analysis for bruise damage detection on white mushrooms (Agaricus bisporus)," Journal of Chemometrics, vol. 22, no. 3-4, pp. 259–267, 2008.

85. T. C. Pearson and D. T. Wicklow, "Detection of corn kernels infected by fungi," Transactions of the ASABE, vol. 49, no. 4, pp. 1235–1245, 2006.

86. S. H. Gordon, B. C. Wheeler, R. B. Schudy, D. T. Wicklow, and R. V. Greene, "Neural network pattern recognition of photoacoustic FTIR spectra and knowledge-based techniques for detection of mycotoxigenic fungi in food grains," Journal of Food Protection, vol. 61, no. 2, pp. 221–230, 1998.

87. N. Berardo, V. Pisacane, P. Battilani, A. Scandolara, A. Pietri, and A. Marocco, "Rapid detection of kernel rots and mycotoxins in maize by near-infrared reflectance spectroscopy," Journal of Agricultural and Food Chemistry, vol. 53, no. 21, pp. 8128–8134, 2005.

88. J. Jin, L. Tang, Z. Hruska, and H. Yao, "Classification of toxigenic and atoxigenic strains of Aspergillus flavus with hyperspectral imaging," Computers and Electronics in Agriculture, vol. 69, no. 2, pp. 158–164, 2009.

89. F. E. Dowell, S. E. Ram, and L. M. Seitz, "Predicting scab, vomitoxin, and ergosterol in single wheat kernels using near-infrared spectroscopy," Cereal Chemistry, vol. 76, no. 4, pp. 573–576, 1999.

90. C. Miralbés, "Prediction chemical composition and alveograph parameters on wheat by near-infrared transmittance spectroscopy," Journal of Agricultural and Food Chemistry, vol. 51, no. 21, pp. 6335–6339, 2003.

91. N. Yang and G. Ren, "Application of near-infrared reflectance spectroscopy to the evaluation of rutin and D-chiro-inositol contents in tartary buckwheat," Journal of Agricultural and Food Chemistry, vol. 56, no. 3, pp. 761–764, 2008.

92. O. V. Brenna and N. Berardo, "Application of near-infrared reflectance spectroscopy (NIRS) to the evaluation of carotenoids content in maize," Journal of Agricultural and Food Chemistry, vol. 52, no. 18, pp. 5577–5582, 2004.

93. S. Kawamura, M. Natsuga, and K. Itoh, "Determination of undried rough rice constituent content using near-infrared transmission spectroscopy," Transactions of the American Society of Agricultural Engineers, vol. 42, no. 3, pp. 813–818, 1999.

94. C. Zhang, Y. Shen, J. Chen, P. Xiao, and J. Bao, "Nondestructive prediction of total phenolics, flavonoid contents, and antioxidant capacity of rice grain using near-infrared spectroscopy," Journal of Agricultural and Food Chemistry, vol. 56, no. 18, pp. 8268–8272, 2008. ·

95. T. L. Stubbs, A. C. Kennedy, and A. Fortuna, "Using NIRS to predict fiber and nutrient content of Dryland cereal cultivars," Journal of Agricultural and Food Chemistry, vol. 58, no. 1, pp. 398–403, 2010. ·

96. P. R. Wiley, G. J. Tanner, P. M. Chandler, and R. S. Anderssen, "Molecular classification of barley (Hordeum vulgare L.) mutants using derivative NIR spectroscopy," Journal of Agricultural and Food Chemistry, vol. 57, no. 10, pp. 4042–4050, 2009.

97. G. Fox and M. Manley, "Hardness methods for testing maize kernels," Journal of Agricultural and Food Chemistry, vol. 57, no. 13, pp. 5647–5657, 2009.

98. M. Manley, P. Williams, D. Nilsson, and P. Geladi, "Near infrared hyperspectral imaging for the evaluation of endosperm texture in whole yellow maize (Zea maize L.) Kernels," Journal of Agricultural and Food Chemistry, vol. 57, no. 19, pp. 8761–8769, 2009. ·

99. S. Kawamura, M. Natsuga, and K. Itoh, "Visual and near-infrared reflectance spectroscopy for rice taste evaluation," Transactions of the American Society of Agricultural Engineers, vol. 40, no. 6, pp. 1755–1759, 1997.

100. F. Davrieux, F. Allal, G. Piombo et al., "Near infrared spectroscopy for high-throughput characterization of shea tree (Vitellaria paradoxa) nut fat profiles," Journal of Agricultural and Food Chemistry, vol. 58, no. 13, pp. 7811–7819, 2010.

101. R. Font, M. del Río, J. M. Fernández-Martínez, and A. de Haro-Bailón, "Use of near-infrared spectroscopy for screening the individual and total glucosinolate contents in indian mustard seed (Brassica juncea L. Czern. & Coss.)," Journal of Agricultural and Food Chemistry, vol. 52, no. 11, pp. 3563–3569, 2004.

102. R. Font, M. del Río, J. M. Fernández, and A. de Haro, "Acid detergent fiber analysis in oilseed brassicas by near-infrared spectroscopy," Journal of Agricultural and Food Chemistry, vol. 51, no. 10, pp. 2917–2922, 2003.

103. S. Kim, H. Park, and G. Choung, "Nondestructive determination of oil content and fatty acid composition in perilla seeds by near-infrared spectroscopy," Journal of Agricultural and Food Chemistry, vol. 55, no. 5, pp. 1679–1685, 2007.

104. G. Hacisalihoglu, B. Larbi, and Mark Settles, "Near-infrared reflectance spectroscopy predicts protein, starch, and seed weight in intact seeds of common bean (Phaseolus vulgaris L.)," Journal of Agricultural and Food Chemistry, vol. 58, no. 2, pp. 702–706, 2010.

105. L. Wang, Q. Wang, H. Liu, L. Liu, and Y. Du, "Determining the contents of protein and amino acids in peanuts using near-infrared reflectance spectroscopy," Journal of the Science of Food and Agriculture, vol. 93, no. 1, pp. 118–124, 2013.

106. L. Yang and Q. Sun, "Recognition of the hardness of licorice seeds using a semi-supervised learning method and near-infrared spectral data," Chemometrics and Intelligent Laboratory Systems, vol. 114, pp. 109–115, 2012.

107. G. Downey, R. Briandet, R. H. Wilson, and E. K. Kemsley, "Near- and mid-infrared spectroscopies in food authentication: coffee varietal identification," Journal of Agricultural and Food Chemistry, vol. 45, no. 11, pp. 4357–4361, 1997.

108. G. Downey, P. McIntyre, and A. N. Davies, "Detecting and quantifying sunflower oil adulteration in extra virgin olive oils from the Eastern Mediterranean by visible and near-infrared spectroscopy," Journal of Agricultural and Food Chemistry, vol. 50, no. 20, pp. 5520–5525, 2002.

109. T. Woodcock, G. Downey, and C. P. O›Donnell, "Confirmation of declared provenance of European extra virgin olive oil samples by NIR spectroscopy," Journal of Agricultural and Food Chemistry, vol. 56, no. 23, pp. 11520–11525, 2008.

110. L. D. Gautz, P. Kaufusi, M. C. Jackson, H. C. Bittenbender, and C. Tang, "Determination of kavalactones in dried kava (Piper methysticum) powder using near-infrared reflectance spectroscopy and partial least-squares regression," Journal of Agricultural and Food Chemistry, vol. 54, no. 17, pp. 6147–6152, 2006. ·

111. A. Zalacain, S. A. Ordoudi, E. M. Díaz-Plaza et al., "Near-infrared spectroscopy in saffron quality control: determination of chemical composition and geographical origin," Journal of Agricultural and Food Chemistry, vol. 53, no. 24, pp. 9337–9341, 2005.

112. K. Tanaka, Y. Kuba, T. Sasaki, F. Hiwatashi, and K. Komatsu, "Quantitation of curcuminoids in curcuma rhizome by near-infrared spectroscopic analysis," Journal of Agricultural and Food Chemistry, vol. 56, no. 19, pp. 8787–8792, 2008.

113. G. Ren and F. Chen, "Simultaneous quantification of ginsenosides in American ginseng (Panax quinquefolium) root powder by visible/near-infrared reflectance spectroscopy," Journal of Agricultural and Food Chemistry, vol. 47, no. 7, pp. 2771–2775, 1999. ·

114. S. Kim, H. Park, and G. Choung, "Nondestructive determination of lignans and lignan glycosides in sesame seeds by near infrared reflectance spectroscopy," Journal of Agricultural and Food Chemistry, vol. 54, no. 13, pp. 4544–4550, 2006.

115. N. Xin, X. Gu, H. Wu, Y. Hu, and Z. Yang, "Discrimination of raw and processed Dipsacus asperoides by near infrared spectroscopy combined with least squares-support vector machine and random forests,"Spectrochimica Acta—Part A, vol. 89, pp. 18–24, 2012.

116. K. Horváth, Z. Seregély, É. Andrássy, I. Dalmadi, and J. Farkas, "A preliminary study using near infrared spectroscopy to evaluate freshness and detect spoilage in sliced pork meat," Acta Alimentaria, vol. 37, no. 1, pp. 93–102, 2008.

117. D. I. Ellis, D. Broadhurst, D. B. Kell, J. J. Rowland, and R. Goodacre, "Rapid and quantitative detection of the microbial spoilage of meat by fourier transform infrared spectroscopy and machine learning,"Applied and Environmental Microbiology, vol. 68, no. 6, pp. 2822–2828, 2002.

118. W. R. Windham, D. P. Smith, B. Park, K. C. Lawrence, and P. W. Feldner, "Algorithm development with visible/near-infrared spectra for detection of poultry feces and ingesta," Transactions of the American Society of Agricultural Engineers, vol. 46, no. 6, pp. 1733–1738, 2003.

119. H. L. A. Tarr, "Flavor of flesh foods," Advances in Chemistry, vol. 56, pp. 190–202, 1969.

120. C. E. Byrne, G. Downey, D. J. Troy, and D. J. Buckley, "Non-destructive prediction of selected quality attributes of beef by near-infrared reflectance spectroscopy between 750 and 1098 nm," Meat Science, vol. 49, no. 4, pp. 399–409, 1998.

121. S. Andrés, I. Murray, E. A. Navajas, A. V. Fisher, N. R. Lambe, and L. Bünger, "Prediction of sensory characteristics of lamb meat samples by near infrared reflectance spectroscopy," Meat Science, vol. 76, no. 3, pp. 509–516, 2007.

122. M. Renerre and R. Labas, "Biochemical factors influencing metmyoglobin formation in beef muscles,"Meat Science, vol. 19, no. 2, pp. 151–165, 1987.

123. M. Renerre, "Factors involved in the discoloration of beef meat," International Journal of Food Science and Technology, vol. 25, no. 6, pp. 613–630, 1990.

124. A. H. A. Van den Oord and J. J. Wesdorp, "Colour rating and pigment composition of beef: relation between colour measurement by reflectance spectrophotometry, subjective colour evaluation, and the relative concentration of oxymyoglobin and ferric myoglobin in chilled retail-packed beef," Journal of Food Technology, vol. 6, no. 1, pp. 15–20, 1971.

125. B. Leroy, S. Lambotte, O. Dotreppe, H. Lecocq, L. Istasse, and A. Clinquart, "Prediction of technological and organoleptic properties of beef Longissimus thoracis from near-infrared reflectance and transmission spectra," Meat Science, vol. 66, no. 1, pp. 45–54, 2004.

126. S. D. Shackelford, T. L. Wheeler, and M. Koohmaraie, "On-line classification of US Select beef carcasses for longissimus tenderness using visible and near-infrared reflectance spectroscopy," Meat Science, vol. 69, no. 3, pp. 409–415, 2005.

127. D. Cozzolino, N. Barlocco, A. Vadell, F. Ballesteros, and G. Gallieta, "The use of visible and near-infrared reflectance spectroscopy to predict colour on both intact and homogenised pork muscle," LWT—Food Science and Technology, vol. 36, no. 2, pp. 195–202, 2003. ·

128. S. Andrés, A. Silva, S. L. Soares-Pereira, C. Martins, A. M. Bruno-Soares, and L. Murray, "The use of visible and near infrared reflectance spectroscopy to predict beef M. longissimus thoracis et lumborum quality attributes," Meat Science, vol. 78, no. 3, pp. 217–224, 2008.

129. J. D. Wood, M. Enser, A. V. Fisher et al., "Fat deposition, fatty acid composition and meat quality: a review," Meat Science, vol. 78, no. 4, pp. 343–358, 2008.

130. N. Prieto, M. E. R. Dugan, O. López-Campos, T. A. McAllister, J. L. Aalhus, and B. Uttaro, "Near infrared reflectance spectroscopy predicts the content of polyunsaturated fatty acids and biohydrogenation products in the subcutaneous fat of beef cows fed flaxseed," Meat Science, vol. 90, no. 1, pp. 43–51, 2012. ·

131. J. R. Anderson, C. Borggaard, A. J. Rasmussen, and L. P. Houmoller, "Optical measurements of pH in meat," Meat Science, vol. 53, no. 2, pp. 135–141, 1999.

132. B. Savenije, G. H. Geesink, J. G. P. van der Palen, and G. Hemke, "Prediction of pork quality using visible near infrared reflectance spectroscopy," Meat Science, vol. 73, no. 1, pp. 181–184, 2006.

133. Y. Liao, Y. Fan, and F. Cheng, "On-line prediction of fresh pork quality using visible/near-infrared reflectance spectroscopy," Meat Science, vol. 86, no. 4, pp. 901–907, 2010.

134. E. Tornberg, M. Wahlgren, J. Brøndum, and S. B. Engelsen, "Pre-rigor conditions in beef under varying temperature- and pH-falls studied with rigometer, NMR and NIR," Food Chemistry, vol. 69, no. 4, pp. 407–418, 2000.

135. D. E. Chan, P. N. Walker, and E. W. Mills, "Prediction of pork quality characteristics using visible and near-infrared spectroscopy," Transactions of the ASAE, vol. 45, no. 5, pp. 1519–1527, 2002.

136. V. Sierra, N. Aldai, P. Castro, P. Osoro, A. Coto-Montes, and M. Oliván, "Prediction of the fatty acid composition of beef by near infrared transmittance spectroscopy," Meat Science, vol. 78, no. 3, pp. 248–255, 2008.

137. M. Prevolnik, M. Candek-Potokar, D. Skorjanc et al., "Predicting intramuscular fat content in pork and beef by near infrared spectroscopy," Journal of Near Infrared Spectroscopy, vol. 13, no. 2, pp. 77–85, 2005. ·

138. N. Barlocco, A. Vadell, F. Ballesteros, G. Galietta, and D. Cozzolino, "Predicting intramuscular fat, moisture and Warner-Bratzler shear force in pork muscle using near infrared reflectance spectroscopy,"Animal Science, vol. 82, no. 1, pp. 111–116, 2006.

139. J. Hong and K. Yasumoto, "Near-infrared spectroscopic analysis of heme and nonheme iron in raw meats," Journal of Food Composition and Analysis, vol. 9, no. 2, pp. 127–134, 1996.

140. N. Prieto, S. Andrés, F. J. Giráldez, A. R. Mantecón, and P. Lavín, "Potential use of near infrared reflectance

spectroscopy (NIRS) for the estimation of chemical composition of oxen meat samples," Meat Science, vol. 74, no. 3, pp. 487–496, 2006.

141. D. Pérez-Marín, E. De Pedro Sanz, J. E. Guerrero-Ginel, and A. Garrido-Varo, "A feasibility study on the use of near-infrared spectroscopy for prediction of the fatty acid profile in live Iberian pigs and carcasses," Meat Science, vol. 83, no. 4, pp. 627–633, 2009.

142. E. De Pedro, N. Núñez, J. García et al., "Implementing near infrared spectroscopy in the Iberian pig industry," OPtions Méditerranéennes, Series A, no. 76, pp. 225–227, 2007.

143. J. García-Olmo, A. Garrido, and E. De Pedro, "Methodological aspects on near infrared analysis of Iberian pig fat using interactance-reflectance fiber optic mode," Journal of Near Infrared Spectroscopy, vol. 6, no. 1–4, pp. A307–A312, 1998.

144. I. González-Martín, C. González-Pérez, J. Hernández-Méndez, and N. Álvarez-García, "Determination of fatty acids in the subcutaneous fat of Iberian breed swine by near infrared spectroscopy (NIRS) with a fibre-optic probe," Meat Science, vol. 65, no. 2, pp. 713–719, 2003.

145. H. J. Swatland, On-Line Evaluation of Meat, Technomic Publishing, Lancaster, UK, 1995.

146. D. Cozzolino, I. Murray, J. R. Scaife, and R. Paterson, "Study of dissected lamb muscles by visible and near infrared reflectance spectroscopy for composition assessment," Animal Science, vol. 70, no. 3, pp. 417–423, 2000.

147. R. A. Lawri, Meat Science, Pergamon Press, Oxford, UK, 1985.

148. M. Mitsumoto, S. Maeda, T. Mitsuhashi, and S. Ozawa, "Near-infrared spectroscopy determination of physical and chemical characteristics in beef cuts," Journal of Food Science, vol. 56, no. 6, pp. 1493–1496, 1991.

149. D. T. Delpy and M. Cope, "Quantification in tissue near-infrared spectroscopy," Philosophical Transactions of the Royal Society B, vol. 352, no. 1354, pp. 649–659, 1997.

150. D. Cozzolino and I. Murray, "Effect of sample presentation and animal muscle species on the analysis of meat by near infrared reflectance spectroscopy," Journal of Near Infrared Spectroscopy, vol. 10, no. 1, pp. 37–44, 2002.

151. I. Murray, "The NIR spectra of homologous series of organic compounds," in Proceedings of the International NIR/NIT Conference, Budapest, Hungary, 1986.

152. G. Downey, J. McElhinney, and T. Fearn, "Species identification in selected raw homogenized meats by reflectance spectroscopy in the mid-infrared, near-infrared, and visible ranges," Applied Spectroscopy, vol. 54, no. 6, pp. 894–899, 2000.

153. D. Lomiwes, M. M. Reis, E. Wiklund, O. A. Young, and M. North, "Near infrared spectroscopy as an on-line method to quantitatively determine glycogen and predict ultimate pH in pre rigor bovine M. longissimus dorsi," Meat Science, vol. 86, no. 4, pp. 999–1004, 2010.

154. N. Prieto, S. Andrés, F. J. Giráldez, A. R. Mantecón, and P. Lavín, "Discrimination of adult steers (oxen) and young cattle ground meat samples by near infrared reflectance spectroscopy (NIRS)," Meat Science, vol. 79, no. 1, pp. 198–201, 2008.

155. J. W. S. Yancey, J. K. Apple, J. F. Meullenet, and J. T. Sawyer, "Consumer responses for tenderness and overall impression can be predicted by visible and near-infrared spectroscopy, Meullenet-Owens razor shear, and Warner-Bratzler shear force," Meat Science, vol. 85, no. 3, pp. 487–492, 2010.

156. B. Park, Y. R. Chen, W. R. Hruschka, S. D. Shackelford, and M. Koohmaraie, "Near-infrared reflectance analysis for predicting beef longissimus tenderness," Journal of Animal Science, vol. 76, no. 8, pp. 2115–2120, 1998.

157. R. Rødbotten, Prediction of beef tenderness by NIR spectroscopy, and examination of the influence of ante- and post-mortem factors on tenderness [Ph.D. thesis], Agricultural University of Norway, Bergen, Norway, 2001.

158. C. Venel, A. M. Mullen, G. Downey, and Troy, "Prediction of tenderness and other quality attributes of beef by near infrared reflectance spectroscopy between 750 and 1100 nm; further studies," Journal of Near Infrared Spectroscopy, vol. 9, no. 3, pp. 185–198, 2001.

159. N. Prieto, D. W. Ross, E. A. Navajas et al., "On-line application of visible and near infrared reflectance spectroscopy to predict chemical-physical and sensory characteristics of beef quality," Meat Science, vol. 83, no. 1, pp. 96–103, 2009.

160. G. H. Geesink, F. H. Schreutelkamp, R. Frankhuizen et al., "Prediction of pork quality attributes from near infrared reflectance spectra," Meat Science, vol. 65, no. 1, pp. 661–668, 2003.

161. R. J. Winger and C. J. Hagyard, "Juiciness—its importance and some contributing factors," in Quality Attributes and Their Measurement in Meat, Poultry and Fish Products, A. M. Pearson and T. R. Dutson, Eds., vol. 9 of Advances in Meat Research, chapter 4, pp. 94–124, Aspen Publisher, Frederick, Md, USA, 1994.

162. K. I. Hildrum, B. N. Nilsen, M. Mielnik, and T. Næs, "Prediction of sensory characteristics of beef by near-infrared spectroscopy," Meat Science, vol. 38, no. 1, pp. 67–80, 1994.

163. K. I. Hildrum, T. Isaksson, T. T. Naes, B. N. Nilsen, M. Rodbotten, and P. Lea, "Near infrared reflectance spectroscopy in the prediction of sensory properties of beef," Journal of Near Infrared Spectroscopy, vol. 3, no. 2, pp. 81–87, 1995.

164. T. Næs and K. I. Hildrum, "Comparison of multivariate calibration and discriminant analysis in evaluating NIR spectroscopy for determination of meat tenderness," Applied Spectroscopy, vol. 51, no. 3, pp. 350–357, 1997.

165. S. D. Shackelford, T. L. Wheeler, and M. Koohmaraie, "Development of optimal protocol for visible and near-infrared reflectance spectroscopic evaluation of meat quality," Meat Science, vol. 68, no. 3, pp. 371–381, 2004.

166. D. Cozzolino, I. Murray, and R. Paterson, "Visible and near infrared reflectance spectroscopy for the determination of moisture, fat and protein in chicken

breast and thigh muscle," Journal of Near Infrared Spectroscopy, vol. 4, no. 1–4, pp. 213–223, 1996.

167. D. Cozzolino, D. De Mattos, and D. Vaz Martins, "Visible/near infrared reflectance spectroscopy for predicting composition and tracing system of production of beef muscle," Animal Science, vol. 74, no. 3, pp. 477–484, 2002.

168. P. H. M. Dian, D. Andueza, M. Jestin, I. N. Prado, and S. Prache, "Comparison of visible and near infrared reflectance spectroscopy to discriminate between pasture-fed and concentrate-fed lamb carcasses," Meat Science, vol. 80, no. 4, pp. 1157–1164, 2008.

169. P. H. M. Dian, D. Andueza, C. M. P. Barbosa, S. Amoureux, M. Jestin, and P. C. F. Carvalho, "Methodological developments in the use of visible reflectance spectroscopy for discriminating pasture-fed from concentrate-fed lamb carcasses," Animal, vol. 1, no. 8, pp. 1198–1208, 2007.

170. D. Cozzolino and I. Murray, "Identification of animal meat muscles by visible and near infrared reflectance spectroscopy," Lebensmittel-Wissenschaft und-Technologie, vol. 37, no. 4, pp. 447–452, 2004.

171. H. B. Ding and R. J. Xu, "Differentiation of beef and kangaroo meat by visible/near-infrared reflectance spectroscopy," Journal of Food Science, vol. 64, no. 5, pp. 814–817, 1999.

172. M. Prevolnik, M. Čandek-Potokar, M. Novič, and D. Škorjanc, "An attempt to predict pork drip loss from pH and colour measurements or near infrared spectra using artificial neural networks," Meat Science, vol. 83, no. 3, pp. 405–411, 2009.

173. G. Downey and D. Beauchêne, "Discrimination between fresh and frozen-then-thawed beef m. longissimus dorsi by combined visible-near infrared reflectance spectroscopy: a feasibility study," Meat Science, vol. 45, no. 3, pp. 353–363, 1997.

174. K. Thyholt and T. Isaksson, "Differentiation of frozen and unfrozen beef using near-infrared spectroscopy," Journal of the Science of Food and Agriculture, vol. 73, no. 4, pp. 525–532, 1997.

175. N. S. Hobson, I. Tothill, and A. P. Turner, "Microbial detection," Biosensors and Bioelectronics, vol. 11, no. 5, pp. 455–477, 1996.

176. M. Kawasaki, S. Kawamura, H. Nakatsuji, and M. Natsuga, "Online real-time monitoring of milk quality during milking by near infrared spectroscopy," in ASABE Meeting Presentation, 2005.

177. H. M. Al-Qadiri, M. Lin, M. A. Al-Holy, A. G. Cavinato, and B. A. Rasco, "Monitoring quality loss of pasteurized skim milk using visible and short wavelength near-infrared spectroscopy and multivariate analysis," Journal of Dairy Science, vol. 91, no. 3, pp. 950–958, 2008.

178. R. Karoui and J. De Baerdemaeker, "A review of the analytical methods coupled with chemometric tools for the determination of the quality and identity of dairy products," Food Chemistry, vol. 102, no. 3, pp. 621–640, 2007.

179. M. Laporte and P. Paquin, "Near-infrared analysis of fat, protein, and casein in cow›s milk," Journal of Agricultural and Food Chemistry, vol. 47, no. 7, pp. 2600–2605, 1999.

180. S. Šašić and Y. Ozaki, "Short-wave near-infrared spectroscopy of biological fluids. 1. Quantitative analysis of fat, protein, and lactose in raw milk by partial least-squares regression and band assignment,"Analytical Chemistry, vol. 73, no. 1, pp. 64–71, 2001.

181. T. Sato, S. Kawano, and M. Iwamoto, "Detection of foreign fat adulteration of milk by near infrared spectroscopic method," Journal of Dairy Science, vol. 73, no. 12, pp. 3408–3413, 1990.

182. Y. R. Chen, R. W. Huffman, B. Park, and M. Nguyen, "Transportable spectrophotometer system for on-line classification of poultry carcasses," Applied Spectroscopy, vol. 50, no. 7, pp. 910–916, 1996.

183. H. Chen and B. P. Marks, "Evaluating previous thermal treatment of chicken patties by visible/near-infrared spectroscopy," Journal of Food Science, vol. 62, no. 4, pp. 753–780, 1997.

184. H. Chen and B. P. Marks, "Visible/near-infrared spectroscopy for physical characteristics of cooked chicken patties," Journal of Food Science, vol. 63, no. 2, pp. 279–282, 1998.

185. Y. R. Chen, B. Park, R. W. Huffman, and M. Nguyen, "Classification of on-line poultry carcasses with backpropagation neural networks," Journal of Food Process Engineering, vol. 21, no. 1, pp. 33–48, 1998.

186. J. McElhinney, G. Downey, and T. Fearn, "Chemometric processing of visible and near infrared reflectance spectra for species identification in selected raw homogenised meats," Journal of Near Infrared Spectroscopy, vol. 7, no. 3, pp. 145–154, 1999.

187. H. Rannou and G. Downey, "Discrimination of raw pork, chicken and turkey meat by spectroscopy in the visible, near- and mid-infrared ranges," Analytical Communications, vol. 34, no. 12, pp. 401–404, 1997.

188. B. Park, S. Yoon, W. R. Windham, K. C. Lawrence, M. S. Kim, and K. Chao, "Line-scan hyperspectral imaging for real-time in-line poultry fecal detection," Sensing and Instrumentation for Food Quality and Safety, vol. 5, no. 1, pp. 25–32, 2011.

189. H. Ding, R.-J. Xu, and D. K. O. Chan, "Identification of broiler chicken meat using a visible/near-infrared spectroscopic technique," Journal of the Science of Food and Agriculture, vol. 79, no. 11, pp. 1382–1388, 1999.

190. M. Lin, M. Al-Holy, M. Mousavi-Hesary, H. Al-Qadiri, A. G. Cavinato, and B. A. Rasco, "Rapid and quantitative detection of the microbial spoilage in chicken meat by diffuse reflectance spectroscopy (600-1100 nm)," Letters in Applied Microbiology, vol. 39, no. 2, pp. 148–155, 2004.

191. Y. Liu and Y. Chen, "Two-dimensional correlation spectroscopy study of visible and near-infrared spectral variations of chicken meats in cold storage," Applied Spectroscopy, vol. 54, no. 10, pp. 1458–1470, 2000.

192. Y. Liu, F. E. Barton II, B. G. Lyon, W. R. Windham, and C. E. Lyon, "Two-dimensional correlation analysis

of visible/near-infrared spectral intensity variations of chicken breasts with various chilled and frozen storages," Journal of Agricultural and Food Chemistry, vol. 52, no. 3, pp. 505–510, 2004.

193. Y. Liu, Y. Chen, and Y. Ozaki, "Two-dimensional visible/near-infrared correlation spectroscopy study of thermal treatment of chicken meats," Journal of Agricultural and Food Chemistry, vol. 48, no. 3, pp. 901–908, 2000.

194. S. K. Stormo, A. H. Sivertsen, K. Heia, and D. Skipnes, "Endpoint temperature of heat-treated surimi can be measured by visible spectroscopy," Food Control, vol. 26, no. 1, pp. 92–97, 2012.

195. S. H. Olsen, Quantification and characterisation of residual blood in fish muscle—impact of slaughtering methods [Ph.D. thesis], University of Tromsø UIT, Tromsø, Norway, 2011.

196. S. H. Olsen, N. K. Sørensen, R. Larsen, E. O. Elvevoll, and H. Nilsen, "Impact of pre-slaughter stress on residual blood in fillet portions of farmed Atlantic cod (Gadus morhua)—measured chemically and by Visible and Near-infrared spectroscopy," Aquaculture, vol. 284, no. 1–4, pp. 90–97, 2008.

197. D. Cozzolino, A. Chree, J. R. Scaife, and I. Murray, "Usefulness of Near-infrared reflectance (NIR) spectroscopy and chemometrics to discriminate fishmeal batches made with different fish species,"Journal of Agricultural and Food Chemistry, vol. 53, no. 11, pp. 4459–4463, 2005.

198. M. Uddin, E. Okazaki, S. Turza, Y. Yumiko, M. Tanaka, and Y. Fukuda, "Non-destructive visible/NIR spectroscopy for differentiation of fresh and frozen-thawed fish," Journal of Food Science, vol. 70, no. 8, pp. C506 C510, 2005.

199. N. Bøknaes, K. N. Jensen, C. M. Andersen, and H. Martens, "Freshness assessment of thawed and chilled cod fillets packed in modified atmosphere using near-infrared spectroscopy," Lebensmittel-Wissenschaft und-Technologie, vol. 35, no. 7, pp. 628–634, 2002.

200. H. Nilsen, M. Esaiassen, K. Heia, and F. Sigernes, "Visible/near-infrared spectroscopy: a new tool for the evaluation of fish freshness?" Journal of Food Science, vol. 67, no. 5, pp. 1821–1826, 2002.

201. G. Olafsdottir, P. Nesvadba, C. Di Natale et al., "Multisensor for fish quality determination," Trends in Food Science and Technology, vol. 15, no. 2, pp. 86–93, 2004. ·

202. M. Lin, M. Mousavi, M. Al-Holy, A. G. Cavinato, and B. A. Rasco, "Rapid near infrared spectroscopic method for the detection of spoilage in rainbow trout (Oncorhynchus mykiss) fillet," Journal of Food Science, vol. 71, no. 1, pp. S18–S23, 2006.

203. H. Zhang and T. Lee, "Rapid near-infrared spectroscopic method for the determination of free fatty acid in fish and its application in fish quality assessment," Journal of Agricultural and Food Chemistry, vol. 45, no. 9, pp. 3515–3521, 1997.

204. B. Gjerde and H. Martens, "Predicting carcass composition of rainbow trout by near-infrared reflectance spectroscopy," Journal of Animal Breeding and Genetics, vol. 104, pp. 137–148, 1987.

205. J. Shimamoto, K. Hasegawa, S. Hattori, Y. Hattori, and T. Mizuno, "Non-destructive determination of the fat content in glazed bigeye tuna by portable near infrared spectrophotometer," Fisheries Science, vol. 69, no. 6, pp. 1247–1256, 2003.

206. J. Shimamoto, K. Hasegawa, K. Ide, and S. Kawano, "Nondestractive determination of the fat content in raw and frozen horse mackerel by near infrared spectroscopy," Nippon Suisan Gakkaishi, vol. 67, no. 4, pp. 717–722, 2001.

207. S. Yamauchi, T. Sawada, and S. Kawano, "Nondestructive determination of fat content in frozen skipjack by near infrared (NIR) spectroscopy with fiber optics in interactance mode," Nippon Suisan Gakkaishi, vol. 65, no. 4, pp. 747–752, 1999.

208. J. Shimamoto, S. Hiratsuka, K. Hasegawa, M. Sato, and S. Kawano, "Rapid non-destructive determination of fat content in frozen skipjack using a portable near infrared spectrophotometer," Fisheries Science, vol. 69, no. 4, pp. 856–860, 2003.

209. J. A. Mathias, P. C. Williams, and D. C. Sobering, "The determination of lipid and protein in freshwater fish using near-infrared reflectance spectroscopy," Aquaculture, vol. 61, no. 3-4, pp. 303–311, 1987.

210. M. H. Lee, A. G. Cavinato, D. M. Mayes, and B. A. Rasco, "Noninvasive short-wavelength near-infrared spectroscopic method to estimate the crude lipid content in the muscle of intact rainbow trout," Journal of Agricultural and Food Chemistry, vol. 40, no. 11, pp. 2176–2180, 1992.

211. M. Lin, A. G. Cavinato, D. M. Mayes et al., "Bruise detection in Pacific pink salmon (Oncorhynchus gorbuscha) by visible and short-wavelength near-infrared (SW-NIR) spectroscopy (600-1100 nm),"Journal of Agricultural and Food Chemistry, vol. 51, no. 22, pp. 6404–6408, 2003.

212. R. Font, M. Del Río-Celestino, D. Vélez, A. De Haro-Bailón, and R. Montoro, "Visible and near-infrared spectroscopy as a technique for screening the inorganic arsenic content in the red crayfish (Procambarus clarkii Girard)," Analytical Chemistry, vol. 76, no. 14, pp. 3893–3898, 2004.

213. J. Gayo and S. A. Hale, "Detection and quantification of species authenticity and adulteration in crabmeat using visible and near-infrared spectroscopy," Journal of Agricultural and Food Chemistry, vol. 55, no. 3, pp. 585–592, 2007.

214. J. Gayo, S. A. Hale, and S. M. Blanchard, "Quantitative analysis and detection of adulteration in crab meat using visible and near-infrared spectroscopy," Journal of Agricultural and Food Chemistry, vol. 54, no. 4, pp. 1130–1136, 2006.

215. M. R. Brown, "Rapid compositional analysis of oysters using visible-near infrared reflectance spectroscopy," Aquaculture, vol. 317, no. 1–4, pp. 233–239, 2011.

216. V. Lebot, A. Champagne, R. Malapa, and D. Shiley, "NIR determination of major constituents in tropical root and tuber crop flours," Journal of Agricultural and Food

Chemistry, vol. 57, no. 22, pp. 10539–10547, 2009.

217. A. Dacal-Nieto, A. Formella, P. Carrión, E. Vazquez-Fernandez, and M. Fernández-Delgado, "Non-destructive detection of hollow heart in potatoes using hyperspectral imaging," in Computer Analysis of Images and Patterns: 14th International Conference (CAIP ‹11), P. Real, D. Díaz-Pernil, H. Molina-Abril, A. Berciano, and W. Kropatsch, Eds., vol. 6855 of Lecture Notes in Computer Science, pp. 180–187.

218. E. Joubert, M. Manley, and M. Botha, "Use of NIRS for quantification of mangiferin and hesperidin contents of dried green honeybush (Cyclopia genistoides) plant material," Journal of Agricultural and Food Chemistry, vol. 54, no. 15, pp. 5279–5283, 2006.

219. S. Bellato, V. Del Frate, R. Redaelli et al., "Use of near infrared reflectance and transmittance coupled to robust calibration for the evaluation of nutritional value in naked oats," Journal of Agricultural and Food Chemistry, vol. 59, no. 9, pp. 4349–4360, 2011.

220. L. R. Schimleck, J. C. Doran, and A. Rimbawanto, "Near infrared spectroscopy for cost effective screening of foliar oil characteristics in a Melaleuca cajuputi breeding population," Journal of Agricultural and Food Chemistry, vol. 51, no. 9, pp. 2433–2437, 2003. ·

221. E. Joubert, M. Manley, B. R. Gray, and H. Schulz, "Rapid measurement and evaluation of the effect of drying conditions on harpagoside content in Harpagophytum procumbens (devil›s claw) root," Journal of Agricultural and Food Chemistry, vol. 53, no. 9, pp. 3493–3502, 2005.

222. H. Schulz, U. H. Engelhardt, A. Wegent, H. Drews, and S. Lapczynski, "Application of near-infrared reflectance spectroscopy to the simultaneous prediction of alkaloids and phenolic substances in green tea leaves," Journal of Agricultural and Food Chemistry, vol. 47, no. 12, pp. 5064–5067, 1999.

223. J. M. Hernández-Hierro, J. Valverde, S. Villacreces et al., "Feasibility study on the use of visible-near-infrared spectroscopy for the screening of individual and total glucosinolate contents in broccoli,"Journal of Agricultural and Food Chemistry, vol. 60, no. 30, pp. 7352–7358, 2012.

224. M. Laasonen, T. Harmia-Pulkkinen, C. L. Simard, E. Michiels, M. Räsänen, and H. Vuorela, "Fast identification of Echinacea purpurea dried roots using near-infrared spectroscopy," Analytical Chemistry, vol. 74, no. 11, pp. 2493–2499, 2002.

225. W. Li, P. Goovaerts, and M. Meurens, "Quantitative analysis of individual sugars and acids in orange juices by near-infrared spectroscopy of dry extract," Journal of Agricultural and Food Chemistry, vol. 44, no. 8, pp. 2252–2259, 1996.

226. Y. C. Chen, H. Zhang, and R. Matsunaga, "Rapid determination of the main organic acid composition of raw Japanese apricot fruit juices using near-infrared spectroscopy," Journal of Agricultural and Food Chemistry, vol. 54, no. 26, pp. 9652–9657, 2006.

227. J. M. González-Sáiz, C. Pizarro, I. Esteban-Díez et al., "Monitoring of alcoholic fermentation of onion juice by NIR spectroscopy: valorization of worthless onions," Journal of Agricultural and Food Chemistry, vol. 55, no. 8, pp. 2930–2936, 2007.

228. H. Cen, Y. He, and M. Huang, "Measurement of soluble solids contents and pH in orange juice using chemometrics and Vis–NIRS," Journal of Agricultural and Food Chemistry, vol. 54, no. 20, pp. 7437–7443, 2006.

229. F. Liu and Y. He, "Use of visible and near infrared spectroscopy and least squares-support vector machine to determine soluble solids content and pH of cola beverage," Journal of Agricultural and Food Chemistry, vol. 55, no. 22, pp. 8883–8888, 2007.

230. C. L. Ng, R. L. Wehling, and S. L. Cuppett, "Method for determining frying oil degradation by near-infrared spectroscopy," Journal of Agricultural and Food Chemistry, vol. 55, no. 3, pp. 593–597, 2007. ·

231. D. Cozzolino, R. G. Dambergs, L. Janik, W. U. Cynkar, and M. Gishen, "Analysis of grapes and wine by near infrared spectroscopy," Journal of Near Infrared Spectroscopy, vol. 14, no. 5, pp. 279–289, 2006.

232. X. Chen and X. Lei, "Application of a hybrid variable selection method for determination of carbohydrate content in soy milk powder using visible and near infrared spectroscopy," Journal of Agricultural and Food Chemistry, vol. 57, no. 2, pp. 334–340, 2009.

233. R. G. Dambergs, A. Kambouris, L. Francis, and M. Gishen, "Rapid analysis of methanol in grape-derived distillation products using near-infrared transmission spectroscopy," Journal of Agricultural and Food Chemistry, vol. 50, no. 11, pp. 3079–3084, 2002.

234. M.-J. Sáiz-Abajo, J.-M. González-Sáiz, and C. Pizarro, "Classification of wine and alcohol vinegar samples based on near-infrared spectroscopy. Feasibility study on the detection of adulterated vinegar samples,"Journal of Agricultural and Food Chemistry, vol. 52, no. 25, pp. 7711–7719, 2004.

235. M. J. Sáiz-Abajo, J. M. González-Sáiz, and C. Pizarro, "Prediction of organic acids and other quality parameters of wine vinegar by near-infrared spectroscopy. A feasibility study," Food Chemistry, vol. 99, no. 3, pp. 615–621, 2006.

236. W. Fan, Y. Shan, G. Li, H. Lv, H. Li, and Y. Liang, "Application of competitive adaptive reweighted sampling method to determine effective wavelengths for prediction of total acid of vinegar," Food Analytical Methods, vol. 5, no. 3, pp. 585–590, 2012.

237. S. W. Bruun, Protein-lipid-carbohydrate interactions and importance to food quality—spectroscopic detection of macromolecular interactions [Ph.D. thesis], Technical University of Denmark, Lyngby, Denmark, 2006.

238. F. Liu, W. Kong, and Y. He, "Quality determination of fruit vinegars using visible/near infrared spectroscopy and least squares-support vector machine," African Journal of Agricultural Research, vol. 7, no. 15, pp. 2395–2404, 2012.

239. D. Cozzolino, H. E. Smyth, and M. Gishen, "Feasibility study on the use of visible and near-infrared spectroscopy together with chemometrics to discriminate between

commercial white wines of different varietal origins," Journal of Agricultural and Food Chemistry, vol. 51, no. 26, pp. 7703–7708, 2003.

240. L. Liu, D. Cozzolino, W. U. Cynkar, M. Gishen, and C. B. Colby, "Geographic classification of Spanish and Australian tempranillo red wines by visible and near-infrared spectroscopy combined with multivariate Analysis," Journal of Agricultural and Food Chemistry, vol. 54, no. 18, pp. 6754–6759, 2006.

241. S. Buratti, D. Ballabio, G. Giovanelli et al., "Monitoring of alcoholic fermentation using near infrared and mid infrared spectroscopies combined with electronic nose and electronic tongue," Analytica Chimica Acta, vol. 697, no. 1-2, pp. 67–74, 2011.

242. H. Yu, H. Lin, H. Xu, Y. Ying, B. Li, and X. Pan, "Prediction of enological parameters and discrimination of rice wine age using least-squares support vector machines and near infrared spectroscopy," Journal of Agricultural and Food Chemistry, vol. 56, no. 2, pp. 307–313, 2008.

243. L. L. Vines, S. E. Kays, and P. E. Koehler, "Near-infrared reflectance model for the rapid prediction of total fat in cereal foods," Journal of Agricultural and Food Chemistry, vol. 53, no. 5, pp. 1550–1555, 2005.

244. Windham, S. E. Kays, and F. E. Barton II, "Effect of cereal product residual moisture content on total dietary fiber determined by near-infrared reflectance spectroscopy," Journal of Agricultural and Food Chemistry, vol. 45, no. 1, pp. 140–144, 1997.

245. S. E. Kays, Windham, and F. E. Barton II, "Prediction of total dietary fiber in cereal products using near-infrared reflectance spectroscopy," Journal of Agricultural and Food Chemistry, vol. 44, no. 8, pp. 2266–2271, 1996.

246. S. E. Kays, F. E. Barton II, W. R. Windham, and D. S. Himmelsbach, "Prediction of total dietary fiber by near-infrared reflectance spectroscopy in cereal products containing high sugar and crystalline sugar,"Journal of Agricultural and Food Chemistry, vol. 45, no. 10, pp. 3944–3951, 1997.

247. S. E. Kays, Windham, and F. E. Barton II, "Prediction of total dietary fiber by near-infrared reflectance spectroscopy in high-fat- and high-sugar-containing cereal products," Journal of Agricultural and Food Chemistry, vol. 46, no. 3, pp. 854–861, 1998.

248. S. E. Kays and F. E. Barton II, "Rapid prediction of gross energy, and utilizable energy in cereal food products using near-infrared reflectance spectroscopy," Journal of Agricultural and Food Chemistry, vol. 50, no. 5, pp. 1284–1289, 2002.

249. S. E. Kays and F. E. Barton II, "Near-infrared analysis of soluble and insoluble dietary fiber fractions of cereal food products," Journal of Agricultural and Food Chemistry, vol. 50, no. 10, pp. 3024–3029, 2002.

250. S. E. Kays and F. E. Barton II, "Energy from fat determined by near-infrared reflectance spectroscopy,"Journal of Agricultural and Food Chemistry, vol. 52, no. 6, pp. 1669–1674, 2004.

251. D. D. Archibald and S. E. Kays, "Determination of total dietary fiber of intact cereal food products by near-infrared reflectance," Journal of Agricultural and Food Chemistry, vol. 48, no. 10, pp. 4477–4486, 2000.

252. R. M. Ozanich, M. L. Schrattenholzer, and J. B. Callis, "Noninvasive determination of moisture and oil content of wheat-flour cookies—near-infrared spectroscopy in the wavelength range 700-1100 nm," inBiosensor Design and Application, vol. 511 of ACS Symposium Series, chapter 12, pp. 137–164, 1992.

253. L. F. de Alencar Figueiredo, F. Davrieux, G. Fliedel et al., "Development of NIRS equations for food grain quality traits through exploitation of a core collection of cultivated sorghum," Journal of Agricultural and Food Chemistry, vol. 54, no. 22, pp. 8501–8509, 2006.

254. S. W. Bruun, I. Søndergaard, and S. Jacobsen, "Analysis of protein structures and interactions in complex food by near-infrared spectroscopy. 1. Gluten powder," Journal of Agricultural and Food Chemistry, vol. 55, no. 18, pp. 7234–7243, 2007.

255. S. W. Bruun, I. Søndergaard, and S. Jacobsen, "Analysis of protein structures and interactions in complex food by near-infrared spectroscopy. 2. Hydrated gluten," Journal of Agricultural and Food Chemistry, vol. 55, no. 18, pp. 7244–7251, 2007.

256. N. Sinelli, E. Casiraghi, and G. Downey, "Studies on proofing of yeasted bread dough using near- and mid-infrared spectroscopy," Journal of Agricultural and Food Chemistry, vol. 56, no. 3, pp. 922–931, 2008.

257. R. M. García-Rey, J. García-Olmo, E. De Pedro, R. Quiles-Zafra, and M. D. Luque de Castro, "Prediction of texture and colour of dry-cured ham by visible and near infrared spectroscopy using a fiber optic probe," Meat Science, vol. 70, no. 2, pp. 357–363, 2005.

258. A. Lucas, D. Andueza, E. Rock, and B. Martin, "Prediction of dry matter, fat, pH, vitamins, minerals, carotenoids, total antioxidant capacity, and color in fresh and freeze-dried cheeses by visible-near-infrared reflectance spectroscopy," Journal of Agricultural and Food Chemistry, vol. 56, no. 16, pp. 6801–6808, 2008.

259. E. S. Çelik, Determination of aroma compounds and exopolysaccharides formation by lactic acid bacteria isolated from traditional yogurts [M.S. thesis], İzmir Institute of Technology, İzmir, Turkey, 2007.

260. M. Navrátil, C. Cimander, and C. Mandenius, "On-line multisensor monitoring of yogurt and filmjölk fermentations on production scale," Journal of Agricultural and Food Chemistry, vol. 52, no. 3, pp. 415–420, 2004.

261. G. Cozzi, J. Ferlito, G. Pasini, B. Contiero, and F. Gottardo, "Application of near-infrared spectroscopy as an alternative to chemical and color analysis to discriminate the production chains of Asiago d›Allevo cheese," Journal of Agricultural and Food Chemistry, vol. 57, no. 24, pp. 11449–11454, 2009.

262. J. Blasco, N. Aleixos, J. Gómez, and E. Moltó, "Citrus sorting by identification of the most common defects using multispectral computer vision," Journal of Food Engineering, vol. 83, no. 3, pp. 384–393, 2007.

263. P. M. Mehl, K. Chao, M. Kim, and Y. R. Chen, "Detection of defects on selected apple cultivars using hyperspectral

and multispectral image analysis," Applied Engineering in Agriculture, vol. 18, no. 2, pp. 219–226, 2002.

264. I. S. Arvanitoyannis and M. Van Houwelingen-Koukaliaroglou, "Implementation of chemometrics for quality control and authentication of meat and meat products," Critical Reviews in Food Science and Nutrition, vol. 43, no. 2, pp. 173–218, 2003.

265. T. Woodcock, C. C. Fagan, C. P. O›Donnell, and G. Downey, "Application of near and mid-infrared spectroscopy to determine cheese quality and authenticity," Food and Bioprocess Technology, vol. 1, no. 2, pp. 117–129, 2008

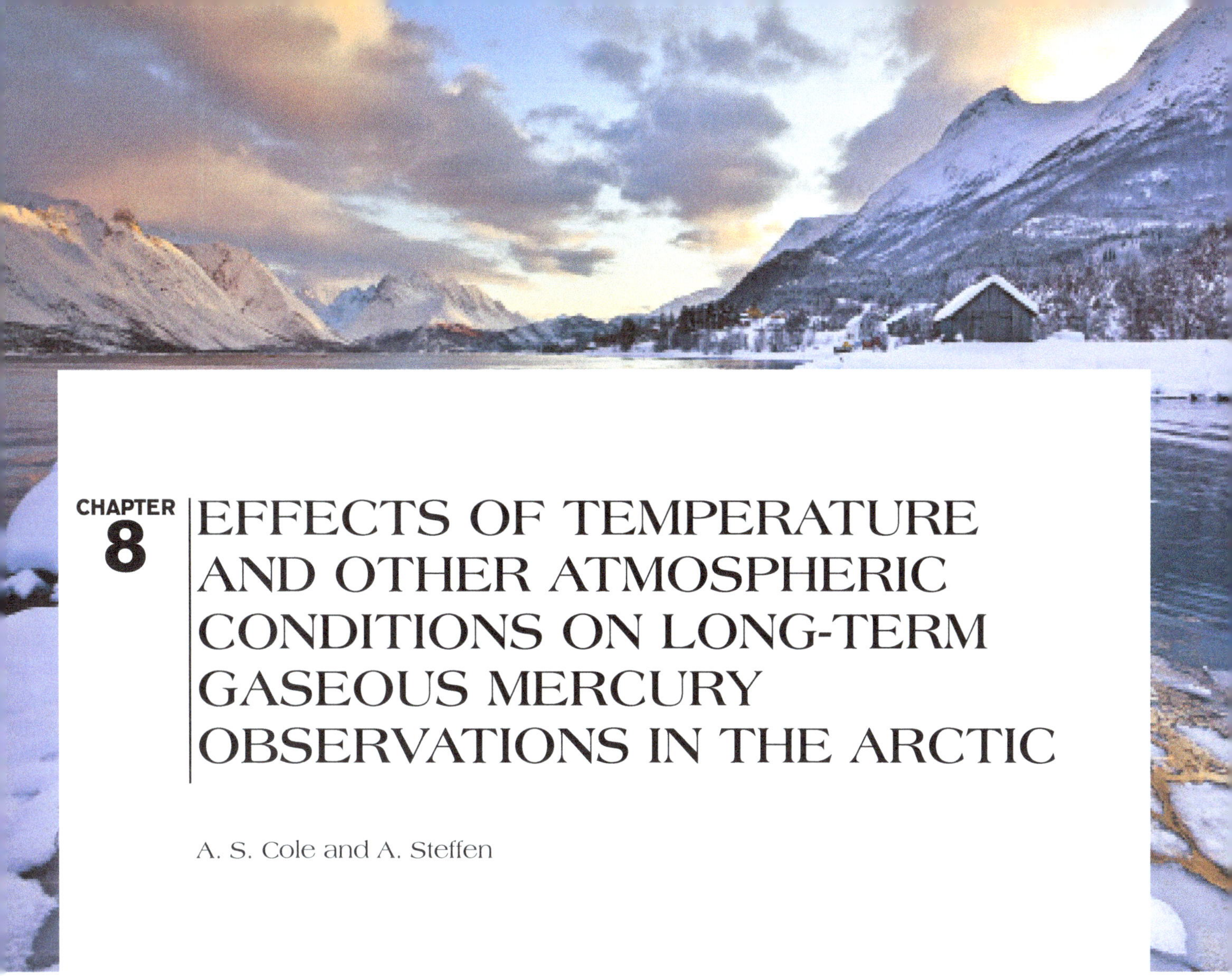

CHAPTER 8 EFFECTS OF TEMPERATURE AND OTHER ATMOSPHERIC CONDITIONS ON LONG-TERM GASEOUS MERCURY OBSERVATIONS IN THE ARCTIC

A. S. Cole and A. Steffen

INTRODUCTION

Mercury in the Arctic is of particular interest for two reasons. From a health perspective, mercury is found in high levels in Arctic wildlife and in some populations despite there being no major mercury sources nearby. For atmospheric scientists, monitoring of atmospheric mercury revealed unusual "atmospheric mercury depletion events" (AMDEs) in which mercury concentrations drop precipitously on timescales of a few hours (Schroeder et al., 1998). Based on research to date, these AMDEs are believed to be due to chemical oxidation by halogen radicals of long-lived gaseous elemental mercury (GEM, or Hg^0) to much more quickly deposited reactive gaseous mercury compounds (RGM) and particle-bound mercury (PHg), and are estimated to contribute as much as 30–55% of the total atmospheric mercury deposited to the Arctic annually (Ariya et al., 2004; Christensen et al., 2004; Skov et al., 2004; Dastoor et al., 2008). As the Arctic undergoes dramatic change due to reduced sea ice and warming temperatures (Cosimo, 2006), it is crucial to understand how climate change may affect the frequency and magnitude of AMDEs in the future in order to predict future inputs of mercury to polar regions

There is already some evidence suggesting that mercury oxidation chemistry is in- fluenced directly and indirectly by climate. Theoretical calculations predict that the net oxidation of Hg^0 to $HgBr_2$ by bromine radicals – a likely oxidation pathway during AMDEs – will be much faster at low temperatures (Goodsite et al., 2004). The origin of high levels of these bromine radicals ("bromine explosions") appears to be heterogeneous reactions with bromide ions (Br^-) in snow, ice and/or aerosol, where sea salt is the original source of bromide (Vogt et al., 1996; Fan and Jacob, 1992). These reactions are likely temperature-dependent, as shown by laboratory experiments on ice (Adams et al., 2002) and predicted by calculations accounting for low-temperature acidification of sea-salt aerosol (Sander et al., 2006). Additionally, bromine explosions, and by extension both ozone depletion events and AMDEs, may be tied to snow and ice characteristics through frost flowers (Kaleschke et al., 2004), first-year sea ice (Simpson et al., 2007), and inland snow pack chemistry (Simpson et al., 2005; Piot and von Glasow, 2008), all of which would be sensitive to climate change. Finally, a short field study at McMurdo, Antarctica, showed a correlation between GEM and air temperature although it was not clear if temperature was the direct cause of the GEM variability (Brooks et al., 2008). However, there has not yet been an analysis directly linking mercury depletion events and temperature during long-term monitoring.

Citation: A. S. Cole and A. Steffen, Effects of temperature and other atmospheric conditions on long-term gaseous mercury observations in the Arctic, http://www.atmos-chem-phys-discuss.net/9/27167/2009/acpd-9-27167-2009.pdf

In addition to the springtime low GEM concentrations, the Arctic regularly experiences late spring and summer GEM concentrations that are significantly higher than the hemispheric background (Steffen et al., 2005). These high values can be analogously defined as atmospheric mercury emission events (AMEEs), however, the mechanism for AMEEs is not well understood. They do tend to coincide with melt conditions (Dommergue et al., 2003), so there is certainly potential for their occurrence to be affected by climate change and therefore influence the future Arctic mercury budget.

Measurements at Alert, Canada, comprise the longest continuous data set on Arctic atmospheric mercury at 14 yr. Although it is generally accepted that climate-driven trends require at least 30 yr of data coverage to be distinguishable from interannual variability, it is likely that areas which experience more rapid change – such as the Arctic – will also see the trend emerge from the noise on shorter timescales. Therefore, an analysis of atmospheric mercury was undertaken in order to determine if there was a statistical change in concentrations from 1995 to 2007. A previous analysis up to 2005 showed no significant long term trend in the annual GEM concentration at Alert (Temme et al., 2007), but in this paper we extend the analysis in time, use an alternative statistical method, and look at seasonal behaviour in order to isolate the springtime AMDE chemistry. In addition, we examine the relationship between AMDEs and meteorological and climate parameters in this extensive data set and a shorter one from Amderma, Russia, in order to assess whether climate change may be expected to influence future springtime mercury deposition to the Arctic ecosystem.

METHODS

Measurements of atmospheric gaseous mercury have been obtained using a commercial Tekran 2537A instrument at Alert, Canada (82.50° N, 62.33° W) since 1995 and at Amderma, Russia (69.72° N, 61.62° E) since 2001. The instruments are set up with a 0.2 µm Teflon filter at the outside inlet (47 mm diameter) followed by an approximately 30-foot sample line heated to 50°C to the back of the instrument, which houses an additional identical Teflon filter. Due to the presence of the two filters, the measured concentrations are assumed to represent gaseous elemental mercury based on results from a previous measurement campaign (Steffen et al., 2002). Two gold collectors are used, alternating between collection and analysis cycles every five or 30 min for continuous monitoring. Hourly averages of these data were calculated for this analysis, requiring 25% completeness. The quality control process used for these data sets was the Environment Canada-developed Research Data Management and Quality Assurance System (RDMQ) and follows the protocol for the Canadian Atmospheric Mercury Measurements Network (CAMNet) (Steffen and Schroeder, 1999). GEM measurements at Alert were only valid for the first half of 2008, therefore this analysis was limited to 1995–2007.

Meteorological data from the Global Atmospheric Watch station at Alert – such as temperature, wind speed, and wind direction – were reported as five-minute means which were then averaged to hourly means, using vector averaging in the case of the wind speed and direction. Hourly relative humidity (up to 2006) and global solar radiation (up to 2003) were obtained by request from Environment Canada's Climate Services Branch (Environment Canada). Temperature, wind speed, wind direction, and relative humidity were reported at Amderma approximately every three hours.

Monthly sea ice area index for the Northern Hemisphere was obtained from the National Snow and Ice Data Center records (Fetterer et al., 2002, updated 2009). To approximate the area of seasonal (first-year) sea ice in the spring months, the sea ice area for the previous September was subtracted from the total sea ice area for the month.

Temporal trends in GEM concentrations by month were calculated using the seasonal Kendall test and seasonal Kendall slope estimator (Gilbert, 1987). This method is an extension of the non-parametric Mann-Kendall test for trend, which is a recom- 5 mended trend test when there are missing values and where the data are not normally distributed – both of which apply to the Alert data set. In the seasonal method, data from each month are treated separately and a slope is estimated for each month using Sen's nonparametric estimator of slope. For the purposes of these calculations, each hourly data point in the month is treated as a replicate measurement. An overall annual trend can be estimated from the monthly trend statistics; however, this estimate is less reliable if the monthly trends are not sufficiently homogeneous. A test for seasonal homogeneity was therefore performed as well (van Belle and Hughes, 1984).

In the past, depletion events were defined by the ambient GEM concentration dipping below 1.0 ng m^{-3} (Steffen et al., 2002). For the purposes of this analysis, the mean background GEM concentration at Alert, as well as its standard deviation, was calculated from all the September–December hourly averages from all years. This time period was chosen to best represent background mercury concentrations that are minimally affected by spring AMDEs, summer AMEEs, or late winter Arctic Haze buildup, as shown by the relatively low variability in measured GEM. An atmospheric mercury depletion event (AMDE) was then defined as a GEM concentration more than 3.9 standard deviations below this mean, corresponding to the 99.99% confidence limit and thereby minimizing the number of false positives in the classification of a measurement as a depletion event. Similarly, mercury emission events (AMEEs) were defined as measurements more than 3.9 standard deviations above the September– December mean. This calculation resulted in standard cutoff values of<1.063 ng m^{-3} for AMDEs and>1.970 ng m^{-3} for AMEEs. These cutoffs implicitly assume a constant September–December GEM concentration, which was acceptable when looking at AMDEs or AMEEs as a function of variables such as temperature, etc. However, when assessing the presence or absence of trends in AMDEs and AMEEs the cutoffs were adjusted for each year by the calculated trend in the September–December means to give values of, respectively, 1.116 ng m^{-3} – (0.0088 ng m^{-3} yr−1)·(N_{1995}) and 2.022 ng m^{-3} – (0.0088 ng m^{-3} yr−1)·(N_{1995}), where N_{1995} is the number of years since $_{1995}$

In order to quantify the combined frequency and strength of depletion (or emission) events, an integrated AMDE quantity was calculated by summing the magnitude of each hourly measurement classified as an AMDE – that is, the difference between the measured value and the cutoff value calculated as described above – and dividing by the total number of hourly measurements, N:

$$\frac{\sum_{i=1}^{N} |C_{AMDEcutoff} - C_i|}{N}, \text{ where } C_i < C_{AMDEcutoff} \quad (1)$$

This integrated AMDE frequency thus provides a measure of both the frequency and intensity of depletion events. The analogous calculation for integrated AMEE frequency was also done, using the data points where $Ci > C_{AMEEcutoff}$. These calculations were performed for measurements grouped by time period (e.g. by month or season), requiring 70% data coverage within the time period, and by meteorological parameters such as a temperature or wind speed range, requiring a minimum of twenty valid hourly measurements. When calculating total springtime integrated AMDEs or AMEEs at Alert, each month was required to have 70% completeness

RESULTS AND DISCUSSION

Trends in atmospheric Hg^0at Alert

The mean GEM concentrations of each month, averaged over 1995–2001 and 2002– 2007, are shown in the top of Fig. 1 for illustrative purposes. Only months with at least 70% data completeness were included in the mean. In the bottom of Fig. 1 are the annual concentration changes calculated using the Sen's slope estimate (based on the seasonal Kendall test for trend) for each month and the 95% confidence limits on the slope. Positive slopes represent an overall increasing trend in the concentration of GEM in that particular month over the period 1995–2007, while negative slopes indicate a decreasing trend. Two major changes in the annual GEM concentration pro- file are clear from this figure: (1) The spring minimum GEM concentration, driven by depletion event chemistry, has shifted from May to April; and (2) there has been a decrease in the background (fall/winter) GEM concentration at Alert. An overall annual trend of -0.0086 ± 0.0014 ng m^{-3} yr^{-1} (−0.56% yr^{-1}) was estimated based on monthly trends in daily mean GEM (which were equivalent to the trends using hourly data but have larger confidence bands), since using hourly data to calculate the annual slopes was not feasible due to the number of data points. The statistical test for seasonal heterogeneity, however, indicates that the monthly trends based on daily or hourly mean GEM values are sufficiently different that the estimated trend for the entire year is not reliable. This can be seen in Fig. 1 as the lack of overlap between error bars of different months. However, monthly trends using a single data point for each month (median monthly GEM) were equivalent to those using hourly or daily values (within error bars), but with much larger statistical uncertainties that resulted in a positive test for seasonal homogeneity and an identical annual trend estimate of −0.0086 ng m^{-3} yr^{-1} with a 95% confidence range of(−0.0134 ng m^{-3} yr^{-1}, −0.0021 ng m^{-3} yr^{-1}). Therefore, we conclude that despite increasing GEM concentrations in May, the overall GEM concentration at Alert is significantly decreasing by approximately 0.6% yr^{-1}. In addition to changes in the frequency and intensity of depletion events by month, resulting in the observed monthly GEM trends, there was a significant decrease in the total springtime (March–June) integrated AMDEs at Alert. However, there was no significant change in integrated AMEEs during the late spring and summer period (May–August).

The decrease in background GEM over this period represents a change in the balance between sources and sinks of atmospheric mercury in the Arctic. One change could be a decrease in global mercury emissions, since emissions were estimated to have changed from 1995 to 2000 by −2%, or −0.4% yr^{-1} (Pacyna et al., 2006), comparable to the trend in ambient GEM concentration at Alert. The trend at Alert is also comparable to the decreasing trend of -0.015 ± 0.003 ng m^{-3} yr^{-1} estimated for the period 1996–2004 at Cape Point, South Africa (Slemr et al., 2008). If decreasing 5 emissions are the cause of decreasing GEM concentrations at Alert, this trend may reverse as predicted emissions increases from Asia offset decreases from other regions. Alternately, an overall decrease in air concentrations could also be due to an increase in the total deposition of mercury out of the air. While springtime depletion events do not appear to have increased at Alert, AMDE deposition in other regions of the Arctic or ongoing deposition in other seasons may be increasing, possibly influenced by decreasing ice cover or changes in the oxidative capacity of the Arctic atmosphere. If this increased deposition is occurring, atmospheric concentrations could continue to decrease while input of inorganic mercury to the ecosystem could increase. While a long-term record of total filterable mercury in one Arctic location – at Resolute, Canada – reports a decrease from 1974 to 2000 of approximately 3% yr^{-1} in the summer and fall, which suggests a decrease in mercury deposition over that period in those seasons (Li et al., 2009), this trend is likely dominated by decreases in the late 1970s and 1980s. More recent and widespread data on Arctic mercury deposition trends is needed to determine if this trend is ongoing and regional.

The springtime minimum in GEM concentration is due to the atmospheric chemistry of AMDEs (Steffen et al., 2008), thus the observed change in the timing of this minimum could be caused by a number of factors that might affect this chemistry. It is thought that meteorology and climate affect the frequency and length of AMDEs, both directly and through the existence of sea ice (Dastoor et al., 2008; Steffen et al., 2002). For example, these events occur primarily at high latitudes where cold temperatures prevail and are likely sustained by stable inversion layers that prevent mixing with low-halogen, higher-mercury air from aloft, as is predicted for ozone depletion events (Lehrer et al., 2004). Therefore, an examination of the relationships between AMDEs and a number of meteorological and climate parameters was performed to determine if any of these

Correlations of depletion and emission events with meteorological and climate parameters

The frequency and intensity of AMDEs was compared with local temperature, wind 5 speed, and wind direction as well as with sea ice area and climate indices. In the case of high frequency measurements such as the local meteorological conditions, data from all years was binned with respect to the meteorological parameter (temperature, wind speed, or wind direction) and the integrated frequency of depletion events was calculated within each bin as described previously. The results are shown for Alert in Fig. 2 and for Amderma in Fig. 3. Also shown are the mean GEM concentrations for each bin, which incorporate both AMDEs and AMEEs.

In both locations, depletion events were less frequent and/or intense at higher temperatures within each month. In addition, at a given temperature, depletion events increased from earlier to later in the spring, at least until AMDEs began to taper off in June. The AMDEs at both locations disappear once the local temperature is above freezing, consistent with what has been previously observed (Lindberg et al., 2002; Steffen et al., 2005). As can also be seen in Figs. 2 and 3, there were fewer AMDEs at Amderma than at Alert, and they began and ended earlier in the spring, consistent with the lower latitude and warmer temperatures there. However, the trend of decreasing AMDEs and increasing GEM concentrations with warmer temperature is the same at both locations. This trend is consistent with what was expected from experimental and modelling results (Adams et al., 2002; Goodsite et al., 2004; Sander et al., 2006), and suggests that increasing temperatures across the Arctic in the future may decrease the frequency or severity of depletion events, at least in the short term. Previous analysis of ozone depletion events (Tarasick and Bottenheim, 2002) suggested a "threshold" surface temperature for severe depletions around −20◦C, below which the generation of bromine radicals – which lead to both ozone and mercury depletion events – is very efficient. The temperature dependence seen here does not exhibit this behaviour, however, the actual temperatures recorded at Alert and Amderma are likely indicative of, but not equal to, the temperatures of the air masses in previous days when the depletion chemistry was likely occurring. As would be expected from the tight correlation between springtime ozone and GEM concentrations, ozone depletion events exhibit similar temperature dependence (not shown), suggesting a common underlying cause – whether halogen radical generation or meteorology. There could also be direct temperature effects in the reactions that oxidize GEM to reactive gas phase mercury (RGM). While experiments in the temperature range 393–448 K showed no temperature dependence of the oxidation of GEM by Br radicals (Grieg et al., 1970), similar studies at the low temperatures seen in the Arctic springtime (e.g. 230–273 K) are lacking, and theoretical calculations of the reaction rate predict an inverse temperature dependence (Goodsite et al., 2004). A detailed comparison of ozone and mercury may lead to a better understanding of the mechanism for the temperature dependence but is complicated by the very different properties of these two species and is, therefore, left for further investigation

The correlation of AMDEs with wind speed was much weaker than that with temperature, as shown in the middle of Figs. 2 and 3. In general, the integrated AMDE frequency in each wind speed bin was not nearly as variable as the AMDE frequency binned by temperature, and a slight peak at around 5 m s^{-1} for all months at Alert was not echoed at Amderma, where the highest values were observed at the lowest wind speeds for April and May. The relationship with wind speed for April and May at Amderma is consistent with depletion events being sustained by more stable (low wind speed), stratified air masses due to strong temperature inversions (Lehrer et al., 2004), but similar behaviour is not seen at Alert. While temperature inversions might be expected to be more prevalent at colder surface temperatures, the fact that the dependence of GEM concentrations on wind speed is much weaker than the dependence on temperature at both locations suggests that the correlation between temperature and depletion events is not primarily due to stable meteorological conditions sustaining mercury-depleted air masses over the measurement site, but is more likely due to temperature-dependent chemistry local to or upwind of the sites. This conclusion is supported by the poor correlation between temperature and local wind speed in these data sets. While previous short-term observations in the Antarctic showed a correlation between local temperature and wind speed during an AMDE period (Brooks et al., 2008), hourly (or three-hourly at Amderma) temperature and wind speed at our sampling sites were weakly correlated in March ($R^2=0.$ at Amderma, $R^2=0.22$ at Alert) and showed no correlation during April and May at either site ($R^2<0.1$).

The relationship between AMDEs and local wind direction was very different between the two sites. At Alert, the integrated AMDE frequency was significantly higher when the wind was approaching from a northerly or northeasterly direction in March, April, and May. This direction corresponds to air masses arriving from the Arctic Ocean and is consistent with previous trajectory studies which found that ozone-depleted air masses (which correlate very well with GEM-depleted air) arriving at Alert in April were most likely to have traversed the Arctic Ocean north of Siberia (Bottenheim and Chan, 2006). Winds from the southwest predominate in the spring due to the "funnelling" of air masses by the topography of Ellesmere Island and therefore represent air that has most recently traversed land and in many cases is descending from the free troposphere. These air masses were not as likely to be depleted in mercury. Although local wind direction is not necessarily a robust representation of the back trajectory of air arriving at Alert, a subset of spring days were chosen to perform 3–10 day back trajectories using the NOAA HYSPLIT model (Draxler and Rolph, 2003) and did show that air masses originating over the Arctic Ocean were generally associated with local winds from the north. Hourly wind direction data, therefore, provided a way to confirm using a large data set (>30 000 points) what was already believed from a more intense trajectory analysis (Bottenheim and Chan, 2006), and extend the comparison to other spring months. In contrast to Alert, AMDEs at Amderma did not exhibit any dependence on local wind direction, suggesting that either they do not consistently originate in one particular region or that the longer-term back trajectories of air arriving at Amderma is not well represented by local wind direction. Therefore, a trajectory-based analysis may be needed in order to identify the source(s) of AMDEs observed at Amderma. The relationship between AMEEs at Alert and temperature, wind speed and direction, and month is illustrated in Fig. 4. The integrated AMEE frequency peaked around 0 ◦C in all months, was highest in June and July, and was overall much smaller than the integrated AMDE frequencies for the year. The same temperature dependence was observed at Amderma (not shown), and shows that high GEM concentrations were most common during melt periods, whether due to emissions from snowmelt or due to a pulse of oceanic emissions during the breakup of sea ice. AMEEs also appeared to depend on wind speed and direction, at least in June–August. A peak in AMEEs at the lowest wind speeds suggests that the surface is a source of GEM during these months, such that low mixing conditions tend to maintain high GEM concentrations. The correlation of summer AMEE frequency with wind direction is similar to what is seen with spring AMDEs, that is, higher values when the wind is from the north or northeast. This suggests that during melt conditions the ocean and/or its snow and ice cover are a source of GEM. This idea is supported by Andersson et al. (2008), who observed from the Arctic Ocean that the sea ice can act as a barrier to GEM evasion and that increased GEM concentrations were observed when the sea ice was broken by the research ship (Andersson et al., 2008). In contrast to temperature, which is reasonably conserved in a transported air mass, other climate-related parameters that are not represented by local data may affect the number of AMDEs observed at a site – for instance, the quantity and type of sea ice encountered by the air mass, or large-scale mixing patterns that influence the source region of air arriving at the site. While these parameters may influence the origin and transport of Hg-poor air masses, it is difficult to compare them with atmospheric mercury concentrations at Alert at high time resolution without knowing the air mass origin and transit time in addition to high-resolution sea ice characteristics. Therefore, in order to address the more general question of whether the interannual variability in AMDEs is related to variability in seasonal or total sea ice, or to large-scale climate patterns that influence transport, monthly averages of sea ice area and climate indices were compared with monthly integrated AMDE frequencies at Alert for each month in which sufficient data were available. The entire period 1995–2007 was used for this comparison. There were not enough months with the required data completeness at Amderma to perform a similar comparison, therefore this analysis is limited to Alert. The results of these comparisons are shown in Table 1. The somewhat significant correlations in March between AMDEs at Alert and both the North Atlantic Oscillation (NAO) Index (p0.2), despite the significant correlations that do exist when looking at hourly measurements (e.g., top of Fig. 2). This suggests that AMDEs in a given month are related to the distribution of temperatures more than the monthly mean temperature. As a result, it would be difficult to predict the effect of future temperature increases on AMDEs, since an increase in the mean monthly temperature could also feasibly be accompanied by a wider distribution of temperatures, and therefore an increase in extremely cold days that are associated with increased AMDE activity. This is discussed further in Sect. 3.3. It is also likely that additional parameters that are not addressed by this analysis are also important drivers of AMDE frequency. For example, while radiative flux is somewhat represented by the julian day (or month, broadly), interannual variability in cloudiness may influence the photochemical generation of halogen radicals. However, this effect would be most important in the region where generation of these radicals take place and would not necessarily relate to radiation levels at Alert. Finally, as previously discussed, it may be that total Northern Hemisphere seasonal sea ice (or Arctic Ocean sea ice, which also showed no relationship to AMDEs) is not a good representation of seasonal ice in the region(s) where AMDEs at Alert originate. Future detailed analysis of higher-resolution sea ice conditions, particularly now that seasonal sea ice area can be more accurately estimated from satellite scatterometer data (Nghiem et al., 2006), may reveal a clearer relationship with AMDEs at Alert once a longer data set is collected.

Finally, to assess the robustness of the relationship between hourly springtime GEM concentration and various parameters, a multiple linear regression was performed and used to predict GEM concentrations for March–June 2008. Temperature alone was able to explain only 16% of the variance in the 1995–2007 data (R^2=0.16), while using both temperature and Julian day (which incorporates the observed dependence on month and serves as a proxy for radiation levels) explained 22% of the variance. Adding wind speed and the cosine of the wind direction (representing the northerly component) as independent variables improved the fit slightly, with this four-parameter model explaining 26% of the variance, but adding solar radiation, sea ice area, and daily NAO index as additional parameters did not improve the fit any further. There fore, a two-parameter fit using temperature and Julian day was chosen to predict 2008 springtime GEM values. The results of this prediction are compared with measured values in Fig. 5. As would be expected based on the total R^2 of the regression (0.22), much of the variance in the measured concentration is lost in the model. However, this simple two-parameter linear fit does exhibit an ability to predict the timing – though not the magnitude – of some of the observed fluctuations in the GEM concentration. This provides an indication that the relationship between temperature, month, and depletion events derived from 1995–2007 data is robust for 2008 as well, though clearly, other parameters that are not accounted for in our analysis account for a large part of the total variance. More sophisticated models of mercury chemistry in the Arctic may possibly be improved, then, by parametrizing the temperature dependence found in this analysis.

Qualitative comparison of trends

Although local temperature has a significant relationship with depletion events, the overall springtime trends in temperature, as shown in Fig. 6, are not completely consistent with springtime GEM trends (Fig. 1). Decreased GEM concentrations in February, March, and April are consistent with decreased temperatures in those months, but the temperature in May has also decreased slightly from 1995 to 2007 while GEM has increased. It is possible that the inconsistency in the May temperature and GEM trends, as well as the lack of correlation in monthly means discussed in Sect. 3.2, is due to the influence of other factors. For example, changes in incoming radiation or winds arriving from the north could be enough to overwhelm the relatively small change in temperature. However, further analysis reveals that despite the overall decreasing trend in May temperature, the percentage of hourly measurements below −°C – i.e. at temperatures that exhibit strong and frequent AMDEs – was 28% for 1995–2001 and only 21% for 2002–2007. In this case, although the mean temperature decreased, the distribution of temperatures was narrower, resulting in fewer low temperature extremes and fewer (or less intense) AMDEs. The same was not true of the temperature distributions in March and April, which were uniformly shifted to lower temperatures during the period, consistent with increased AMDE frequency. Therefore, the shift in the annual GEM minimum shown in Fig. 1 is consistent with changes in air temperature over the period 1995–2007, though contributions from other factors are also likely

CONCLUSIONS

Atmospheric elemental mercury measurements at Alert, Canada, show a significant decrease in local GEM concentration over the period 1995–2007 in every month except May, resulting in an annual trend of −0.0086±0.0014 ng m^{-3} yr^{-1}. This is the first time 5 that a significant trend has been reported at this long-term monitoring site. It is also shown that mercury depletion events have shifted to earlier in the spring during the last thirteen years. The results presented here provide additional information about the conditions for both low and high concentrations of atmospheric mercury based on multi-year high resolution observations. A month-by-month analysis revealed a robust correlation between local temperature and depletion events at both Alert and Amderma, Russia. While this correlation could indicate a potential decrease in AMDEs with increasing temperatures in the future, until the mechanism for the temperature effect is known this correlation is primarily useful as a guide to refining the parametrization of AMDEs in models. We must be very cautious in extending that relationship into future Arctic conditions that deviate significantly from what has been experienced in recent history.

Other parameters, such as local wind speed, total sea ice area, and climate indices, did not correlate with AMDEs at either location, with the exception of a correlation between the March Polar/Eurasian Teleconnection Pattern and AMDEs at Alert that may be due to higher inputs of atmospheric mercury from midlatitudes during years with a weak polar vortex. Wind direction was an important factor in both depletion and emission events observed at Alert but not at Amderma. Further research into the mechanism of the temperature effect in AMDEs and AMEEs would be valuable for providing models with the tools to predict the effect of climate change on Arctic mercury. For example, while the good correlation between AMDEs and ozone depletion events suggests that the temperature dependence seen in AMDEs is due to a common factor, such as higher levels of halogen radicals at lower temperatures, temperature-dependent laboratory studies of mercury oxidation reactions would quantify (or rule out) any effect of temperature on mercury concentrations separate from those of ozone. Differences in the temperature of maximum integrated AMDE frequency at Alert and Amderma, as well as the relationship between AMDEs at measurement sites and differences in regional ice conditions, could be investigated using a trajectory study similar to what has been done for ozone (Bottenheim and Chan, 2006) that incorporates the past temperature history of air parcels. Finally, the origin of AMEEs is a current knowledge gap, and it is hoped that the dependence on meteorological conditions reported here may motivate further research into their source and mechanism

ACKNOWLEDGEMENTS

The authors would like to thank A. Konoplev, P. Lee, and C. Scherz for their contributions to mercury data collection and analysis; the Global Atmospheric Watch program for meteorological data at Alert; numerous personnel at the GAW Alert laboratory and Canadian Forces Station Alert; and J. Bottenheim and D. Durnford for valuable discussions. Funding for this project was provided by the Northern Contaminants Program of Indian and Northern Affairs Canada, Environment Canada, the Government of Canada Program for International Polar Year, and the Arctic Monitoring and Assessment Program.

Table 1: Pearson's correlation coefficients (R), and statistical significance level (p), expressed as R (p), for monthly integrated AMDEs at Alert with monthly climate-related parameters. The number of months in the period 1995–2007 with sufficient GEM data are reported as N.

	March (N=9)	April (N=9)	May (N=9)	June (N=10)
Northern Hemisphere sea ice area	−0.54 (0.14)	−0.27 (0.48)	0.52 (0.15)	0.46 (0.18)
Northern Hemisphere seasonal sea ice area (estimated)	0.14 (0.72)	−0.05 (0.89)	−0.25 (0.52)	0.21 (0.57)
Arctic Oscillation Index	0.44 (0.24)	−0.11 (0.78)	−0.35 (0.36)	0.04 (0.90)
North Atlantic Oscillation Index	−0.61 (0.08)	−0.34 (0.37)	0.50 (0.18)	−0.07 (0.85)
Polar/Eurasia Teleconnection Pattern (PET)	0.84 (0.005)	0.14 (0.37)	−0.45 (0.23)	0.05 (0.90)
Mean monthly temperature	−0.44 (0.24)	−0.02 (0.96)	−0.41 (0.27)	0.43 (0.22)

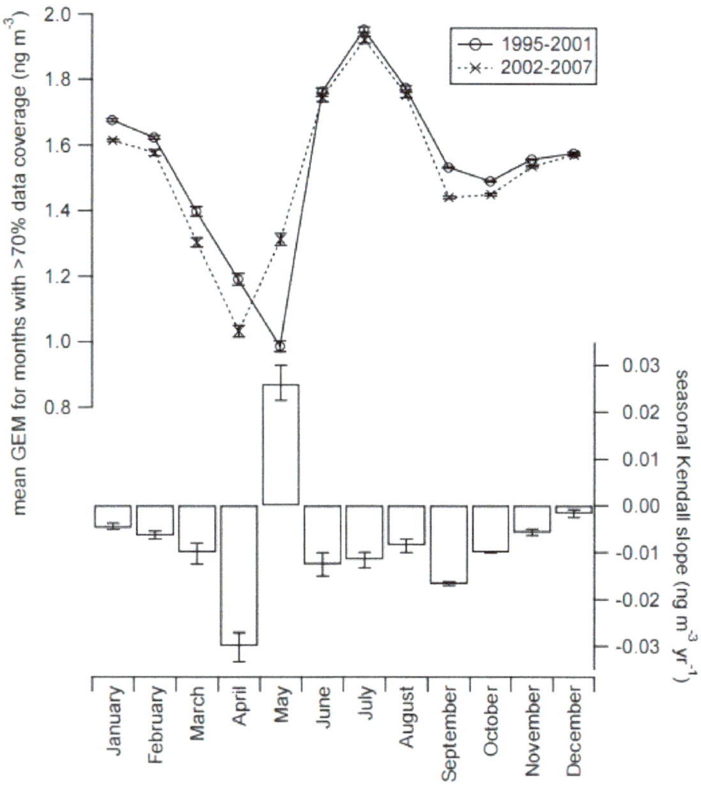

Figure 1: Seasonal profiles of gaseous elemental mercury concentrations at Alert for 1995–2001 and 2002–2007 (top); change in concentration with time calculated for each month by Sen's estimator of slope (bottom).

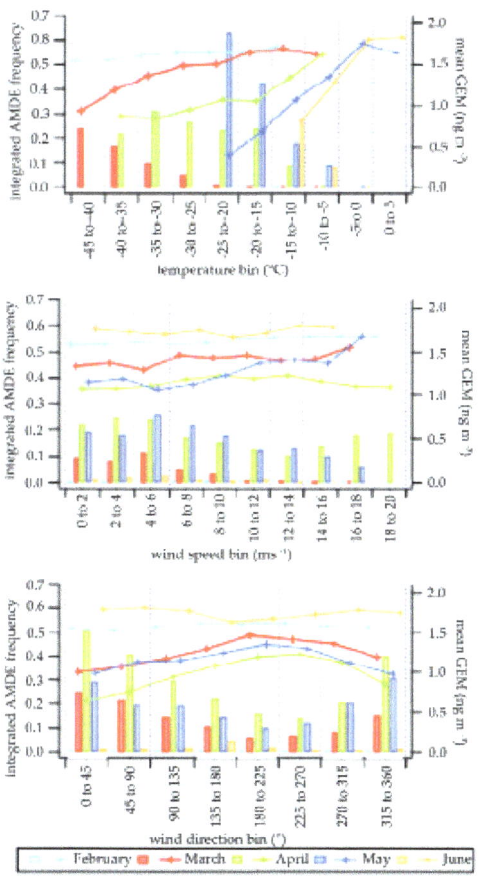

Figure 2: Integrated AMDE frequency (coloured bars) and mean gaseous elemental mercury (lines), as a function of temperature, wind speed, wind direction and month for Alert data 1995–2007. For reference, springtime frequency histograms of the observed meteorological parameters are included in gray.

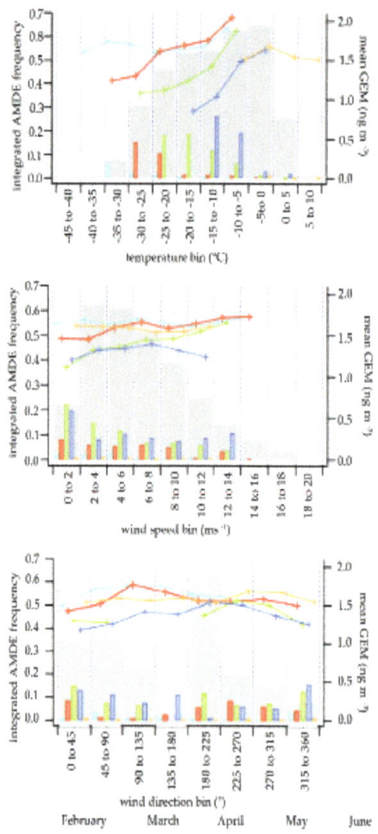

Figure 3: Integrated AMDE frequency (coloured bars) and mean gaseous elemental mercury (lines), as a function of temperature, wind speed, wind direction and month for Amderma data 2001–2007. For reference, springtime frequency histograms of the observed meteorological parameters are included in gray.

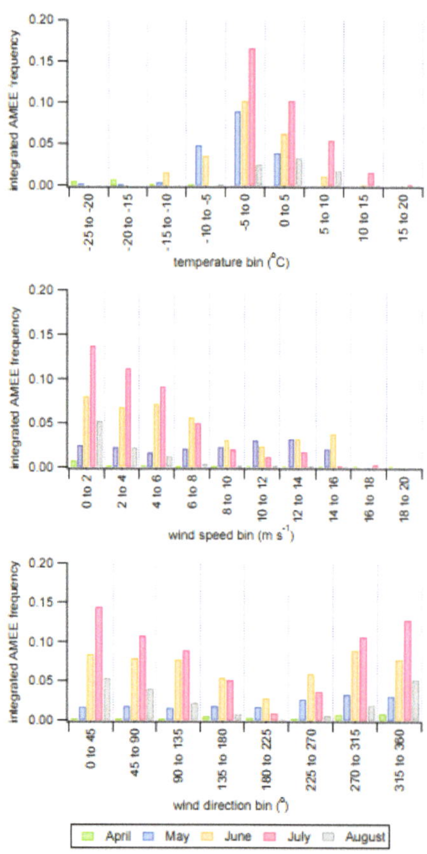

Figure 4: Integrated AMEE frequency as a function of temperature, wind speed, wind direction and month for Alert data 1995–2007. Note the change in scale from Figs. 2 and 3.

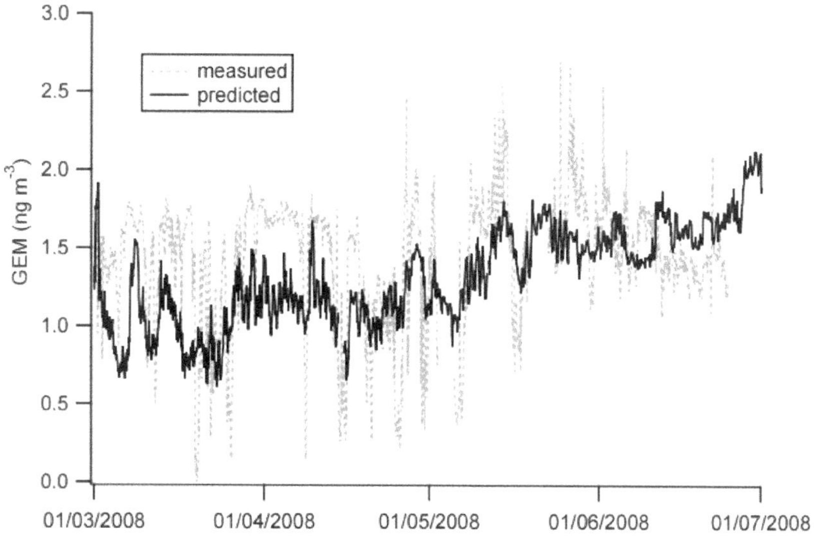

Figure 5: Comparison of measured hourly GEM concentrations at Alert for March–June 2008 with those predicted based on a multiple linear regression fit of 1995–2007 GEM concentrations to local temperature and Julian day.

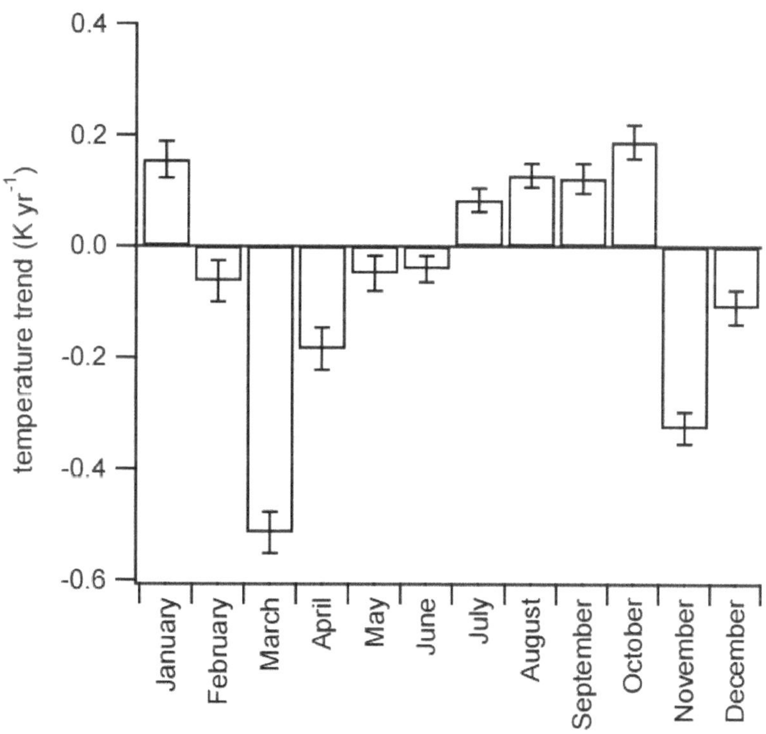

Figure 6: 1995–2007 trends in local temperature at Alert calculated for each month by Sen's estimator of slope.

REFERENCES

1. Adams, J. W., Holmes, N. S., and Crowley, J. N.: Uptake and reaction of HOBr on frozen and dry NaCl/ NaBr surfaces between 253 and 233 K, Atmos. Chem. Phys., 2, 79–91, 2002, http://www.atmos-chem-phys.net/2/79/2002/.
2. Andersson, M. E., Sommar, J., Gardfeldt, K., and Lindqvist, O.: Enhanced concentrations of dissolved gaseou
3. Ariya, P., Dastoor, A., Amyot, M., Schroeder, W., Barrie, L., Anlauf, K., Raofie, F., Ryzhkov, A., Davignon, D., Lalonde, J., and Steffen, A.: The Arctic: A sink for mercury, Tellus B, 56, 397–403, 2004
4. Bottenheim, J. and Chan, E.: A trajectory study into the origin of spring time Arctic boundary layer ozone depletion, J. Geophys. Res.-Atmos, 111, D19301, doi:10.1029/2006JD007055, 2006.
5. Brooks, S., Lindberg, S., Southworth, G., and Arimoto, R.: Springtime atmospheric mercury speciation in the McMurdo, Antarctica coastal region, Atmos. Environ., 42, 2885–2893, 2008.
6. Christensen, J. H., Brandt, J., Frohn, L. M., and Skov, H.: Modelling of mercury in the Arctic 5 with the Danish Eulerian Hemispheric Model, Atmos. Chem. Phys., 4, 2251–2257, 2004, http://www.atmos-chem-phys.net/4/2251/2004/.

7. Cosimo, J. C.: Arctic warming signals from satellite observations, Weather, 61, 70–76, 2006.

8. Dastoor, A. P., Davignon, D., Theys, N., van Roozendael, M., Steffen, A., and Ariya, P. A.: Modeling dynamic exchange of gaseous elemental mercury at polar sunrise, Environ. Sci. Technol., 42, 5183–5188, 2008.

9. Dommergue, A., Ferrari, C., Gauchard, P.-A., Boutron, C. F., Poissant, L., Pilote, M., Jitaru, P., and Adams, F. C.: The fate of mercury species in a sub-arctic snowpack during snowmelt, Geophys. Res. Lett., 30, 1621, doi:10.1029/2003GL017308, 2003.

10. Draxler, R. R. and Rolph, G. D.: HYSPLIT (HYbrid Single-Particle Lagrangian Integrated Trajectory) Model access via NOAA ARL READY Website (http://www.arl.noaa.gov/HYSPLIT.php). Silver Spring, MD, NOAA Air Resources Laboratory, 2003.

11. Environment Canada. Climate Services: http://www.climate.weatheroffice.ec.gc.ca/, access: July 2008.

12. Fan, S.-M. and Jacob, D. J.: Surface ozone depletion in Arctic spring sustained by bromine reactions on aerosols, Nature, 359, 522–524, 1992.

13. Fetterer, F., Knowles, K., Meier, W., and Savoie, M. Sea Ice Index: http://nsidc.org/data/g02135. html, access: 2009 Boulder, CO, National Snow and Ice Data Center, 2002, updated 2009.

14. Gilbert, R. O.: Statistical Methods for Environmental Pollution Monitoring, Van Nostrand Reinhold Company, New York, 204–240 pp., 1987

15. Goodsite, M. E., Plane, J. M. C., and Skov, H.: A theoretical study of the oxidation of Hg^0 25 to $HgBr_2$ in the troposphere, Environ. Sci. Technol., 38, 1772–1776, 2004.

16. Grieg, G., Gunning, H. E., and Strausz, O. P.: Reactions of metal atoms. II. The combination of mercury and bromine atoms and the dimerization of HgBr, J. Chem. Phys., 52, 3684–3690, 10.1063/1.1673544, 1970.

17. Kaleschke, L., Richter, A., Burrows, J. P., Afe, O., Heygster, G., Notholt, J., Rankin, A. M., Roscoe, H. K., Hollwedel, J., Wagner, T., and Jacobi, H.-W.: Frost flowers on sea ice as a source of sea salt and their influence on tropospheric halogen chemistry, Geophys. Res. Lett., 31, L16114, doi:16110.11029/12004GL020655., 2004.

18. Lehrer, E., Honninger, G., and Platt, U.: A one dimensional model study of the mechanism of ¨ halogen liberation and vertical transport in the polar troposphere, Atmos. Chem. Phys., 4, 2427–2440, 2004, http://www.atmos-chem-phys.net/4/2427/2004/.

19. Li, C., Cornett, J., Willie, S., and Lam, J.: Mercury in Arctic air: The long-term trend, Sci. Total Environ., 407, 2756–2759, 2009.

20. Lindberg, S. E., Brooks, S., Lin, C. J., Scott, K. J., Landis, M. S., Stevens, R. K., Goodsite, M. E., and Richter, A.: Dynamic oxidation of gaseous mercury in the Arctic troposphere at polar sunrise, Environ. Sci. Technol., 36, 1245–1256, 2002.

21. Nghiem, S. V., Chao, Y., Neumann, G., Li, P., Perovich, D. K., Street, T., and Clemente-Colon, P.: Depletion of perennial sea ice in the East Arctic Ocean, Geophys. Res. Lett., 33, L17501, doi:10.1029/2006GL027198, 2006.

22. Pacyna, E. G., Pacyna, J. M., Steenhuisen, F., and Wilson, S.: Global anthropogenic mercury emission inventory for 2000, Atmos. Environ., 40, 4048–4063, 2006.

23. 5 Piot, M. and von Glasow, R.: The potential importance of frost flowers, recycling on snow, and open leads for ozone depletion events, Atmos. Chem. Phys., 8, 2437–2467, 2008, http://www.atmos-chem-phys.net/8/2437/2008/.

24. Sander, R., Burrows, J., and Kaleschke, L.: Carbonate precipitation in brine – a potential trigger for tropospheric ozone depletion events, Atmos. Chem. Phys., 6, 4653–4658, 2006, http://www.atmos-chem-phys.net/6/4653/2006/.

25. Schroeder, W. H., Anlauf, K. G., Barrie, L. A., Lu, J. Y., Steffen, A., Schneeberger, D. R., and Berg, T.: Arctic springtime depletion of mercury, Nature, 394, 331–332, 1998.

26. Sharma, S., Andrews, E., Barrie, L. A., Ogren, J. A., and Lavoue, D.: Variations and sources of ´ the equivalent black carbon in the high Arctic revealed by long-term observations at Alert and Barrow: 1989–2003, J. Geophys. Res.-Atmos., 111, D14208, doi:10.1029/2005JD006581, 2006.

27. Simpson, W. R., Alvarez-Aviles, L., Douglas, T. A., and Sturm, M.: Halogens in the coastal snow pack near Barrow, Alaska: Evidence for active bromine air-snow chemistry during springtime, Geophys. Res. Lett., 32, L04811, doi:10.1029/2004GL021748, 2005.

28. Simpson, W. R., Carlson, D., Honninger, G., Douglas, T. A., Sturm, M., Perovich, D., and ¨ Platt, U.: First-year sea-ice contact predicts bromine monoxide (BrO) levels at Barrow, Alaska better than potential frost flower contact, Atmos. Chem. Phys., 7, 621–627, 2007, http://www.atmos-chem-phys.net/7/621/2007/

29. Skov, H., Christensen, J. H., Heidam, N. Z., Jensen, B., Wahlin, P., and Geernaert, G.: Fate of elemental mercury in the Arctic during atmospheric depletion episodes and the load of atmospheric mercury to the Arctic, Environ. Sci. Technol., 38, 2373–2382, 2004.

30. Slemr, F., Brunke, E.-G., Labuschagne, C., and Ebinghaus, R.: Total gaseous mercury concentrations at the Cape Point GAW station and their seasonality, Geophys. Res. Lett., 35, L11807, doi:10.1029/2008GL033741, 2008.

31. Steffen, A. and Schroeder, W. Standard operating procedures manual for total gaseous mercury measurements: Canadian Atmospheric Mercury Measurements Network (CAMNet), Toronto, Canada, Environment Canada, 1999

32. Steffen, A., Schroeder, W., Bottenheim, J., Narayan, J., and Fuentes, J.: Atmospheric mercury concentrations: measurements and profiles near snow and ice surfaces in the Canadian Arctic during Alert 2000, Atmos. Environ., 36, 2653–2661, 2002.

33. Steffen, A., Schroeder, W., Macdonald, R., Poissant, L., and Konoplev, A.: Mercury in the Arctic atmosphere: An analysis of eight years of measurements of GEM at Alert (Canada) and a comparison with observations

at Amderma (Russia) and Kuujjuarapik (Canada), Sci. Total Environ., 342, 185–198, 2005.

34. Steffen, A., Douglas, T., Amyot, M., Ariya, P., Aspmo, K., Berg, T., Bottenheim, J., Brooks, S., Cobbett, F., Dastoor, A., Dommergue, A., Ebinghaus, R., Ferrari, C., Gardfeldt, K., Goodsite, M. E., Lean, D., Poulain, A. J., Scherz, C., Skov, H., Sommar, J., and Temme, C.: A synthesis of atmospheric mercury depletion event chemistry in the atmosphere and snow, Atmos. Chem. Phys., 8, 1445–1482, 2008, http://www.atmos-chem-phys.net/8/1445/2008/.

35. Tarasick, D. W. and Bottenheim, J. W.: Surface ozone depletion episodes in the Arctic and Antarctic from historical ozonesonde records, Atmos. Chem. Phys., 2, 197–205, 2002, http://www.atmos-chem-phys.net/2/197/2002/.

36. van Belle, G. and Hughes, J. P.: Nonparametric tests for trend in water quality, Water Resour. Res., 20, 127–136, 1984.

37. Vogt, R., Crutzen, P. J., and Sander, R.: A mechanism for halogen release from sea-salt aerosol in the remote marine boundary layer, Nature, 383, 327–330, 1996.

CHAPTER 9 CHEMOMETRICS IN FINGERPRINTING BY MEANS OF THIN LAYER CHROMATOGRAPHY

Łukasz Komsta

INTRODUCTION

Exactly 40 years ago, in 1971, Swante Wold used the term "chemometrics" for the first time in his grant application. These 40 years of continuous development of computer abilities and analytical chemistry methods made chemometrics accessible to any interested scientist. The chemometric techniques allow the researcher to compare and explore complex and multivariate data by a proper projection methods [1].

The chemometric exploration of chromatographic data is a well-established topic in the case of HPLC equipped with different detection techniques [2–4]. The application of this approach is very wide, from ion chromatography of mineral waters [5], through oil spills [6], metabolic profiling [7], different kinds of food [8], petroleum oils [9], and ending in pharmaceutical and herbal investigations [10–16]. Although such approach needs some chemometric (mathematical) knowledge, it is more and more often preferred because of the lack of the need of identifying every peak or, at least significant peaks. The chromatogram is treated not as the set of any peaks, but as an unique multivariate signal. Therefore, there is an increasing interest in chemometric chromatographical fingerprinting.

Citation: Łukasz Komsta, "Chemometrics in Fingerprinting by Means of Thin Layer Chromatography," Chromatography Research International, doi:10.1155/2012/893246

Although thin-layer chromatography is a well-established technique and its pharmaceutical and herbal applications are very wide (see e.g., books of Macek [17] and Waksmundzka-Hajnos et al. [18]), the fingerprinting is done here mainly in a nonchemometric way. The most often used approach relies on identifying peaks and comparing their height, area, or intensity. There is still lack of comprehensive fingerprinting TLC approaches based on the full chemometric processing of collected data. The reason for this cannot be definitely justified; it is most probably caused by narrow specializing of thin-layer chromatographers, some fear of chemometrics, and lack of cooperation between TLC chromatographers and chemometricians.

Therefore, the purpose of this paper is to review and summarize all possibilities in TLC fingerprinting and settle some starting point for any TLC chromatographer, which would be interesting in introducing chemometrics to his/her research.

FINGERPRINTING: CLASSICAL AND BY MEANS OF CHEMOMETRICS

The classical fingerprinting by TLC is done visually. The chromatographic system is optimized to separate and visualize (e.g., by different colors) all interesting components from particular herbal sample. Next, the samples belonging to different herbal species are compared and the differences in particular peaks are manually notified. Several peaks can be then proposed as markers, that is, substances responsible for differences between discriminated classes.

This allows the chromatographer to analyze an unknown sample and—based on absolute or relative intensity of marker peaks—to distinguish between particular species or chemotaxonomic groups. The fingerprints can be done with several TLC systems simultaneously, each system allowing to see and compare the markers found in another group of phytochemicals found in the sample.

The main disadvantage of classical fingerprinting is its subjectivity. The comparison of chromatograms is mainly based on a subjective (not objective) method, mainly chromatographer perception. The chemometric approach, although more difficult, is based only on objective mathematical methods.

The chemometric fingerprinting treats the chromatogram as a unique signal, without a need to identify and interpret the peaks. The chromatogram is represented as a vector—a serie of numbers with a particular sampling rate. Therefore, each chromatogram can be treated as one point in a multidimensional space, where the number of dimensions is defined by the signal length. As the chromatograms consist of tens, hundreds, or thousands of points, the high multidimensionality does not allow the chemometrician to imagine such point; however, the spaces of higher dimensions have analogous to 3D space metric properties (angles, distances, hyperplanes, etc.).

Considering the chromatograms as multivariate points, one could measure a similarity/dissimilarity between any two of them in different manner. The most often used measure is Euclidean distance (which is analogous to distance between points in 3D space), and angular distance (which is the angle between vectors pointed from origin to the two points). The first one is sensitive to absolute values of the peaks, the second one works in a correlation-like manner and compares only shape of the chromatograms. Based on the similarity/dissimilarity or correlation matrix one can perform a number of unsupervised chemometric techniques, which will be explained below.

CREATING FINGERPRINTS

In the case of HPLC, the chromatography equipment allows a researcher to export a chromatogram as a serie of numbers, because the chromatogram is already digitized during its development. In the case of TLC, the plate must be scanned by densitometer or photographed after development. In a general way, three approaches can be considered.(1)Scanning a plate with a densitometer.(2)Taking a photo of the plate and extracting univariate, densitogram-like signal along a particular track.(3)Taking a photo of whole plate, when 2D TLC is performed.

The densitometric scanning can be performed in ultraviolet wavelength or visual wavelength, before or after applying a developing agent. The scanning can be performed with one wavelength, several wavelengths, or in multiwavelength mode. The last one results in the data analogous to DAD chromatogram, but the main disadvantage is a time needed to scan the track. Only the newest densitometers have DAD built in, and can collect many wavelengths during one pass of the scanner. The plate can be scanned two times, before and then after development by visual agent, densitograms can be then combined together with one matrix with more channels.

In the case of videoscanning, one can take a color photo at 254 nm light, 366 nm light, or visible light. Each photo has three channels (RGB), so there is a theoretical possibility to collect 9 channels of information for the same plate. However, one must not move the plate under the camera during taking subsequent pictures to preserve compatibility of spatial information. For this reason, it is almost impossible to move the plate out of videoscanner, develop it by some agent, and then scan it again: obtained photo will not exactly match the previous one and the spatial difference can make the chemometric processing almost impossible (or at least the results will be incorrect). It could be possible with a special holder, always holding the plate at an exact position, but to the best knowledge of the author, no commercially available videoscanning camera is equipped with it.

The images of the plate can be collected not only by special videoscanning equipment—in many cases a popular digital camera is also an appropriate hardware. However, it must be done in repetitive conditions, in the same place, with the same ambient light, and with the camera fixed in one position with tripod. The univariate tracks can be then obtained and exported from the photos by an open source software, such as ImageJ [19].

CHEMOMETRIC PRETREATMENT

Before applying the chemometric techniques, proper pretreatment of the signals is a necessity. The typical order of preprocessing is as follows: normalization, denoising, and baseline removal, followed by warping/registering.

The normalization step is not mandatory and there is no consensus when it is obligatory—sometimes it makes the results better, sometimes even worse. Therefore it is recommended to compare the results with and without normalization. The most often method of signal normalization is the standard normal variate (SNV) method [20]. It is a centering to zero mean and scaling to unit variance of each signal itself (independent of scaling or centering used in chemometric methods themselves).

Denoising

The denoising step is also not mandatory and it should be performed only in the case of visible noise. However, the densitograms and videoscans have quite a high noise level in general. The signal processing science knows many methods of denoising and smoothing of the signals. The simplest one is Savitzky-Golay smoothing filter (which can be used also with simultaneous differentiation) [21]. The filter parameters should be optimized against the correlation of filtered noise [22]. The other most important method is the wavelet shrinkage [23,24]. The other ones, such as digital filters, Fourier denoising, or Whittaker smoother are not so often used.

The optimum denoising method is different in the case of different noise properties—noise spectrum (and autocorrelation) and homo/heteroscedastcity. The densitometer noise is in most cases pink and heteroscedastic whereas the CCD camera noise is white and homoscedastic. Therefore the recommended methods are different. The reader can be referred to two papers with comprehensive comparison of denoising techniques in the case of densitograms [25] and univariate videoscans [26]. In the first case, denoising is more difficult and the Savitzky-Golay filter or Haar wavelet shrinkage with soft thresholding is a good choice. In the second case, the Haar mother wavelet with soft denoising and any decomposition level larger than 1 was the best choice.

There is also another report [27] on image processing techniques on whole images (when the 2D chromatogram is processed and one cannot extract a univariate path). Several other kind of filters (averaging, circular, Gaussian, median, and Wiener) can be used here regardless of 2D version of previous techniques. This study recommends use of median filter with the window width optimized against filtered noise autocorrelation. The wavelet shrinkage at high decomposition level performs also very well.

Baseline Removal

The baseline removal process is almost always mandatory. If this step is not applied, a baseline variation causes a random variance greater than the one caused by chromatogram variability. The real trends in data are then completely obscured by baseline trends; therefore one cannot make any real conclusion without baseline removal.

If the univariate signals are considered (densitograms, univariate videoscans), the reader is referred to a comprehensive study and review of current baseline removal methods [28]. Many approaches exists in the literature and the methods of choice could be asymmetric penalized least squares (e.g., in airPLS implementation) or quantile regression.

In the case of 2D videoscans (images), the simple algorithms, such as rollerball method [29] are sufficient to filter the smooth baseline trends caused by inhomogenous illumination or spraying. The more advanced methods, used mainly in the case of 2D gel electropherograms [30] are not necessary in the case of TLC [29].

Warping

An another significant problem is the random shift of the chromatographic peaks. This phenomenon cannot be fully avoided and the variability caused by these shifts can also obscure the variability between the peaks themselves. This problem is extensively discussed on univariate chromatograms in [31]. The most often used approach is called correlation optimized warping (COW) whereas dynamic time wariping (DTW), Fuzzy Warping or several other ideas are also present in the literature [32–34].

Regardless of the method used, there is a need that all aligned fingerprints contain a set of common peaks to be warped correctly and the target chromatogram (the reference one to be aligned) is also carefully selected [35].

The only one way to bypass the alignment/warping procedure is to use so-called Gram-matrix in the case of multichannel chromatograms [36]. In TLC this would mean multiwavelength DAD-like signals. However, the time is the main drawback here in the case of classical densitometer.

In the case of image processing, a warping step (called also in the image context "registering") is a little bit more difficult. The example approach [29] requires a registering done by special curves called splines, against the correlation. The random variation of the spot location in 2D TLC is very significant and correct processing of such fingerprints are almost impossible without this step.

CHEMOMETRICS ITSELF: UNSUPERVISED AND SUPERVISED TECHNIQUES

The chemometric techniques applied to TLC fingerprints can be unsupervised or supervised. The mainly used unsupervised techniques [1, 37] are principal component analysis [38] with its robust variant [39] independent component analysis [40], and cluster analysis, including dendrograms [41]. In the case of multimode data, such as multichannel fingerprints, the PARAFAC method can be also considered [42].

The unsupervised approaches perform analysis taking into the account some objective criteria, such as dissimilarity or variance in the data matrix. PCA and ICA find directions in the datasets, therefore after applying this technique on the "calibration set" one can store the directions (expressed as loading vectors) and make a further projections of any unknown future sample. In the case of dendrograms, one can add the further samples to dendrogram, because adding another object does not change shape of the dendrogram. In all the cases, one should point a significant attention for outlying objects, as they can disturb data analysis completely [43].

The supervised techniques are used as multivatiate calibration methods [44], or as discriminative methods [45], when data do not cluster against any objective criterion. Although there is a possibility to construct a regression model with a number of first latent variables of PCA and such approach is called principal component regression (PCR), the analysis of data matrix is then based only on variance and, therefore, is unsupervised. Due to this limitation, the most often used linear supervised method is the partial least squares (PLS), known also as projection to latent structures [46, 47]. It finds the directions in the data matrix, which maximize the covariance between matrix itself and predicted variable, and then constructs a regression model with several first such directions. There are also some extensions-generalizations of the method, such as cyclic subspace regression [48] or continuum regression [49]. In the case of outlying object, there may be a need to introduce their robust variants [50]. The linear methods are sufficient in many cases; however, sometimes there is a need to use nonlinear ones. The PLS method can be nonlinearized by applying the radial basis function concept [51]. The neural networks are also often used [52].

CONCLUSION

Although the chemometric approach in TLC fingerprinting requires some prior mathematical and informatical knowledge, there are many publications which can be treated as starting points for interested TLC chromatographers. Tracing the citations inside these papers, one can gain all the knowledge for chemometric treatment for TLC data. However, it is always a nice idea to consult an idea with some experienced chemometrician. Due to increasing interest in chemometrics, we can expect an increasing interest in chemometric evaluation of all TLC data.

REFERENCES

1. M. Daszykowski, B. Walczak, and D. L. Massart, "Projection methods in chemistry," Chemometrics and Intelligent Laboratory Systems, vol. 65, no. 1, pp. 97–112, 2003.
2. J. H. Christensen, J. Mortensen, A. B. Hansen, and O. Andersen, "Chromatographic preprocessing of GC-MS data for analysis of complex chemical mixtures," Journal of Chromatography A, vol. 1062, no. 1, pp. 113–123, 2005.
3. Y. Ni, L. Zhang, J. Churchill, and S. Kokot, "Application of high performance liquid chromatography for the profiling of complex chemical mixtures with the aid of chemometrics," Talanta, vol. 72, no. 4, pp. 1533–1539, 2007.
4. Y. Ni, Y. Peng, and S. Kokot, "Fingerprinting of complex mixtures with the use of high performance liquid chromatography, inductively coupled plasma atomic emission spectroscopy and chemometrics,"Analytica Chimica Acta, vol. 616, no. 1, pp. 19–27, 2008.
5. E. Blicharska, Ł. Komsta, R. Kocjan, A. Gumieniczek, and A. Wiśniewska, "Chemometric processing of ion chromatograms application to comparative analysis of polish bottled mineral and spring waters,"Polish Journal of Environmental Studies, vol. 19, no. 5, pp. 1071–1075, 2010.
6. J. H. Christensen and G. Tomasi, "Practical aspects of chemometrics for oil spill fingerprinting," Journal of Chromatography A, vol. 1169, no. 1-2, pp. 1–22, 2007.
7. T. Gröger and R. Zimmermann, "Application of parallel computing to speed up chemometrics for GC × GC-TOFMS based metabolic fingerprinting," Talanta, vol. 83, no. 4, pp. 1289–1294, 2011.
8. K. A. Aliferis, P. A. Tarantilis, P. C. Harizanis, and E. Alissandrakis, "Botanical discrimination and classification of honey samples applying gas chromatography/mass spectrometry fingerprinting of headspace volatile compounds," Food Chemistry, vol. 121, no. 3, pp. 856–862, 2010.
9. I. Eide and K. Zahlsen, "A novel method for chemical fingerprinting of oil and petroleum products based on electrospray mass spectrometry and chemometrics," Energy and Fuels, vol. 19, no. 3, pp. 964–967, 2005.
10. X. H. Fan, Y. Y. Cheng, Z. L. Ye, R. C. Lin, and Z. Z. Qian, "Multiple chromatographic fingerprinting and its application to the quality control of herbal medicines," Analytica Chimica Acta, vol. 555, no. 2, pp. 217–224, 2006.
11. Y. Hu, Y. Z. Liang, B. Y. Li, X. N. Li, and Y. P. Du, "Multicomponent spectral correlative chromatography applied to complex herbal medicines," Journal of Agricultural and Food Chemistry, vol. 52, no. 26, pp. 7771–7776, 2004.
12. F. Gong, Y. Z. Liang, Y. S. Fung, and F. T. Chau, "Correction of retention time shifts for chromatographic fingerprints of herbal medicines," Journal of Chromatography A, vol. 1029, no. 1-2, pp. 173–183, 2004.
13. F. Gong, B. T. Wang, F. T. Chau, and Y. Z. Liang, "Data preprocessing for chromatographic fingerprint of herbal medicine with chemometric approaches," Analytical Letters, vol. 38, no. 14, pp. 2475–2492, 2005.
14. Y. Liang, P. Xie, and F. Chau, "Chromatographic fingerprinting and related chemometric techniques for quality control of traditional Chinese medicines," Journal of Separation Science, vol. 33, no. 3, pp. 410–421, 2010.
15. Y.-Z. Liang, P.-S. Xie, and K. Chan, "Chromatographic fingerprinting and metabolomics for quality control of TCM," Combinatorial Chemistry and High Throughput Screening, vol. 13, no. 10, pp. 943–953, 2010.
16. Y. Z. Liang, P. S. Xie, and K. Chan, "Perspective of chemical fingerprinting of Chinese herbs," Planta Medica, vol. 76, no. 17, pp. 1997–2003, 2010.
17. K. Macek, Pharmaceutical Applications of Thin-Layer and Paper Chromatography, Elsevier, 1972.
18. M. Waksmundzka-Hajnos, J. Sherma, and T. Kowalska, Thin Layer Chromatography in Phytochemistry, Taylor-Francis, 2010.
19. http://rsbweb.nih.gov/ij/.
20. R. J. Barnes, M. S. Dhanoa, and S. J. Lister, "Standard normal variate transformation and de-trending of near-infrared diffuse reflectance spectra," Applied Spectroscopy, vol. 43, no. 5, pp. 772–777, 1989.
21. A. Savitzky and M. J. E. Golay, "Smoothing and differentiation of data by simplified least squares procedures," Analytical Chemistry, vol. 36, no. 8, pp. 1627–1639, 1964.
22. G. Vivó-Truyols and P. J. Schoenmakers, "Automatic selection of optimal Savitzky-Golay smoothing,"Analytical Chemistry, vol. 78, no. 13, pp. 4598–4608, 2006.
23. B. K. Alsberg, A. M. Woodward, and D. B. Kell, "An introduction to wavelet transforms for chemometricians: a time-frequency approach," Chemometrics and Intelligent Laboratory Systems, vol. 37, no. 2, pp. 215–239, 1997.
24. B. Walczak and D. L. Massart, "Wavelets—something for analytical chemistry?" Trends in Analytical Chemistry, vol. 16, no. 8, pp. 451–462, 1997.
25. Ł. Komsta, "A comparative study on several algorithms for denoising of thin layer densitograms,"Analytica Chimica Acta, vol. 641, no. 1-2, pp. 52–58, 2009.
26. Ł. Komsta, "Suppressing the charged coupled device noise in univariate thin-layer videoscans: a comparison of several algorithms," Journal of Chromatography A, vol. 1216, no. 12, pp. 2548–2553, 2009.
27. Ł. Komsta, "Dealing with charged-coupled device noise in thin-layer videodensitometry. Optimization of several image-denoising techniques," Acta Chromatographica, vol. 21, no. 3, pp. 355–367, 2009.
28. Ł. Komsta, "Comparison of several methods of chromatographic baseline removal with a new approach based on quantile regression," Chromatographia, vol. 73, no. 7-8, pp. 721–731, 2011.

29. Ł. Komsta, T. Cieśla, A. Bogucka-Kocka, A. Józefczyk, J. Kryszeń, and M. Waksmundzka-Hajnos, "The start-to-end chemometric image processing of 2D thin-layer videoscans," Journal of Chromatography A, vol. 1218, no. 19, pp. 2820–2825, 2011.
30. M. Daszykowski, I. Stanimirova, A. Bodzon-Kulakowska, J. Silberring, G. Lubec, and B. Walczak, "Start-to-end processing of two-dimensional gel electrophoretic images," Journal of Chromatography A, vol. 1158, no. 1-2, pp. 306–317, 2007.
31. M. Daszykowski, Y. Vander Heyden, C. Boucon, and B. Walczak, "Automated alignment of one-dimensional chromatographic fingerprints," Journal of Chromatography A, vol. 1217, no. 40, pp. 6127–6133, 2010.
32. B. Walczak and W. Wu, "Fuzzy warping of chromatograms," Chemometrics and Intelligent Laboratory Systems, vol. 77, no. 1-2, pp. 173–180, 2005.
33. V. Pravdova, B. Walczak, and D. L. Massart, "A comparison of two algorithms for warping of analytical signals," Analytica Chimica Acta, vol. 456, no. 1, pp. 77–92, 2002.
34. A. M. van Nederkassel, M. Daszykowski, P. H. C. Eilers, and Y. V. Heyden, "A comparison of three algorithms for chromatograms alignment," Journal of Chromatography A, vol. 1118, no. 2, pp. 199–210, 2006.
35. M. Daszykowski and B. Walczak, "Target selection for alignment of chromatographic signals obtained using monochannel detectors," Journal of Chromatography A, vol. 1176, no. 1-2, pp. 1–11, 2007.
36. M. Daszykowski, R. Danielsson, and B. Walczak, "No-alignment-strategies for exploring a set of two-way data tables obtained from capillary electrophoresis-mass spectrometry," Journal of Chromatography A, vol. 1192, no. 1, pp. 157–165, 2008.
37. M. Daszykowski, "From projection pursuit to other unsupervised chemometric techniques," Journal of Chemometrics, vol. 21, no. 7–9, pp. 270–279, 2007.
38. S. Wold, K. Esbensen, and P. Geladi, "Principal component analysis," Chemometrics and Intelligent Laboratory Systems, vol. 2, no. 1–3, pp. 37–52, 1987.
39. C. Croux and A. Ruiz-Gazen, "A fast algorithm for robust principal components based on projectrion pursuit," in Proceedings of the Computational Statistics (COMPSTAT ‹96), Physica, Heidelberg, Germany, 1996.
40. A. Hyvarinen, J. Karhunen, and E. Oja, Independent Component Analysis, John Willey and Sons, New York, NY, USA, 2001.
41. N. Bratchell, "Cluster analysis," Chemometrics and Intelligent Laboratory Systems, vol. 6, pp. 105–125, 1987.
42. R. Bro, "PARAFAC. Tutorial and applications," Chemometrics and Intelligent Laboratory Systems, vol. 38, no. 2, pp. 149–171, 1997.
43. M. Daszykowski, K. Kaczmarek, Y. Vander Heyden, and B. Walczak, "Robust statistics in data analysis—a review. Basic concepts," Chemometrics and Intelligent Laboratory Systems, vol. 85, no. 2, pp. 203–219, 2007.
44. H. Martens and T. Naes, Multivariate Calibration, John Willey and Sons, 1989.
45. M. Barker and W. Rayens, "Partial least squares for discrimination," Journal of Chemometrics, vol. 17, no. 3, pp. 166–173, 2003.
46. S. Wold, M. Sjostrom, and L. Eriksson, "PLS—a basic tool of chemometrics," Chemometrics and Intelligent Laboratory Systems, vol. 58, pp. 109–130, 2001.
47. J. Hinkle and W. Rayens, "Partial least squares and compositional data: problems and alternatives,"Chemometrics and Intelligent Laboratory Systems, vol. 30, no. 1, pp. 159–172, 1995.
48. J. H. Kalivas, "Cyclic subspace regression with analysis of the hat matrix," Chemometrics and Intelligent Laboratory Systems, vol. 45, no. 1-2, pp. 215–224, 1999.
49. S. De Jong and R. W. Farebrother, "Extending the relationship between ridge regression and continuum regression," Chemometrics and Intelligent Laboratory Systems, vol. 25, no. 2, pp. 179–181, 1994.
50. M. Daszykowski, Y. Vander Heyden, and B. Walczak, "Robust partial least squares model for prediction of green tea antioxidant capacity from chromatograms," Journal of Chromatography A, vol. 1176, no. 1-2, pp. 12–18, 2007.
51. B. Walczak and D. L. Massart, "The radial basis functions—partial least squares approach as a flexible non-linear regression technique," Analytica Chimica Acta, vol. 331, no. 3, pp. 177–185, 1996.
52. B. Walczak, "Neural networks with robust backpropagation learning algorithm," Analytica Chimica Acta, vol. 322, no. 2, pp. 21–29, 1996.

CHAPTER 10

ETHANOL PRODUCTION FROM WASTE POTATO MASH BY USING SACCHAROMYCES CEREVISIAE

Gulten Izmirlioglu and Ali Demirci

Keywords: Enzyme hydrolysis, response surface method, bio-ethanol, Saccharomyces cerevisiae, fermentation, waste potato mash

INTRODUCTION

Renewable energy attracts attention for the protection of the environment and supplies our energy needs by reducing dependence on petroleum and non-renewable energy sources. Bio-ethanol, which is one of the energy sources, is known to be a potential alternative to petroleum-derived fuels and has the potential to meet the increasing demand for energy for industrial processes, heating and transportation [1]. In 2009, 19,534.99 millions of gallon of ethanol were produced worldwide [2]. The U.S. and Brazil are the two major countries and produced 10,600.00 and 6,577.89 million gallon of ethanol in 2009, respectively [2]. However, the U.S. is still the largest petroleum consumer in the world and consumed 18.8 million barrels per day of petroleum products in 2011. Advanced bio-fuels (may include ethanol derived from cellulose, sugar or starch, or waste material, including crop residue, other vegetative waste material, animal waste, and food waste) production under The U.S. Renewable Fuel Standard (RFS) scenario could reduce U.S. petroleum import by approximately $23 billion in 2016 according to the report for "U.S. Economic Impact of Advanced Biofuels Production: Perspectives to 2030" [3].

Sugar cane, as a raw material, is used for 60% of global ethanol production, however; corn is the main raw material of ethanol production in the United States (90%) [1]. These carbon sources are high value products as a food source [4,5]. Potato is another high value crop as a food source and currently, utilized 60% frozen, 14% fresh, 13% chip, 13% dehydrated, and 1% potato seedin the US [6]. Today, the U.S. produces about 460 billion pound of potatoes annually on 1.3 million acres with an approximate value of 3$ billion. During processing of potato, some of the potatoes wasted, for example 18% in the potato chips industry and 5%–20% in potato cultivation [7,8] Keystone Potato Products, LLC (Hegins, PA) manufactures potato products including potato flakes with the capacity of 7,700 kg/h, out of 900 kg skin and 225 kg low quality waste mash potato per h (Personal communication with Keith Masser). Therefore, the waste from potato industry can be utilized as growth media for the fermentation processes. The wastes of potato industry are currently being utilized as animal feed [9]. However, they can be consumed for ethanol production without requiring drying process. Waste of potato industry could be an economical carbon source for ethanol fermentation in the industry, because of high starch content of potato wastes. Waste potato mash is a waste of potato flakes process and composed of approximately 80% moisture.

Citation: Gulten Izmirlioglu and Ali Demirci, Ethanol Production from Waste Potato Mash by Using Saccharomyces Cerevisiae, doi:10.3390/app2040738

Hydrolysis is a process of breaking down starch (amylopectin and amylose) into fermentable sugars and is needed before the fermentation. Hydrolysis is carried out at high temperature (90–110 °C); however, at low temperatures, it is possible and can contribute to energy savings [10]. To convert starch into the fermentable sugars, either acid hydrolysis or enzymatic hydrolysis needs to be performed. Each has their own set of advantages and disadvantages for use. The limitations of acid hydrolysis can be by-products inhibition on growth of yeast (such as 5-hydroxymethylfurfural (5-HMF)), neutralization before fermentation and expensive constructional material due to corrosion risks [11]. On the other hand, high prices of enzymes play a crucial role when feasibility is of concern. Enzyme hydrolysis is chosen even though high cost of enzymes and initial investment because of high conversion yield of glucose [11]. Amylases (α-amylase, β-amylase, and glucoamylase) are employed for hydrolysis of starchy materials. Although amylases are derived from plants, animals, and microorganisms, microbial amylases are in use commonly [12]. α-Amylase (endo-1,4-α-D-glucan glucohydrolase, EC 3.2.1.1) hydrolyses the 1,4-α-D-glucosidic linkages in the linear amylase chain, randomly. However, glucoamylase (exo-1,4-α-D-glucan glucohydrolase, EC 3.2.1.3) cleaves the 1,6-α-linkages at the branching points of amylopectin as well as 1,4-α-linkages [13].

Response Surface Methodology (RSM) is a combination of mathematical and statistical techniques and used for the modeling and analysis of problems in which a response of interest is influenced by several variables and the objective is to optimize this response [14]. Furthermore, Box-Behnken is a sub-category of response surface design and is used to study the quadratic effect of factors after identifying the significant factors using screening factorial experiments. The advantage of this method is a reduced number of samples and replicates (12 edges for three factors with three levels; for a total of 12 data points) and the center of the factor space (center point is replicated three times, in total 15 data points), whereas the full factorial design has 27 data points (3 factors × 3 levels × 3 replicates). The three variables (X1, X2, and X3) can predict what the response of product production (Y) is.

Ethanol production from waste potato is a relatively new topic and limited research has been conducted about the utilization of potato waste for ethanol production. Fadel [7] and Liimatainen *et al.* [8] showed that different wastes of potato industry can be a carbon source for yeast during alcohol fermentation by studying waste from potato chips industry (98.67% total carbohydrate) and different potato cultivations (starch content in a range of 11.2% to over 19.3%), respectively. Fadel [7] reported that the highest alcohol concentration (13.2% *v*/*v*) was achieved after 24 h at 34 °C in a medium contained 25% *w*/*v*glucose with initial pH level of 5, using 0.075% urea as the sole nitrogen source, 0.05% orthophosphoric acid; 8% inoculum size (*v*/*v*) and agitation rate of 100 rpm.

Therefore, this study was undertaken to further investigation of optimum liquefaction and saccharification conditions specifically for waste potato mash while evaluating fermentation parameters, such as pH, inoculum size and nitrogen sources for ethanol production from waste potato mash.

EXPERIMENTAL SECTION

Microorganism and Medium

Saccharomyces cerevisiae (ATCC 24859) was obtained from the American Type Culture Collection (Manassas, VA, USA). To prepare inoculum, *S. cerevisiae* was grown in medium composed of 20 g/L of glucose (Domino Sugar, Domino Foods Inc., Yankers, NY, USA), 6 g/L of yeast extract (Difco, Sparks, MD, USA), 0.3 g/L of $CaCl_2 \cdot 2H_2O$, 4 g/L of $(NH_4)_2SO_2$, 1 g/L of $MgSO_4 \cdot 7H_2O$, and 1.5 g/L of KH_2PO_4 at 30 °C for 24 h. In order to maintain viability, the culture was stored at 4 °C and sub-cultured biweekly, whereas stock cultures were kept in 20% glycerol at −80 °C.

Waste Potato Mash

Waste potato mash was obtained from Keystone Potato Products (Hegins, PA, USA), which manufactures potato flakes commercially. There were a variety of potatoes used throughout the study, including Frito-lay FL 1833, Atlantis, and Russet Burbank. The starch content of waste potato mash was 17%–24%. Waste potato mash was stored at −20 °C until use without any pretreatment.

Enzymes

α-Amylase (EC 3.2.1.1) for liquefaction and amyloglucosidase (EC 3.2.1.3) for saccharification were used. These enzymes were manufactured by Novozyme Corporation and distributed by Sigma-Aldrich (Saint Louis, MO, USA). Their activity or concentration was reported as 18.8 mg protein/mL for α-amylase and 300 Unit/mL for amyloglucosidase by the manufacturer.

Hydrolysis of Starch

Experimental Design

Response Surface Methodology (RSM) was employed to optimize three experimental factors for optimization of liquefaction and saccharification of waste potato mash. RSM with a three-factor, three-level Box–Behnken design [15] was used to optimize the response as % loss in non-dissolved solid for three variables.

The actual factor levels corresponding to coded factor levels are shown in Table 1 and Table 2. The ranges of factor levels for experimental design were selected based on our preliminary studies (unpublished data). The optimal conditions for maximum loss in non-dissolved solid and glucose conversion were estimated by statistical analysis using Minitab (Version 13.3; Minitab Inc., State College, PA, USA). The coefficients in the second-order polynomial were calculated by multiple regression analysis on the experimentally obtained data.

Table 1: Levels of factors chosen for the experimental design of liquefaction.

Factors	Symbols	Coded Levels		
		-1	0	1
Temperature (°C)	X_1	50	72.5	95
Enzyme dose (mL)	X_2	0.2	0.6	1
Dry weight of WPM * (g/100 mL)	X_3	1	5.5	10

* Waste potato mash.

Table 2: Levels of factors chosen for the experimental design of saccharification.

Factors	Symbols	Coded Levels		
		-1	0	1
Temperature (°C)	X_1	30	45	60
Enzyme dose (mL)	X_2	0.2	0.6	1
Time (h)	X_3	24	48	72

Liquefaction

Various concentrations of waste potato mash slurries were prepared and pH of the slurry was adjusted to 6.5 by 1 N NaOH. The mixture was agitated at 120 rpm agitation in a shaker water bath for 3 h at temperature chosen by the design based on the preliminary study. Also, 0.2, 0.6, and 1 mL of α-amylase solution, and 1, 5.5, and 10 g dry weight of waste potato mash/100 mL of DI water in 250 mL flasks were evaluated to determine the optimum combination based on the Box-Behnken design. Initial and final samples were taken and analyzed for non-dissolved solid in the mixture to determine optimum condition to yield the maximum loss in dry weight and non-dissolved solid ratio. After optimum combination was obtained, three run was performed at this combination for verification.

The %loss (Equation (1)) in dry weight for liquefaction was chosen as response variable to optimize conditions of liquefaction and calculated as follows:

$$\%\text{ loss} = \frac{\text{Solid reduced in waste potato mash}}{\text{Total solid in waste potato mash}} \times 100 \qquad (1)$$

Saccharification

Liquefied slurry, which was processed at the determined optimum liquefaction condition, was used for saccharification. Box-Behnken RSM design was constructed by the Minitab Statistical Software (Version 13.3; Minitab Inc., State College, PA, USA). Various volumes of enzyme solution (1, 0.6, and 0.2 mL of amyloglucosidase solution) and various temperatures (30, 45, and 60 °C) were evaluated to determine the optimum combination based on the Box-Behnken design as well as three different incubation times (24, 48, and 72 h). Also, agitation was set up at 120 rpm in a shaker water bath. For verification, saccharification at determined optimum combination was performed in triplicate. Samples were taken at the end of the process and analyzed for glucose concentration to determine optimum temperature-time-enzyme concentration combination after 20-fold dilution without any other sample preparation step.

After optimization of hydrolysis, enzyme concentration and dry weight levels increased with same ratio at the same temperature to obtain more glucose. As a result, a linear relation was obtained with a 0.9985 R^2.

Fermentation Media

The base-line fermentation (glucose/yeast extract) medium contained 50 g/L of glucose, 6 g/L of yeast extract (Difco Sparks, MD) 0.3 g/L of $CaCl_2{\cdot}2H_2O$, 4 g/L of $(NH_4)_2SO_2$, 1 g/L of $MgSO_4{\cdot}7H_2O$, and 1.5 g/L of KH_2PO_4 per liter of deionized water. For waste potato mash studies, hydrolyzed waste potato mash was used as the carbon source instead of glucose. The rest of the ingredients were kept the same as baseline fermentation medium. Furthermore, four different alternative nitrogen sources were evaluated to investigate an economical substitute of yeast extract. Poultry meal, hull and fines mix, feather meal, and meat and bone meal obtained from Griffin Industries, Inc. (Butler, KY, USA) as dry powders were used as alternative nitrogen sources to replace yeast extract in the medium at the same concentration (6 g/L).

Ethanol Fermentation

Sartorious Biostat B Plus Bioreactors (Allentown, PA, USA) with 2.5 L vessel (working volume of 1.5 L) equipped with pH, temperature, and agitation controls were used. Temperature was maintained at 30 °C and agitation was maintained at 400 rpm. pH was controlled at 5.5 by adding 4 N NaOH or 4 N H_2SO_4. Inoculum was grown for 24 h at 30 °C. After inoculation, 48 h fermentation was carried out and samples were taken every two hour first 12 h of fermentation whereas every 6 h during the remaining of the fermentation time.

Analysis

Microbial Cell Population

The spiral plating method was used to determine cell population by using a spiral auto-plater (Model 4000, Spiral Biotech, Norwood, MA, USA) and Q-count software (Version 2.1; Spiral Biotech, Norwood, MA, USA). Collected samples were serially diluted 0.1% peptone water and spiral plated on potato dextrose agar (Difco, MD, USA). After 24 h incubation at 30 °C, Q-count software (Version 2.1; Spiral Biotech) was used for enumeration. Results were indicated as log10 CFU/mL.

Glucose and Ethanol

Samples were analyzed for glucose and ethanol concentrations by using YSI 2700 Analyzer (Yellow Springs, OH, USA). YSI analyzer included specific membranes for each of the components as well as system buffers and calibration solutions. One milliliter of samples was diluted by 20 fold to bring the concentration of either ethanol or glucose in the range and then analyzed by YSI Analyzer.

Moisture Analysis

To determine moisture of waste potato mash, samples were

weighed and drying process was carried out at 105 °C in an oven for 48 h until weight of samples stabilized.

Non-Dissolved Solid Analysis

The liquefaction slurry was centrifuged at 4,000 rpm for 30 min. After the supernatant is discarded, the distilled water was used to wash the solid and the washing process was repeated two times. Then the mixture was dried in an oven at 105 °C for 24 h. The solid was weighed and percentage loss was calculated with the respect to the initial slurry as the non-dissolved solid (Equation (1)).

Statistical Analysis

Statistical analysis was conducted to test significant differences between each treatment. In particular, the two-sample t test, ANOVA, and Dunnett tests were used to test for significant differences between the mean of production rate and growth rate values of each treatment. Since controlled and uncontrolled pH were compared, the two-sample t-test was chosen, whereas ANOVA was chosen for inocula sizes and nitrogen sources due to number of parameters [16]. The Dunnett test was used to determine whether any of nitrogen sources yields statistically different than yeast extract. The level of significance was set at 0.05. Statistical analysis was performed using Minitab Statistical Software (Version 13.3; Minitab Inc., State College, PA, USA).

RESULTS AND DISCUSSION

Liquefaction and Saccharification

By applying multiple regression analysis on the experimentally determined data, the regression coefficients were estimated and the following second-order polynomial equation (Equation (2)) was obtained using Minitab software:

$$Y = 117.901 - 1.359X_1 - 33.643X_2 - 1.559X_3 - 33.708X_2^2 - 0.219X_3^2 + 8.742X_2X_3 \quad (2)$$

The lowest loss in non-dissolved solids of waste potato mash observed at the combination of 50 °C, 0.6 mL of α-amylase, and 10 g dry waste potato mash with a 43.7% loss. On the other hand, the maximum loss in non-dissolved solids of waste potato mash observed at the combination of 95 °C, 0.6 mL of α-amylase, and 1 g dry waste potato mash with a 79% loss (Table 3). Also, %loss is compared for 1 and 10 g dry waste potato mash while temperature and enzyme concentrations were kept same to determine the enzyme-substrate relation.

From counterplots, it can be seen that higher dry waste potato mash require higher levels of enzymes (Data not shown). When counterplots are interpreted, it can be seen that increasing temperature (more than 95 °C) might have resulted higher % loss, but at that time enzyme would not be active. By using the results of Box-Behnken design, Box-Behnken optimizer was used to determine optimum conditions of liquefaction, and then an optimum parameter combination suggested as 95 °C, 1 mL of α-amylase, and 4.04 g dry weight of waste potato mash per 100 mL of deionized water, with a prediction of 68.8% loss in dry weight. Verification for this optimum combination was performed with three replications and 74 ± 0.05% of loss in non-dissolved components was obtained, which is very close the estimated value by the RSM optimizer.

Table 3: Box–Behnken design matrix for the three factors and their experimental results for liquefaction.

Run	Factors			% Loss	
	X_1	X_2	X_3	Experimental	Predicted
1	50	0.2	5.5	52.1	60.95
2	72.5	0.2	10	58.07	39.68
3	72.5	0.6	5.5	60.09	57.48
4	95	0.6	10	52.1	62.39
5	95	1	5.5	65.13	56.27
6	95	0.6	1	79	69.47
7	72.5	0.2	1	72.41	73.84
8	72.5	1	10	54.37	52.93
9	50	1	5.5	50.03	41.93
10	72.5	1	1	5.77	24.15
11	72.5	0.6	5.5	60.05	57.48
12	95	0.2	5.5	65.59	73.68
13	72.5	0.6	5.5	52.32	57.48
14	50	0.6	10	43.73	53.25
15	50	0.6	1	61.84	51.54

Waste potato mash slurry, 4.04 g dry weight of waste potato mash per 100 mL of deionized water, was liquefied at 95 °C after addition of 1 mL of α-amylase solution, which was determined as optimum combination for liquefaction process. This liquefied slurry obtained at the optimum liquefaction conditions was used for saccharification. By applying multiple regression analysis on the experimentally determined data, the regression coefficients were estimated as shown in the following second-order polynomial equation (Equation (3)) was obtained using Minitab software:

$$Y = 31.0208 - 0.8X_1 + 16.375X_2 - 0.0278X_3 + 1.9792X_2^2 - 0.2417X_1X_2 - 0.1667X_2X_3 \quad (3)$$

Results of saccharification were summarized as glucose concentration, because saccharification is a conversion of liquefied starch to glucose (Table 4). Levels of temperature were 30, 45, and 60 °C, whereas 0.2, 0.6, and 1 mL of amyloglucosidase solutions were compared to find out optimum combination for saccharification. Moreover, time was another factor, which also had three levels; 24, 48, and 72 h. Minimum glucose (19.2 g/L) obtained at 30 °C, 0.2 mL of amyloglucosidase, and 48 h combination. Maximum glucose concentration was found at 60 °C, 72 h and 0.6 mL of amyloglucosidase with 30.7 g/L glucose concentration. Although longer saccharification might result in higher glucose conversion, this wasn't recommended due to high energy costs. By using the results of Box-Behnken design, Box-Behnken optimizer was used to determine optimum conditions of saccharification, an optimum parameter combination suggested as 60 °C, 0.8 mL of amyloglucosidase, and 72 h. Validation for this optimum combination was performed with three replications and 34.9 g/L glucose was obtained, which is very close the estimated value by the optimizer.

It is reported that the highest dextrose equivalent of 94% was achieved after acid hydrolization of fresh potato tubers [11]. Another study showed that 825.1 mg/g glucose content can be obtained when 100 units of α-amylase and 80 units of amyloglucosidase to hydrolyze potato pulp with enzymes from Sigma [17]. Arapoglou *et al.* [18] reported that 19.37 g/L of total amount of sugars was achieved after acidic hydrolysis of potato peel waste. Studies show that glucose conversions of starchy materials are subject to characteristic of raw materials, type of enzyme, dosage of enzyme, temperature, and pH.

Waste potato mash is a potential fermentation medium for industrial ethanol fermentation. Although a promising glucose concentration (34.9 g/L) was generated by enzyme treatment, glucose concentration was still far away from desired glucose level for typical ethanol fermentation. The amount of waste potato mash and enzyme were increased simultaneously while ratio obtained from Box-Behken RSM optimization was kept constant. To enhance the conversion yield, amount of enzyme and waste potato mash were increased with linear approach and evaluated. The reason for increasing only the amount of enzymes and waste potato mash and holding the time and temperature is that it was seen that enzyme and amount of waste potato mash have a significant effect on the glucose conversion yield from the results of Box-Behnken RSM. There was a linear regression between glucose concentration and amounts of enzyme and dry weight of waste potato mash with a 0.9985 R2 (data not shown). Based on the average of three replications, 16.16 g dry weight/100 mL DEIONIZED WATER was the best choice with 93.5 g/L glucose yield at the end of saccharification as well as viscosity of the mixture, which allows a reasonable agitation.

Table 4: Box–Behnken design matrix for the three factors and their experimental results for saccharification.

Run Order	Factors			Glucose (g/L)	
	X_1	X_2	X_3	Experimental	Predicted
1	45	1	5.572	20.7	22.1
2	45	0.6	48	24.5	22.9
3	45	0.6	48	22.4	22.9
4	30	0.6	24	22.7	22.4
5	60	0.6	24	21.1	22.2
6	60	1	48	28.2	26.4
7	30	1	48	24	23.7
8	45	1	24	21.7	22.3
9	60	0.6	72	30.7	31
10	45	0.2	72	26	25.4
11	30	0.2	48	19.2	20.9
12	45	0.2	24	20.6	19.2
13	60	0.2	48	29.2	29.4
14	45	0.6	48	22	22.9
15	30	0.6	72	20.7	19.5

Effect of pH on Ethanol Production

Ethanol fermentation was evaluated at two different pH profiles to determine the effect of pH: uncontrolled pH and controlled pH at 5.5. Figure 1 shows the cell population and concentrations of glucose and ethanol in the fermentation broth with controlled pH at 5.5 and uncontrolled pH. The results clearly indicate that a higher growth rate for biomass was obtained with the controlled pH at 5.5 (0.496 log CFU/mL/h) than uncontrolled pH (0.289 log CFU/mL/h).

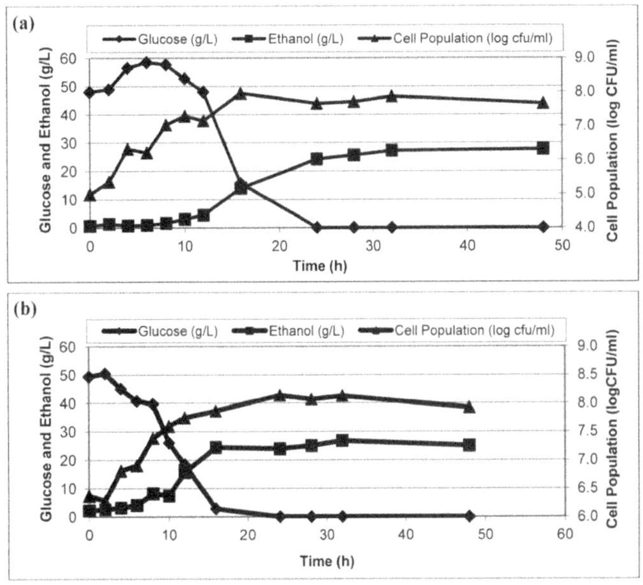

Figure 1: Glucose and ethanol, and biomass concentrations at pH 5.5 (**a**) and uncontrolled pH (**b**).

Furthermore, the maximum ethanol concentration and production rates were 27.7 g/L and 5.47 g/L/h, respectively at controlled pH 5.5, whereas 22.75 g/L and 2.22 g/L/h were obtained at uncontrolled pH. Yields of the fermentations (product produced /substrate consumed) were similar; 46.23% and 44.55% for controlled and uncontrolled pH profiles, respectively. For controlled pH, glucose concentration of the fermentation medium began to increase during the first few hours of fermentation instead of reduction, which might be due to saccharification was still in progress (Figure 1). Therefore, glucose conversion continued simultaneously during earlier stage of the fermentation, and after 24 h of fermentation, glucose was completely consumed. This increase of glucose enhances ethanol yield and growth of biomass.

Yields of the fermentations (product produced/substrate consumed) were similar; 46.23% and 44.55% for controlled and uncontrolled pH profiles, respectively. However, it was difficult to determine the exact amount of consumed glucose because saccharification was still being performed in the earlier stage of fermentation. Therefore, determination of optimum pH was dependent on growth rate, production rate, and ethanol concentrations, which indicated that controlled pH at 5.5 is better for ethanol fermentation.

It was reported that high ethanol production was obtained by using initial pH 5.0 to 6.0 [7], which was in agreement with the results of this study. It was also shown that no ethanol production exists lower than pH 4.0 [19]. During the uncontrolled pH treatment, however, the minimum observed pH was 4.1. Turhan *et al.* [20] reported that maximum ethanol yield, maximum growth rate, and biomass concentration were obtained at pH 5.5 on carob as a medium for ethanol production. Therefore, pH 5.5 was found to be the best pH level and therefore, used for the rest of the study.

Effect of Inoculum Size on Ethanol Production

Three different inoculum sizes (1%, 3%, and 5% (*v*/*v*)) were investigated to determine the effect of inoculum size on kinetic parameters of ethanol fermentation from waste potato mash. Figure 2 shows the ethanol production (g/L), glucose consumption (g/L), and the cell population (log CFU/mL) over 48 h fermentation periods for all cases. The maximum ethanol productivity (6.48 g/L/h) and maximum growth rate (0.3 log CFU/mL/h) were obtained with 3% inoculation, which produced 30.99 g/L ethanol.

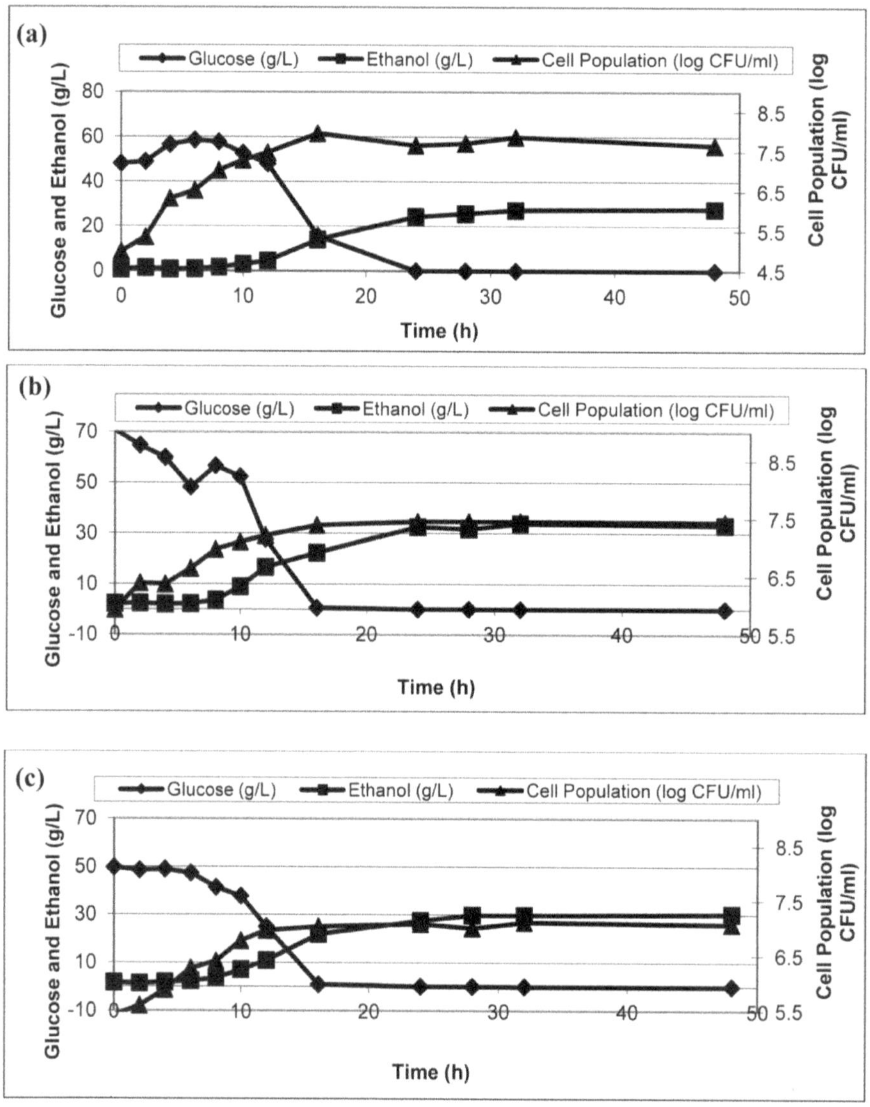

Figure 2: Ethanol, glucose, and cell population in the fermentation broth with different inoculum size; 1% (**a**), 3% (**b**), and 5% (**c**).

Although 5% inoculum sizes demonstrated a higher ethanol yield; growth rate, production rate and consumption rate were lower than parameters of 3% inoculum size. Among 1, 3, and 5% inoculum sizes, 3% was chosen to be the optimum inoculum by comparing production rate, maximum growth rate and produced ethanol. The highest production rate, growth rate, and produced ethanol were obtained for 3% inoculum size as 6.48 g/L/h, 0.3 log CFU/mL/h, and 30.99 g/L, respectively. There was no statistically significant difference in mean production rate among the inoculum sizes ($p > 0.05$), however, growth rates were statistically different for inoculum sizes ($p < 0.05$).

The results demonstrated that there is an increase of ethanol yield up to 3%, however 5% inoculum causes a decrease of kinetic parameters for ethanol fermentation by *S. cerevisiae*. Fadel [7] reported that ethanol production increases by inoculum up to 4%. Furthermore, it was reported that 3% inoculum size was the optimum for ethanol production from carob [20]. Therefore, 3% inoculum size was also suggested to be the optimum level for ethanol by this study.

Effect of Nitrogen Sources

In order to find more economical and efficient alternative nitrogen sources, ethanol fermentations were performed in hydrolyzed waste potato mash media by using poultry meal, hull and fines mix, feather meal, and meat and bone instead of yeast extract as the nitrogen source. Figure 3 shows ethanol, glucose and cell populations using four different nitrogen sources in fermentation medium.

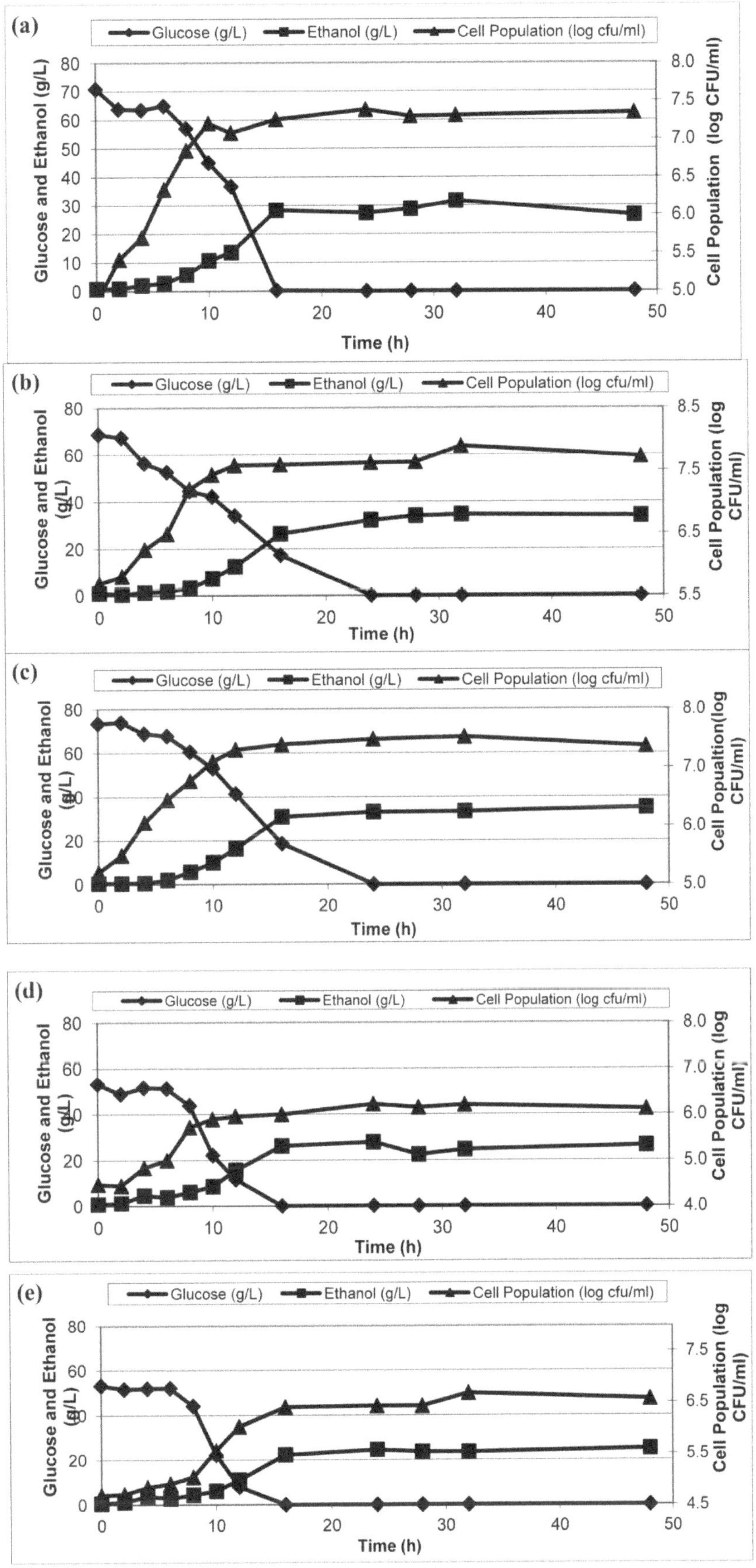

Figure 3: Ethanol, glucose and cell populations in the fermentation broth with different nitrogen sources: Yeast extract (**a**), Feather meal (**b**), Poultry meal (**c**),Meat bone meal (**d**), Hull and fines mix (**e**).

Although the maximum ethanol yield was attained from poultry meal as 35 g/L ethanol, growth rate, production rate, and consumption rate of poultry meal fermentation (0.211 log CFU/mL/h, 3.2 g/L/h, and −5.35 g/L/h, respectively) were lower compared to yeast extract (growth rate 0.28 log CFU/mL/h; production rate 3.68 g/L/h; and consumption −7.07 g/L/h). The second highest ethanol yield (32 g/L ethanol concentration in final broth) was observed with feather meal, which also had very good results for growth rate (0.28 log CFU/mL/h), production rate (3.59 g/L/h), and consumption rate (5.06 g/L/h). Final ethanol concentrations for hull and fines mix and meat bone meal were less than yeast extract with 24.59 and 25.54 g/L ethanol, respectively, whereas 30.80 g/L ethanol obtained while yeast extract was nitrogen source. Moreover, these two nitrogen sources fell behind in growth and production rates (0.194 log CFU/mL/h and 1.97 g/L/h and 0.13 log CFU/mL/h and 2.36 g/L/h, respectively). Overall, the maximum growth rate was observed with yeast extract and feather meal (both 0.28 log CFU/mL/h). However, maximum production rate was obtained with yeast extract (3.68 g/L/h) with a 30.8 g/L ethanol production.

Although poultry meal has the highest ethanol yield, it has lower growth and production rates compared to yeast extract and feather meal. Statistical analysis showed that there is a significant difference among production rates of different nitrogen sources ($p < 0.05$). Compared to yeast extract by Dunnett Test and it was concluded that no nitrogen source produces ethanol significantly different than yeast extract ($p > 0.05$).

According to the results, it can be concluded that all of the animal by-products investigated in this study provide nitrogen for growth of yeast in ethanol fermentation. However, a limitation of the use of alternative nitrogen sources was that these animal-by-products were not 100% soluble, maybe due to partial hydrolysis during manufacturing of these animal by-products, which caused plugging in the tubes of the reactors.

A comparison of poultry meal, hull and fines mix, feather meal and meat bone meal for ethanol production on carob extract by using *S. cerevisiae* was reported by Turhan *et al.* [20]. In their study, maximum production rate and ethanol yield among four alternative nitrogen sources were determined by addition of meat bone meal, whereas none of the other evaluated nitrogen sources reached the yield of ethanol and growth rate when yeast extract was added. Graf-Sirakaya and Demirci [21] also studied the effect of nitrogen sources for ethanol fermentation with S. cerevisiae in a chemically defined media to evaluate seven different animal by-products as a nitrogen source instead of yeast extract. For these studies, yeast extract was compared with poultry meal, hulls and fines mix, beef stock, blood meal, feather meal, pork stock and meat and bone meal [21]. Pork stock and meat and bone meal were reported as giving the two highest ethanol yields as alternative nitrogen sources, however, the second highest production rate after yeast extract was obtained with poultry meal [21]. These animal by-products could be alternative nitrogen sources for ethanol production instead of yeast extract although all of them had lower production rates than the yeast extract.

CONCLUSIONS

In this study, optimal liquefaction and saccharification combinations of three important growth parameters, temperature, enzyme dose, dry weight for liquefaction and temperature-enzyme dose, and time for saccharification were generated using RSM with Box–Behnken design. Optimum combinations were 95 °C, 1 mL of α-amylase, and 4.04 g dry weight of waste potato mash per 100 mL of deionized water and 60 °C, 0.8 mL of amyloglucosidase, and 72 h for liquefaction and saccharification, respectively. The final glucose conversion under optimum medium was 34.9 g/L. Furthermore, higher glucose conversion was obtained by increasing amounts of dry-weight and enzyme by keeping ratio constant. Maximum glucose (137 g/L) was obtained from 24.24 dry-weight/100 mL DIW, however 16.16 g dry-weight/100 mL DIW was promising for ethanol fermentation process when agitation is of concern.

Overall, ethanol fermentation from waste potato mash was found to perform better at pH of 5.5, and 3% inoculum size. The maximum production rate was obtained at pH 5.5, 30 °C, 400 rpm agitation, and 3% inoculum size with a 30.99 g/L ethanol production. By combining optimum conditions, 35 g/L ethanol was produced at pH 5.5 with an inoculum size of 3% when the nitrogen source was poultry meal. However, kinetic parameters of this fermentation were relatively low in comparison with yeast extract and feather meal. Feather meal yielded 32.9 g/L ethanol with 0.28 log CFU/mL/h growth rate and 3.59 g/L/h production rate, which are comparable kinetic parameters to yeast extract. It is clearly indicated that waste potato mash can be an effective fermentation medium for production of ethanol under conditions of controlled pH at 5.5, inoculum size of 3% and supplementation of the nitrogen source. In conclusion, waste potato mash was found promising carbon source for ethanol fermentation with alternate nitrogen sources.

ACKNOWLEDGMENTS

The work was supported in-part by Pennsylvania Agricultural Experiment Station and Turkish Ministry of Education by providing scholarship to Gulten Izmirlioglu. The authors gratefully acknowledge Keystone Potato Products (Hegins, PA, USA) and Griffin Industries, Inc. (Butler, KY, USA) for supplying waste potato mash and alternative nitrogen sources, respectively.

REFERENCES

1. Balat, M.; Balat, H.; Oz, C. Progress in bioethanol processing. *Prog. Energ. Combust.* **2008**, *34*, 551–573. [Google Scholar] [CrossRef]
2. Renewable Fuels Association. Available online: http://ethanolrfa.org/pages/World-Fuel-Ethanol-Production (accessed on 12 April 2012).
3. Biotechnology Industry Organization. Available online: http://www.cleantechsandiego.org/reports/EconomicImpactAdvancedBiofuels.pdf (accessed on 11 October 2012).
4. Nalley, L.; Hudson, D. The potential viability of biomass ethanol as a renewable fuel source: A discussion. Mississippi State University, Starkville, MS, USA. Staff Report 2003-003.
5. United States Department of Agriculture. Available online: http://www.usda.gov/oce/reports/energy/EthanolSugarFeasibilityReport3.pdf (accessed on 12 October 2012).
6. National Potato Council. Available online: http://

nationalpotatocouncil.org/index.php/potato-facts/ (accessed on 12 October 2012).

7. Fadel, M. Alcohol production from potato industry starchy waste. *Egypt J. Microbiol.* **2000**, *35*, 273–287.

8. Liimatainen, H.; Kuokkanen, T.; Kaariainen, J. Development of Bio-Ethanol Production from Waste Potatoes. In Proceedings of the Waste Minimization and Resources Use Optimization Conference, Oulu, Finland, 2004; Pongracz, E., Ed.; Oulu University Press: Oulu, Finland, 2004; pp. 123–129.

9. Yamada, S.; Shinomiya, N.; Ohba, K.; Sekikawa, M.; Oda, Y. Enzymatic hydrolysis and ethanol fermentation of by-products from potato processing plant. *Food Sci. Technol. Res.* **2009**, *15*, 653–658.

10. Sanchez, O.J.; Cardona, C.A. Trends in biotechnological production of fuel ethanol from different feedstocks. *Bioresource Technol.* **2008**, *99*, 5270–5295.

11. Tasic, M.B.; Konstantinovic, B.V.; Lazic, M.L.; Veljkovic, V.B. The acid hydrolysis of potato tuber mash in bioethanol production. *Biochem. Eng. J.* **2009**, *43*, 208–211.

12. Kunamneni, A.; Singh, S. Response surface optimization of enzymatic hydrolysis of maize starch for higher glucose production. *Biochem. Eng. J.* **2005**, *27*, 179–190.

13. Pandey, A.; Nigam, P.; Soccol, C.R.; Soccol, V.T.; Singh, D.; Mohan, R. Advances in microbial amylases. *Biotechnol. Appl. Biochem.* **2000**, *31*, 135–152.

14. Montgomery, D.C. *Design and Analysis of Experiments: Response Surface Method and Designs*; John Wiley and Sons, Inc.: New Jersey, NJ, USA, 2005.

15. Box, G.E.P.; Behnken, D.W. Some new three level designs for the study of quantative variables. *Technometrics* **1960**, *2*, 455–475.

16. Montgomery, D.C. *Design and Analysis of Experiments*, 7th ed; John Wiley & Sons, Inc.: New Jersey, NJ, USA, 2009.

17. Liu, Q. A study of enzymatic hydrolysis of starch in potato pulp. *J. Food Sci.* **2002**, *67*, 2113–2117.

18. Arapoglou, D.; Varzakas, T.; Vlyssides, A.; Israilides, C. Ethanol production from potato peel waste. *Waste Manag.***2010**, *30*, 1898–1902.

19. Graves, T.; Narendranath, K.; Dawson, R.; Power, R. Effect of pH and lactic or acetic acid on ethanol productivity by*Saccharomyces cerevisiae* in corn mash. *J. Ind. Microbiol. Biotechnol.* **2006**, *33*, 469–474.

20. Turhan, I.; Demirci, A.; Karhan, M. Ethanol production from carob extract by *Saccharomyces cerevisiae*. *Bioresource Technol.* **2010**, *101*, 5290–5296.

21. Graf-Sirakaya, B.; Demirci, A. Industrial grade media evaluation for ethanol fermentation with *Saccharomyces cerevisiae*. In Proceedings of The Huck Institutes of Life Sciences' Crossover Symposium on Bioenergy, Pennsylvania State University, University Park, PA, USA, 4–5 September 2004.

CHAPTER 11

INTEGRATED ELECTRICITY PLANNING COMPRISE RENEWABLE ENERGY AND FEED-IN TARIFF

Ho Wai Shin and Haslenda Hashim

Keywords: Renewable energy, feed in-tariff, electricity generation, processing residue, general algebraic, modeling system, generation mix, emission reduction

INTRODUCTION

Increasing Renewable Energy (RE) share had been the goal is many countries over the world, however, due to the high cost of RE compared to conventional resources (fossil fuel), the development of RE into the generation mix is rather slow. In order to promote and increase RE development, Feed-in Tariff (FiT) had been introduced and since, been implemented in 63 jurisdictions worldwide rendering it as the most effective policy at stimulating rapid development of RE (Klein et al., 2006; Couture and Gagnon, 2010).

FiT policies are designed to offer guaranteed prices for fixed periods of time for electricity produced from RE depending on its type of technology, size of installation, quality of the resource and other variables (Couture and Gagnon, 2010; Mendonca, 2007). The remuneration of FiT can be classified as either dependent or independent from the actual electricity market price (Klein et al., 2006; Couture and Gagnon, 2010). Market-independent FIT policies are generally known as fixed-price policies, since they offer a fixed or minimum price for electricity from RE delivered to the grid while market-dependent FIT policies are generally known as premium price policies, or feed-in premiums, since a premium payment is added above the market price (Mendonca et al., 2009).

In order to increase the development of RE in Malaysia, Malaysia Government had proposed FiT to be launched by the mid of 2011 emphasizing on solar PV, biomass, biogas and mini-hydro. The FiT introduced is under the classification of an independent FiT policy covering for RE up to a maximum capacity of 30 MW, while different rate is set for different range of RE size (Chua et al., 2011). In order to ensure a stable funding for FiT, Pusat Tenaga Malaysia (PTM) had introduce a RE fund which collects 2% of consumers electricity bill to fund for the incentive under FiT. The fund is mainly used to equalize the price between non-renewable and renewable sources of energy (Chua et al., 2011).

This study developed a Mixed Integer Linear Programming (MILP) model for Islander Malaysia to plan an optimum fleet-wide electricity generation mix from various sources incorporating RE and FiT using General Algebraic Modeling System (GAMS).

Citation: Ho Wai Shin and Haslenda Hashim, Integrated Electricity Planning Comprise Renewable Energy and Feed-In Tariff, ISSN 1941-7020

MATERIALS AND METHODS

The model is formulated with the objective function to minimize the cost of power plants (Mirzaesmaeeli et al., 2010). In this model, it consists of 4 types of electricity generation, existing Fossil Fuel (FF), new FF, new RE and new RE biomass (separated from other RE due to various fuel sources). The capital cost for existing FF power plant is assumed to be paid off and thus, it is omitted from the objective function. An additional retrofitting cost is included for fuel switching (coal to natural gas)

Superstructure

The superstructure incorporates the 4 types of electricity generation. FF power plant (existing and new) consists of Natural Gas Combined Cycle power plant (NGCC), natural gas combustion turbine (NGCT) and Pulverized Coal (PC). New RE power plant consists of Municipal Solid Waste (MSW), Landfill Gas (LFG), biogas, hydro, Large solar Photovoltaic (LPV) and small Solar Photovoltaic (SPV). While, new RE biomass power plant consists of Biomass Combined Cycle (BCC) and Biomass Bubbling Fluidized Bed (BBFB) with 3 different fuels (empty fruit bunch, EFB, microfiber and kernel). An illustration of the superstructure is as shown in Fig. 1

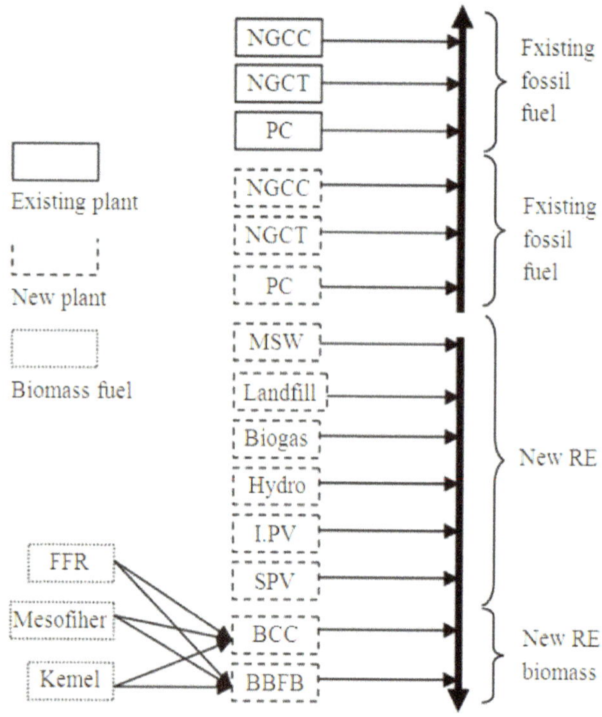

Figure 1: Superstructure for existing and new technology.

Objective function

The objective function is formulated as the total cost subtract with the total remuneration from FiT. The costing included in the objective function is the cost of existing FF power plant (fixed and variable Operating and Maintenance (O&M) cost), fuel cost of existing FF power plant, cost of new FF power plant (capital, fixed and variable O&M cost), fuel cost of new FF power plant, cost of new RE (MSW, LFG, biogas, hydro, LPV and SPV) power plant (capital, fixed and variable O&M cost), cost of new biomass power plant, fuel cost of biomass, retrofitting cost for retrofitting burner to utilize coal fuel to natural gas. The remuneration on the other hand, includes the FiT for each RE Eq. 1:

$$
\begin{aligned}
\min f(i,j,k,l,m,n,) = {} & \underbrace{\sum_i \sum_{k\in FF} \sum_{n\in FC} E^{FF}_{ik} C^{FC}_{km}}_{\text{(Cost of existing FF)}} + \\
& \underbrace{\sum_i \sum_{k\in FF} E^{FF}_{ik} C^{FF}_{k}}_{\text{(Fuel cost of existing FF)}} + \underbrace{\sum_{k\in FF} \sum_{m\in FF} E^{FF}_{k} C^{FC}_{km}}_{\text{Cost of new FF}} + \\
& \underbrace{\sum_{k\in FF} E^{FF}_{k} C^{FF}_{k}}_{\text{(Fuel cost of new FF)}} + \underbrace{\sum_j \sum_{k\in RE} \sum_{m\in RE} E^{RE}_{jkn} C^{RE}_{km}}_{\text{(Cost of new RE)}} \\
& \underbrace{\sum_j \sum_{k\in BI} \sum_l \sum_{m\in BI} \sum_n E^{BI}_{jkln} C^{BI}_{kln}}_{\text{(Cost of new RE biomass)}} + \\
& \underbrace{\sum_j \sum_{k\in BI} \sum_l \sum_{m\in BI} \sum_n E^{BI}_{jkln} P^{BI}_{kl}}_{\text{(Fuel Cost of new RE biomass)}} + \\
& \underbrace{\sum_i \sum_{j\in FF} E^{FF}_{ij} C^{FF}_{ij}}_{\text{(Retrofitting cost)}} - \underbrace{\sum_j \sum_{k\in RE} \sum_n E^{RE}_{jkn} FIT^{RE}_{kn}}_{\text{(Feed-in tariff for RE)}} - \\
& \underbrace{\sum_j \sum_{k\in BI} \sum_l \sum_n \sum_n E^{BI}_{jkln} FIT^{BI}_{kl}}_{\text{(Feed-in tariff for RE biomass)}} +
\end{aligned}
\qquad (2)
$$

Indices:

i = Existing fossil fuel plants

j = RE power plants

k = Type of power plants

l = Biomass fuel

m = Costing

n = Range for FiT

Sets:

FF = Fossil Fuel

FC = Existing Fossil Fuel Costing (Without Capital Cost)

RE = Renewable Energy

BI = Biomass RE

Scalars:

AT = Average Electricity Tariff

REF = Percentage contributed to RE Fund

Red = Percentage reduction of CO_2

CO_2^c = Current CO_2 emission (kMetric Tonne)

Parameters:

C_{km} = Cost 'm' for type of power plant 'k'

C^{BI}_{klm} = Cost 'm' for type power plant 'k' using biomass fuel 'l' operating with biomass 'k'

P_k = Price of fuel for type of power plant 'k'

P^{BI}_{kl} = Price of biomass fuel 'l' for type power plant 'k'

FiT^{BE}_{n} = Feed-in Tariff for RE with range 'n'

FiT^{BI}_{n} = Feed-in Tariff for BI with range 'n'

R_{ij} = Retrofitting cost for power plant 'i' to type power plant 'k'

X_i = Capacity of existing fossil fuel plant

Cap_i = Capital cost of existing fossil fuel plant

V^{MSW} = Conversion rate of MSW (kg) to energy

V^{Lfill} = Conversion rate of MSW (kg) to energy (landfill)

B_i = Biomass Fuel Availability

CO^{FF}_{2k} = CO_2 emission of fossil fuel

Integer variables:

E_{ik} = Electricity generation of existing fossil fuel power plant 'i' for type of power plant 'k'

E_k – Electricity generation for type of power plant 'k'

E_{jkn} = Electricity generation of RE power plant 'j' for type of power plant 'k' with range of FiT 'n'

C^{Bio}_{ikln} = Electricity generation of RE power plant 'j' for type of power plant 'k' using biomass fuel 'l' with range of FiT 'n'

Continuous variables:

A = Adjusted cost ($)

COE = Cost of Electricity ($/kWh)

Constraints:

Annual electricity demand constraint. The summation of electricity generation from all sources (FF and RE) must be equal or greater than the required demand as shown by Eq. 2:

$$\sum_i \sum_{k\in FF} E^{FF}_{ik} + \sum_{k\in FF} E^{FF}_{k} + \sum_j \sum_{k\in RE} \sum_n E^{RE}_{jkn} + \sum_j \sum_{k\in BI} \sum_l \sum_n E^{BI}_{jkln} \geq \text{Demand (MW)} \quad (2)$$

Existing FF power plant constraints

The electricity generation E^{FF}_{ij} must be equal or less Than X_i. Meanwhile, X_i, represents the capacity of each existing FF power plant. The formula is as shown by Eq. 3:

$$\sum_{k\in FF} E^{FF}_{ik} \leq x_i \text{ (MW)}, \quad \forall i \in FF \quad (3)$$

Renewable energy constraint

Electricity generation from renewable energy, E_{jkn} must be equal or less than the availability as shown by Eq. 4:

$$E_{jkn} \leq \text{RE availability (MW)}, \quad \forall k\in \text{bijogas, SPV, LPV} \quad (4)$$

MSW/landfill constraint

There are two technologies to convert waste to energy, direct utilization or capturing landfill gas from decomposing waste. However, since both have different conversion rate (mass to energy), v^k. The summation of E^{MSW}_{jkn}/V^{MSW} and $E^{Lfill}_{jkn}/V^{Lfill}$ must be equal or less than the available mass.

The formula is shown by Eq. 5:

$$\frac{E^{MSW}_{jkn}}{V^{MSW}} + \frac{E^{Lfill}_{jkn}}{V^{Lfill}} \leq \text{MSW Availability (kg)} \quad (5)$$

Biomass constraint

Electricity generation, E^{BI}_{jkln} from biomass must be equal or less than the available fuel resources as shown by Eq. 6:

$$\sum_j \sum_{k\in BI} \sum_n E^{BI}_{jkln} \leq B_l \text{ (MW)}, \quad \forall k \quad (6)$$

CO_2 emission constraint

To meet the carbon emission reduction targets, the summation of fossil fuel emission must be equal or less than the reduction requirement as shown by Eq. 7:

$$\sum_i \sum_{k\in FF} E^{FF}_{ik} CO^{FF}_{2k} + \sum_{k\in FF} E^{FF}_{k} CO^{FF}_{2k} \leq (1-\text{Red}) CO^{c}_{2} \quad (7)$$

RE fund constraint

FiT remuneration must be less or equal to the RE fund as shown by Eq. 8:

$$\sum_j \sum_{k\in BI} \sum_l \sum_n E^{BI}_{jkln} FiT^{BI}_{kn} + \sum_j \sum_{k\in BI} \sum_n E^{RE}_{jkn} FiT^{RE}_{kn} \leq \left(\begin{array}{l} \sum_j \sum_{k\in FF} E^{FF}_{ik} + \sum_{k\in FF} E^{FF}_{k} + \\ \sum_j \sum_{k\in RE} \sum_n E^{RE}_{jkn} + \sum_j \sum_{k\in BI} \sum_l \sum_n E^{BI}_{jkln} \end{array} \right) \times AT \times REF \quad (8)$$

Upper and lower Boundaries

As specific ranges are given for FiT, boundaries are set for the capacity of RE electricity generation. The formula is shown through Eq. 9a-9i. For illustration, Eq. 9a-9i given in the range for the FiT in Malaysia.

Range 'n1':

$0 \leq E_{jkn} \leq 10 \; \forall$ biomass, MSW (9a)

$0 \leq E_{jkn} \leq 1 \; \forall$ SPV (9b)

$0 \leq E_{jkn} \leq 4 \; \forall$ biogas, LFG (9c)

Range 'n2':

$10 < E_{jkn} \leq 20 \; \forall$ biomass, MSW (9d)

$1 < E_{jkn} \leq 10 \; \forall$ LPV (9e)

$4 < E_{jkn} \leq 10 \; \forall$ biogas, LFG (9f)

Table 1: Costing, carbon emission and RE availability Malaysia, 2010 (EIA, 2010)

Power plant	Capital cost[1] ($/MW)	Fixed O and M ($/MW)	Variable O and M ($/MW)	Fuel price, P_i ($/MW)	CO_2 (kMetric tonne)	Existing power plant (MW)	RE availability (MW)
NGCT	974000	6980	128772.0	203159.50	5.04	226	-
NGCC	978000	14390	30046.8	132003.31	3.28	893	-
PC	2521000	23370	37230.0	223856.63	7.20	2100	-
MSW	3860000	100500	43800.0	0.00	-11.51	-	717.17*2
Landfill	8232000	373760	72970.8	0.00	-11.51	-	717.17*2
Biogas	8232000	373760	72970.8	0.00	-11.51	-	47.00
Hydro	3076000	13440	0.0	0.00	0.00	-	0.00
LPV*3	4755000	16700	0.0	0.00	0.00	-	∞
SPV*3	6050000	26040	0.0	0.00	0.00	-	∞
BCC.EFB	7894000	338790	145766.4	129777.78	0.00	-	426.00
BCC.Fiber	7894000	338790	145766.4	126818.68	0.00	-	215.00
BCC.Kernel	7894000	338790	145766.4	153873.33	0.00	-	297.00
BBFB.EFB	3860000	100500	43800.0	141897.90	0.00	-	424.00
BBFB.Fiber	3860000	100500	43800.0	138662.45	0.00	-	215.00
BBFB.Kernel	3860000	100500	43800.0	168243.77	0.00	-	297.00

*1: The capital cost is amortized over a period of 20 years with an interest rate of 15%; *2: MSW and landfill gas power plant share the same source (kMetric Tonne). The conversion rate from waste to energy for MSW, v^{xsw} is 0.1437 (MW/kMetric Tonne) and landfill, $v^{Landfill}$ is 0.0739 (MW/kMetric Tonne); *3: Due to the intermittency of solar energy, 1 MW capacity of photovoltaic is insufficient to meet the actual demand of 1 MW of electricity. With an average sunlight of up to 6 h/day, the cost of solar PV would be 4 times higher to sufficiently meet the required demand

Table 2: Feed-In Tariff for various RE (PTM, 2010; Hashim and Ho, 2011)

RE utilisation	Years	RM/kWh	Degression (%)
Solar PV			
< 1 MW	21	1.14	8.0
> 1 MW < 10 MW	21	0.95	8.0
> 10 MW < 30 MW	21	0.85	8.0
Bonus for BIPV	21	0.25	8.0
Biomass			
< 10 MW	16	0.31	0.5
> 10 MW < 20 MW	16	0.29	0.5
> 20 MW < 30 MW	16	0.27	0.5
Bonus for gasification	16	0.02	0.5
Bonus for MSW	16	0.10	1.8
Biogas			
< 4 MW	16	0.32	0.5
> 4 MW < 10 MW	16	0.30	0.5
> 10 MW < 30 MW	16	0.28	0.5
Bonus for landfill	16	0.08	0.5

Range 'n3':

$20 < E_{jkn} \leq 30 \; \forall$ biomass, MSW (9g)

$10 < E_{jkn} \leq 30 \; \forall$ LPV (9h)

$10 < E_{jkn} \leq 30 \; \forall$ biogas, LFG (9i)

Case study

This model is constructed to provide a road map for Islander Malaysia to implement RE in the total energy generation mix of 1997 MW by 2015 with a carbon emission reduction ranging from 10-40%. As of current, 5 power plants consisting of 1 NGCT power plant located in Pair Gudang (i-1), 3 NGCC power plant, all located in Pasir Gudang (i-2-4) and 1 PC power plant located in Tanjung Bin (i-5), summing up to a capacity of 3219 MW. Several assumptions were drawn in order to simplify the model. The list of assumption is as listed below:

- Currently, the energy generation in Iskandar Malaysia is generated from natural gas (78.8- 1119% MW) and remaining is generated from coal (21.2-287% MW) Malaysia, 2010
- Current cost of electricity if 3.85 cents/kWh.
- Biomass resource is only from palm oil residue
- The only feasible source to produce biogas is from Palm Oil Mill Effluent (POME)
- All RE is considered to be carbon neutral.
- Average electricity tariff is taken to be $ 0.0833 kWh^{-1}

To increase the validity of the model, several data were collected directly from Iskandar Malaysia, PTM and U.S. Energy Information Administration (EIA, 2010) as shown in Table 1 and 2.

Other data used in this model includes, retrofitting cost, R^{FF}_{ik} to retrofit PC to NGCC power plant ($ 23676.79/MW) and current carbon emission, $CO2^C$ (6133.93 kMetric Tonne) Malaysia, 2010. To calculate the COE, the omitted capital cost of existing FF power plant would have to be included. The equation to adjust the total cost and to calculate the cost of electricity is as shown by Eq. 10 and 11:

$$A = f(i, j, k, l, m, n) + \sum_i \sum_{k \in FF} E_{ik}^{FF} Cap_i \quad (10)$$

$$COE = \frac{A(\$)}{\text{Total energy generated (kWh)}} \quad (11)$$

RESULTS

Since the United Nations Climate Change Conference 2009 (COP15), Malaysia had set a target to reduce carbon emission by 40% as announced by the sixth and current Prime Minister of Malaysia, Dato' Seri Najib Razak.

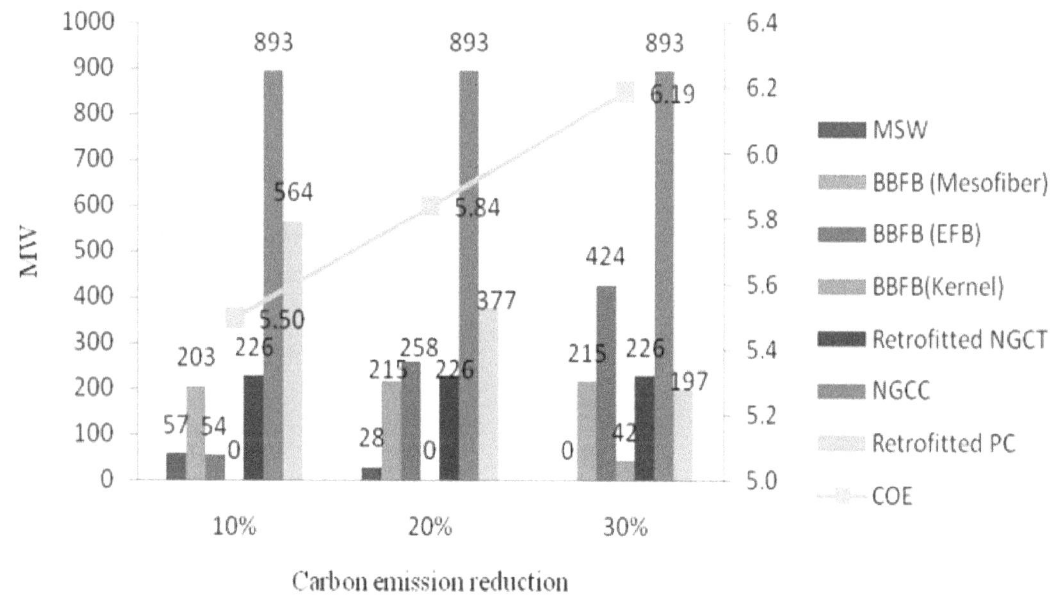

Figure 2: Effect of carbon emission reduction on COE

This section discussed the findings from sensitivity analysis conducted on the model exploring the effects of carbon emission targets toward the cost of electricity and participation of RE. The model includes two methods for carbon reduction, fuel-switching (from coal to natural gas) and implementation of RE. The result of this study is as shown in Fig. 2.

DISCUSSION

Referring to Fig. 2, it can be seen that fuelswitching of PC power plant to NGCC power plant is required for all cases as it proves to be cost effective and a good solution to carbon emission reduction. However, fuel-switching alone could not ultimately reduce the carbon emission and in order to achieve further reduction RE is required. To achieve 10% carbon reduction, a total of 314 MW of RE is required, consisting of 2 MSW power plant adding up to a total of 57 MW (range: 21-30 MW), 27 BBFB (mesofiber) power plant with a total of 203 MW (range: 21-30 MW) and 2 BBFB (EFB) power plant with total of 54 MW (range: 21-30 MW). Two main factors contributes to the selected choice are low costing of these technologies and low FiT (constrained due to RE funding) compare to the other RE (higher range is selected due to lower FiT). While mesofiber and EFB are chosen above kernel as the fuel for BBFB is primarily due lower fuel cost.

To achieve higher carbon emission reduction (20%), the total participation of RE in the generation mix have to be increase by additional of 187 MW. Amount of MSW decrease from 57-MW, BBFB (Mesofiber) increases from 203-215 MW and BBF (EFB) increases from 54-258 MW (all power plants are of the higher range: 21-30 MW). The decrease of MSW participation of due to insufficient funding from RE fund to sustained a stable remuneration. As mesofiber biomass fuel approaches its availability limit, EFB is then used. With 2% of electricity bill contributed to RE fund, it is insufficient to provide enough remuneration to achieve a carbon reduction of 40%, the maximum reduction it could achieve is only 29.62%. Under these scenario, the participation of RE is up to 681 MW, with RES only from biomass (BBFB) with 215 MW from mesofiber, 424 MW from EFB and 42 MW from kernel. Even with incentives from FiT, the cost of RE is still higher than the cost of conventional FF power plant and thus, the COE increases gradually as the carbon emission reduction increases. The COE increases as much as 60.78% hitting a value of 6.19 cents/kWh.

In order to increase the carbon reduction up to 40%, the funding toward RE fund have to be increase, this can be achieve by increasing the percentage collected from electricity bills for RE fund by 0.57% achieving a total contribution of 2.57% or increase the average price of electricity from $ 0.0833/kWh to $ 0.107/kWh (as much as 28.45%). Under this new settings, the contribution from RE is yet again, only from biomass with a total of 875 MW (mesofiber-215 MW, EFB-424 MW and kernel-236 MW). The COE then increases to 6.5 cents/kWh (increment of 68.83%) with all the power plant in the range from 21-30 MW.

CONCLUSION

A MILP model for the optimal planning of gridconnected electricity generation schemes has been developed for IM to meet a specified RE mix target by considering the factor of cost, FiT and carbon reduction. The results indicated that, the selection of type of RE power plant is mainly driven by the cost, FiT, RE fund and the availability of RES.

With FiT introduced, in an investor point of view, it would be more profitable to select the RE with higher remuneration and higher rate of return; however that itself cannot be used as the main criteria in selection of RE projects. To ensure a stable economic, the authority should introduce a quota system which promotes the suitable RE project to be given priority. One method to decide on the priority depends very much on the availability and cost of the resource and the RE fund availability

In this study, to achieve a 40% reduction of carbon emission, the model had selected 875 MW of RE entirely from biomass resources (mesofiber -215 MW, EFB-424 MW and kernel-236 MW) while the remaining demand is met by NGCC. This indicates that focus on biomass resources should be given priority in Malaysia as Malaysia has an abundant amount of biomass especially from palm oil residue. In this scenario, the COE increased by 68.83% to 6.5 cents/kWh. Out of the FF power plants, NGCC is selected due to its low cost, high efficiency and low carbon emission.

ACKNOWLEDGMENT

The authors would also like to thanks Ministry Of Higher Education (MOHE) and University Teknologi Malaysia (UTM) for providing the fund under the GUP research grant of vot number Q.J130000.7125.01H52 for this research.

REFERENCES

1. Chua, S.C., T.H. Oh and W.W. Goh, 2011. Feed-in tariff outlook in Malaysia. Renew. Sustain. Energy Rev., 15: 705-712. DOI: 10.1016/j.rser.2010.09.009
2. Couture, T. and Y. Gagnon, 2010. An analysis of feed-in tariff remuneration models: Implications for renewable energy investment. Energy Policy, 38: 955-965. DOI: 10.1016/j.enpol.2009.10.047
3. EIA, 2010. Updated capital cost estimates for electricity generation plants. EIA, US.
4. Hashim, H. and W.S. Ho, 2011. Renewable energy policies and initiatives for a sustainable energy future in Malaysia. Renew. Sustain. Energy Rev., 15: 4780-4787. DOI: 10.1016/j.rser.2011.07.073
5. Klein, A., A. Held, M. Ragwitz, G. Resch and T. Faber, 2006. Evaluation of different feed-in tariff design options-Best practice paper for the International Feed-In Cooperation. Institute System and Innovation Research.
6. Mendonca, M., 2007. Feed-in Tariffs: Accelerating the Deployment of Renewable Energy. EarthScan, London, ISBN-10: 9781844074662, pp: 150.
7. Mendonca, M., S. Lacey and F. Hvelplund, 2009. Stability participation and transparency in renewable energy policy: Lessons from Denmark and the United States. Policy Soc., 27: 379-398. DOI: 10.1016/j.polsoc.2009.01.007
8. Mirzaesmaeeli, H., A. Elkamel, P.L. Douglas, E. Croiset and M. Gupta, 2010. A multi-period optimization model for energy planning with CO2 emission consideration. J. Environ. Manage., 91: 1063-1070. DOI: 10.1016/j.jenvman.2009.11.009
9. PTM, 2010. Malaysia's 2011 Proposed solar, biomass, biogas and hydro tariffs. Pusat Tenaga Malaysia.

CHAPTER 12

STATISTICAL OPTIMIZATION OF FERMENTATION CONDITIONS FOR CELLULASE PRODUCTION FROM PALM OIL MILL EFFLUENT

Jamal I. Daoud and Md. Zahangir Alam

Keywords: Palm oil mill effluent, cellulase enzyme, bioconversion, statistical optimization

INTRODUCTION

Currently, Malaysia produces 15 million tonnes of crude palm oil per year from which about 12 million tonnes of palm oil is exported, indicates about 52% of the total world production (Wu et al., 2009). The process to extract oil from the Fresh Fruit Bunch (FFB) requires large amount of water, mainly for sterilizing the fruits and oil clarification, resulting in the discharge of organic, non-toxic wastewater known as Palm Oil Mill Effluent (POME). The quantity of POME produced is about 60% for every tonne of FBB processed. Thus, about 18-19.5 tonnes effluent (POME) h^{-1} is generated from the milling process of an average of 30 tonnes FBB h^{-1} (Rashid et al., 2009).

Several techniques have been developed in order to treat the highly biodegradable POME. Ponding, anaerobic and aeration systems are the most adopted treatment processes practiced by more than 85% of the palm oil mills in the country (Ma and Ong, 1985). The drawbacks of these systems are the requirement of a large land area and the system suffers from control and maintenance problems, biogas generation caused in air pollution (Ahmad et al., 2003). The production of this effluent always contributes an environmental problem such as the generation of methane during its anaerobic treatment and the production of high COD (Yacob et al., 2005). It is estimated that POME contains 95-96% water, 0.6-0.7% oil, 4-5% total solids, 2-4% suspended solids, pH 4.7, BOD-25000 mg L^{-1} , COD-50000 mg L^{-1} and total nitrogen-750 mg L^{-1} (Rashid et al., 2009). The carbohydrate and other nutrients content in POME effluent enable it to serve as substrate for the production of cellulolytic enzymes by liquid state bioconversion besides its treatment.

Cellulolytic enzyme production has attracted a worldwide attention due to the possibility of using this enzyme complex for conversion of abundantly available renewable lignocellulosic biomass for production of carbohydrates for numerous industrial applications including bioethanol (Gadjil et al., 1995; Xia and Cen, 1999). Economical production of cellulases is key for feasible bioethanol production from lignocellulosic biomass using cellulase-based processes. To date the production of cellulase has been widely studied in submerged culture processes (liquid state) but the relatively high cost of enzyme production has hindered the industrial applications through various expensive media are used to produce such enzymes (Reczey et al., 1996). The aim of

Citation: Jamal I. Daoud and Md. Zahangir Alam , Statistical Optimization of Fermentation Conditions for Cellulase Production from Palm Oil Mill Effluent, DOI : 10.3844/ajessp.2010.66.70.

this study was to determine the fermentation conditions by applying the statistical technique utilizing Palm Oil Mill Effluent (POME) as new substrate for the production of cellulase enzyme.

MATERIALS AND METHODS

Fermentation media

The major substrate/media used in this study was Palm Oil Mill Effluent (POME). The POME of 5-6% w/w of TSS (pH 4.8) was collected from oil palm industry named Seri Ulu Langat Palm Oil Mill Sdn. Bhd., Dengkil, and Selangor, Malaysia. The POME concentration of 0.5-2% w/w of TSS was prepared by removing the excess water. The final pH of POME was recorded. The POME was supplemented with co-substrate of 1-3% w/w of wheat flour as available nutrients for microbe throughout the study (Alam et al., 2003)

Microbial strain and its inoculum preparation

Culture of Trichoderma harzianum was obtained from lab stock, Bioenvironmental Engineering Lab, IIUM. The fungus was maintained on Potato Dextrose Agar (PDA) plates and subcultured once in a month. The fungus was cultured onto the agar, incubated at 32°C until the entire plate was covered by fungus. After seven days, 100 mL sterilized water was poured onto the surface of four agar plates containing the spore culture. The spores on the surface were gently scraped with sterilized glass rod. The spore (3.2×10^5) suspension of Trichoderma harzianum was then filtered into a 250 mL Erlenmeyer flask and collected for further experiments.

Experimental procedure for cellulase production

The co-substrate of 1-3% (w/v) of wheat flour was added into POME with 0.5-2.0% (w/v) of TSS. The pH of the mixture was adjusted to 4-6 by adding HCl to increase acidity or NaOH to increase alkalinity. The mixture was sterilized at 121°C for 30 min. The inoculum of 1-3% with (3.2×10^5) spores was added into the flask and shaken in the shaker at agitation speed of 100-250 rpm. The pH and temperature was set at 4-6 and 30-35°C respectively during fermentation process. Sampling was done after four days of fermentation for the analysis of cellulase enzymes.

Table 1: Levels of experimental factors for optimization

Factor	Low (-1)	Center (0)	High (+1)
Temperature°C	30.0	32.50	35
Substrate conc (%)	0.5	1.25	2
Co-substrate conc (%)	1.0	2.00	3
Agitation, rpm	100.0	175.00	250
pH	4.0	5.00	6
Inoculums size (%)	1.0	2.00	3

Experimental design and optimization

Fractional factorial design with six center points was performed in order to determine the optimal fermentation conditions for the production of cellulase enzyme by Trichoderma harzianum. The experiments were conducted according to the designated two-level factorial design. In this design, six factors (parameters) with three levels (low, high and medium) were considered for determining the total number of experiments for optimization of cellulase production by liquid state bioconversion. The experiment was design using MinitabTM statistical software. The factors with their levels are shown in Table 1.

Statistical and analytical analysis

Statistical software, MinitabTM was used to analyze regression model of experimental data. The value of F-test, p-test, t-test and R^2 were identified to evaluate the model as well as to determine the optimum conditions. Cellulase activity assay was carried out by the method suggested by Ghose (1987).

RESULTS

The statistical optimization approach using fractional factorial design was used to study the linear, quadratic and interactive effects of various parameters on higher cellulase production by Trichoderma harzianum. The observed (experimental) and predicted results for cellulase enzyme production were obtained with different conditions is shown in Table 2. The highest cellulase activity (13.44 FPU mL^{-1}) was observe at run number 23, where the factors were found to be temperature 30°C, substrate concentration 2% and co-substrate concentration 3%, pH 4, inoculum size 3% and agitation 250 rpm. The experimental result was then analyzed by regression analysis, which gave the following regression Eq. 1 of the levels of cellulase produced (FPU mL^{-1}) as a function of temperature (x_1), substrate concentration (x_2), co-substrate concentration (x_3), inoculums size (x_4), pH (x_5) and agitation (x_6):

Table 2: Result using two level fractional factorial design and six center points showing observed and predicted response (cellulase)

Run	Temperature (°C)	Subs conc. (%)	Co-s conc. (%)	pH	Inoculum size (%)	Agitation (rpm)	Cellulase, FPU mL^{-1} Experimental	Predicted
1	30.0	0.5	1	4	1	100	3.277	2.673
2	35.0	0.5	1	4	1	250	2.643	1.941
3	30.0	2.0	1	4	1	250	2.643	2.194
4	35.0	2.0	1	4	1	100	7.400	6.848
5	30.0	0.5	3	4	1	250	2.696	2.118
6	35.0	0.5	3	4	1	100	2.273	1.540
7	30.0	2.0	3	4	1	100	0.000	-0.490
8	35.0	2.0	3	4	1	250	3.859	3.243
9	30.0	0.5	1	6	1	250	0.000	-0.578
10	35.0	0.5	1	6	1	100	3.753	2.984
11	30.0	2.0	1	6	1	100	0.000	-0.486
12	35.0	2.0	1	6	1	250	1.797	1.225
13	30.0	0.5	3	6	1	100	10.360	9.678
14	35.0	0.5	3	6	1	250	5.550	4.749
15	30.0	2.0	3	6	1	250	12.421	11.900
16	35.0	2.0	3	6	1	100	0.000	-0.676
17	30.0	0.5	1	4	3	250	1.850	1.349
18	35.0	0.5	1	4	3	100	5.286	4.661
19	30.0	2.0	1	4	3	100	0.000	-0.362
20	35.0	2.0	1	4	3	250	4.863	4.431
21	30.0	0.5	3	4	3	100	6.343	5.861
22	35.0	0.5	3	4	3	250	2.326	1.750
23	30.0	2.0	3	4	3	250	13.443	13.048
24	35.0	2.0	3	4	3	100	0.439	-0.034
25	30.0	0.5	1	6	3	100	7.559	7.063
26	35.0	0.5	1	6	3	250	2.114	1.568
27	30.0	2.0	1	6	3	250	3.330	2.994
28	35.0	2.0	1	6	3	100	1.216	0.774
29	30.0	0.5	3	6	3	250	2.167	1.673
30	35.0	0.5	3	6	3	100	5.920	5.263
31	30.0	2.0	3	6	3	100	0.000	-0.374
32	35.0	2.0	3	6	3	250	6.449	5.993
33	32.5	1.5	2	5	2	175	1.586	0.836
34	32.5	1.5	2	5	2	175	1.691	0.836
35	32.5	1.5	2	5	2	175	1.427	0.836
36	32.5	1.5	2	5	2	175	1.269	0.836
37	32.5	1.5	2	5	2	175	1.004	0.836
38	32.5	1.5	2	5	2	175	1.163	0.836

Table 3: ANOVA for polynomial model derived from experimental data

Source	DF	SS	MS	F	P
Regression	31	419.100	13.519	237.40	0.000
Residual error	6	0.324	0.057		
Total	37	419.442			

R^2: 0.999; R^2-adj: 0.995

SS: Sum of Squares; DF: Degree of Freedom; MS: Mean Square

$$\text{Cellulase (FPU mL}^{-1}\text{)} = 155-17.0x_1-89.4x_2+50.4x_3+59.2x_4+77.2x_5+0.0841x_6+0.382x_1^2+2.66x_1x_2-1.69x_1x_3-1.79x_1x_4-2.30x_1x_5-0.00450x_1x_6+9.82x_2x_3+2.31x_2x_4+0.830x_2x_5+0.326x_2x_6-7.91x_3x_4-6.61x_3x_5+0.0204x_3x_6-12.9x_4x_5-0.0979x_4x_6+0.00218x_5x_6 \quad (1)$$

The ANOVA from the analysis are shown in Table 3. The results showed that the coefficient of determination (R^2) was 0.999 and R^2 -adj was 0.995 which ensured satisfactory adjustment of the quadratic model to the experimental data. The value of adjusted determination coefficient is also very high indicating a high significance of the model. The F-value is a measure of variation of the data about the mean. The high F-value and very low probability ($p>F = 0.0000$) indicates that the present model is in good prediction of the experimental result. The t-and p-values for the single, quadratic and interactive effects on the response are shown in Table 4 to evaluate the significance levels ($p<0.5$or $p<0.5$)

DISCUSSION

The t-value and probability value (p-value) serves as a tool for checking the significant of each of the coefficient.

Table 4: t-value and p-value for each coefficient of the parameters and their interactions

Predictor	Coefficient	t	P
Constant	15535.0000	7.07	0.000
Temp, x_1	-17.0000	-14.43	0.000
Subs conc, x_2	-89.4100	-19.41	0.000
Co-s conc, x_3	50.3520	14.81	0.000
pH, x_4	59.1670	25.92	0.000
Inoculum, x_5	77.2390	24.52	0.000
Agitation, x_5	0.0841	2.00	0.092
x_1^2	0.3825	22.32	0.000
x_1x_2	2.6573	18.81	0.000
x_1x_3	-1.6900	-16.20	0.000
x_1x_4	-1.7880	-25.53	0.000
x_1x_5	-2.3010	-23.82	0.000
x_1x_6	-0.0045	-3.49	0.013
x_2x_3	9.8247	13.40	0.000
x_2x_4	2.3059	3.14	0.020
x_2x_5	0.8301	1.13	0.301
x_2x_6	0.3262	33.36	0.000
x_3x_4	-7.9104	-14.38	0.000
x_3x_5	-6.6121	-12.02	0.000
x_3x_6	0.0204	2.79	0.032
x_4x_5	-12.8990	-23.45	0.000
x_4x_6	0.0979	-13.35	0.000
x_5x_6	0.0022	3.88	0.008

The pattern of interactions between the variables is indicated by these coefficients. The larger the magnitude of t-test value and smaller the p-value indicates the high significance of the corresponding coefficient. The variable with low probability levels contribute to the model, whereas the other can be neglected and eliminated from the model. The p-value suggest that the coefficient for linear effect of temperature, x_1, substrate concentration, x_2, cosubstrate concentration, x_3, inoculums size, x_4 and pH, x_5, are most significant with value of p

CONCLUSION

The results show that the optimum conditions for maximum cellulase production (about 14 FPU mL^{-1}) using palm oil mills effluent by liquid state bioconversion was found to be the temperature, 30°C; substrate concentration, 2% (w/v); co-substrate concentration, 3% (w/v); pH, 4; inoculum size, 3% (w/v) and agitation, 250 rpm. The optimization study of fermentation conditions using two-level fractional factorial design would enhance the production of enzymes at different levels using POME as substrate. The lab-scale study on cellulase production from POME as major substrate might give the basic information of further development for large scale production.

ACKNOWLEDGMENT

The researchers thank to the Ministry Of Higher Education (MOHE) for approval a research grant under Fundamental Research Grant Scheme (FRGS 0207-49). The co-operation of Seri Ulu Langat Palm Oil Mill Sdn. Bhd., Dengkil, Selangor, by supplying the experimental samples is acknowledged as well.

REFERENCES

1. Ahmad, A.L., S. Ismail and S. Bhatia, 2003. Water recycling from Palm Oil Mill Effluent (POME) using membrane technology. Desalination, 157: 87-95. DOI: 10.1016/S0011-9164(03)00387-4
2. Alam, M.Z., A. Fakhru'l-Razi and A.H. Molla, 2003. Biosolids accumulation and biodegradation of domestic wastewater treatment plant sludge by developed liquid state bioconversion process using a batch fermenter. Water Res., 37: 3569-3578. DOI: 10.1016/S0043-1354(03)00260-4
3. Gadjil, N.J., H.F. Oaginawala, T. Chakarbartu and P. Khanna, 1995. Enhanced cellulase production by a mutant of Trichoderma reseei. Enz. Microb. Technol., 17: 942-946. DOI: 10.1016/0141- 0229(94)00131-A
4. Ghose, T.K., 1987. Measurement of cellulase activities. Pure Applied Chem., 59: 257-268. DOI: 10.1351/pac198759020257
5. Ma, A.N. and A.S.H. Ong, 1985. Pollution control in palm oil mills in Malaysia. J. Am. Oil Chem. Soc., 62: 261-266. DOI: 10.1007/BF02541389
6. Rashid, S.S., M.Z. Alam, M.I.A. Karim and H.M. Salleh,

2009. Management of palm oil mill effluent through production of cellulases by filamentous fungi. World J. Microbiol. Biotechnol., 25: 2219-2226. DOI: 10.1007/s11274-009-0129-9

7. Reczey, K., Z.S. Szengyel, R. Eklund and G. Zacchi, 1996. Cellulase production by T. reesei. Bioresour. Technol., 57: 25-30. DOI: 10.1016/0960- 8524(96)00038-7

8. Wu, T.W., A.W. Mohammad, J.M. Jahim and N. Anuar, 2009. A holistic approach to managing palm Oil Mill Effluent (POME): Biotechnological advances in the sustainable reuse of POME, Biotechnol. Adv., 27: 40-52. DOI: 10.1016/j.biotechadv.2008.08.005

9. Xia, L. and P. Cen, 1999. Cellulase production by solid state fermentation on lignocellulosic waste from the xylose industry. Process Biochem., 34: 909-912. DOI: 10.1016/S0032-9592(99)00015-1

10. Yacob, S., M.A. Hassan, Y. Shirai, M. Wakisaka and S. Subash, 2005. Baseline study of methane emission from open digesting tanks of palm oil mill effluent treatment. Chemosphere, 59: 1575-1581. DOI: 10.1016/j.chemosphere.2004.11.040

CHAPTER 13

PROCESS SIMULATION OF OXY-COMBUSTION FOR MAXIMIZATION OF ENERGY OUTPUT USING ASPEN PLUS

Xiao Zhang, Subhodeep Banerjee, Ling Zhou, Ramesh Agarwal

Keywords: Carbon capture; Process simulation; Chemical-looping combustion; Maximum energy output; Optimization; Scaled-up simulation.

INTRODUCTION

Coal-fired power plants contribute to significant CO_2 emissions; this reality has driven research in recent years on investigation of combustion processes that can capture CO_2 with reduced energy penalty. One technology that is showing great promise for high-efficiency low-cost carbon capture is the Chemical Looping Combustion (CLC) process. In contrast to other methods for CO_2 separation from flue gas such as oxy-combustion, chemical absorption, and physical adsorption, the CLC is an advanced technology that creates and captures an almost pure and concentrated CO_2 stream with relatively less energy requirement [1, 2]. Several theoretical and experimental studies have demonstrated the potential of CLC to capture almost pure CO_2 very efficiently [3-6]. CLC employs a dual fluidized bed system with circulating fluidized bed process where an oxygen carrier (OC) is used as a bed material providing the oxygen for combustion in the fuel reactor. The reduced OC is then transferred to a second bed and reoxidized by the atmospheric air [7-9] in an air reactor before it is returned to the fuel reactor to complete the loop. Because of the absence of air in the fuel reactor, the combustion products are not diluted by other gases (e.g., N_2), resulting in high purity CO_2 available at the outlet of the fuel reactor. Thus, the CLC process for power generation provides a sequestration-ready CO_2 stream directly after combustion, without the need for using costly gas separation techniques to purify CO_2 from the flue stream. CLC therefore holds significant promise as a next generation combustion technology due to its potential for pre-capturing almost pure CO_2 emission with very limited effect on the efficiency of the power plant. ASPEN Plus is a process simulation software that simulates chemical processes at system level using basic engineering relationships such as mass and energy balance, and multi-phase and chemical reaction models. It consists of flow sheet simulations to calculate stream properties such as flow rate and mass composition given various chemical processes and operating conditions. In this paper, a system level model of CLC process is developed to conduct parametric studies for optimal energy output.. These studies provide valuable insight into the design and operating conditions required in an industrial-scale CLC plant and to assess the feasibility of deploying CLC as an economically viable solution for electricity generation and carbon capture.

Citation: Zhang, Subhodeep Banerjee, Ling Zhou, Ramesh Agarwal, Process simulation of oxy-combustion for maximization of energy output using ASPEN plus, ISSN 2076-2895 (Print)

VALIDATION OF THE CLC PROCESS SIMULATION WITH EXPERIMENT

The CLC process simulation in ASPEN Plus was validated against the experimental work of Sahir et al [10]. The physical and chemical properties of the Colombian coal used as the solid fuel in the experiment are summarized in Table 1.

Table 1: Physical and chemical properties of Colombian coal

Proximate Analysis (wt. %)				Ultimate Analysis (wt. %)					Energy
Moisture	Volatile matter	Fixed carbon	Ash	C	H	N	S	O	LHV (MJ/kg)
3.3	37.0	54.5	5.2	80.7	5.5	1.7	0.6	11.5	29.1

The schematic of the flow sheet for this simulation is shown in Figure 1. The coal is first pulverized and dried before it is pressurized and introduced into a shell gasifier to be partially oxidized to form syngas. The molar ratio of steam and carbon is maintained at unity for the process model. The syngas composition at the gasifier outlet is 34.5% CO, 50.3% H2, 12.3% H_2O, and 2.4% CO_2. The syngas is converted completely to CO_2 and H2O in the fuel reactor while the Fe2O3 in the oxygen carrier is reduced to Fe_3O_4. The oxygen carrier material used is a mixture of 60 wt. % Fe_2O_3 and 40 wt. % inert Al_2O_3 as support. The outflow from the fuel reactor is a concentrated stream of H_2O and CO_2. After condensing the stream, high purity CO_2 is obtained. The reduced oxygen carrier is fed into the air reactor where the oxidation reaction takes place with an 80% conversion of Fe3O_4 to Fe_2O_3.

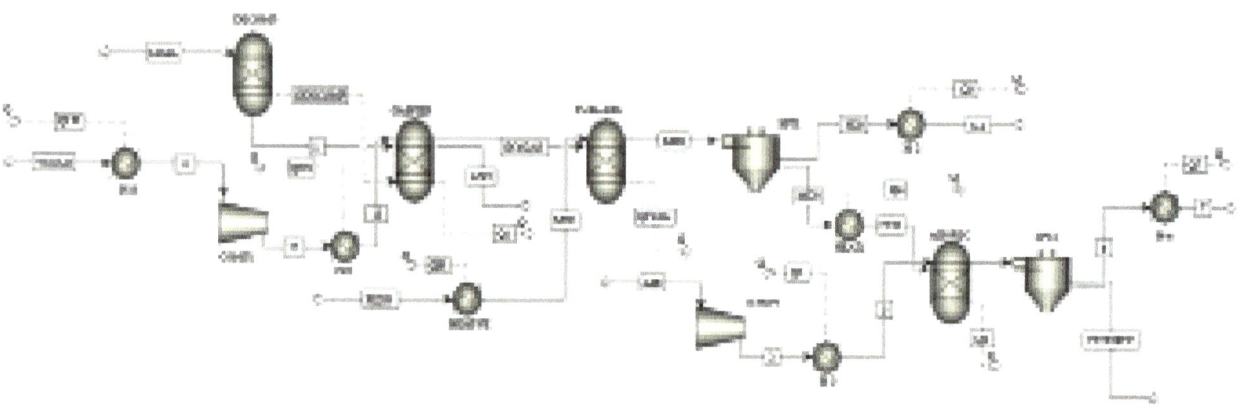

Figure 1: Flow sheet of the CLC model in ASPEN Plus

The various process models used in the ASPEN Plus shown in flow sheet in Figure 1 are summarized in Table 2. The coal devolatilization is defined by the RYIELD reactor, followed by the gasification of coal represented by the RGIBBS reactor. Another RGIBBS reactor defines the actual syngas combustion and the corresponding reduction of the oxygen carrier. These blocks together represent the fuel reactor. The flow sheet within the ASPEN Plus simulation package cannot model this entire reaction with one reactor. As a result, the fuel reactor simulation is broken down into several different reactor simulations. The air reactor is also modeled as an RGIBBS reactor.

Table 2: Process models used in different parts of the CLC process in ASPEN Plus

Name	Model	Function	Reaction formula
DECOMP	RYIELD	Coal devolatilization	Coal → volatile matter + char
BURN	RGIBBS	Gasification	Char + volatile matter → CO_2 + H_2O
FUEL-R	RSTOIC	Carrier reduction reaction	$3Fe_2O_3 + CO/H_2 \rightarrow 2Fe_3O_4 + CO_2/H_2O$
AIR-R	RSTOIC	Carrier oxidation reaction	$4Fe_3O_4 + O_2 \rightarrow 6Fe_2O_3$

For the purpose of validation, the energy balance of the CLC process model was analyzed using the input values from the experiment of Sahir et al [10]. The input values and the energy requirements for the various units and streams in Figure 1 are presented in Table 3; this will be referred to as the baseline case in rest of the paper. Energy is consumed mainly in the compressor processes. Compressed air is required in the air reactor to regenerate Fe_2O_3 from Fe_3O_4. The air compressor for the combustor compresses air to 18atm. Another compressor is used to compress the steam for the gasifier. There is a large amount of energy produced in the air reactor, but the fuel reactor needs to be supplied with energy. This is because the net heat work in the fuel reactor is the summation of the heat work from the DECOMP, GASIFER, and FUEL-R blocks. Although FUEL-R produces energy because of the combustion of syngas, the combined energy requirement of DECOMP and GASIFIER are more than the energy produced in FUELR. Summing the energy requirements of each individual stream, the total energy obtained from the CLC process is 554.2 kW.

Table 3: Input values and energy balance for the baseline case corresponding to the experiment of Sahir et al.[10]

Input values	Coal (kg/h)	100
	Steam (kg/h)	140
	Air Flow Rate (kg/h)	71
	Temperature of Fuel Reactor (°C)	950
	Temperature of Air Reactor (°C)	935
	Fe_2O_3 flow in the Fuel Reactor (kg/h)	5921
	Al_2O_3 in the System (kg/h)	3951
	Particle Density (kg/m³)	3200
Energy Balance (kW)	Fuel reactor	-161.8
	Air reactor	688.0
	Cool air reactor exhaust	135.4
	Cool flue gas	148.3
	Cool OC for air reactor	40.9
	Reheat OC for fuel reactor	-42.7
	Heat steam	-69.8
	Heat air	-184.1
	Net	554.2

The results shown in Table 3 for the baseline case with a coal feed rate of 100 kg/h are in excellent agreement with those reported by Sahir et al [10]. These calculations validate our CLC model developed in ASPEN Plus.

INVESTIGATION OF THE EFFECT OF VARIOUS PARAMETERS ON THE ENERGY OUTPUT OF THE CLC PROCESS SIMULATION

With the successful validation of the process simulation of the CLC experiment of Sahir et al in the previous section, the ASPEN Plus simulation is expanded to consider the effect of varying the air flow rate and the oxygen carrier feeding rate. Additional scaled-up simulations are also conducted to determine these effects on an industrial scale power plant.

Effect of varying the air flow rate on energy output of baseline case with 100 kg/h coal feeding rate

The recent paper of Mukherjee et al. [11] suggests that it is favorable to operate the air reactor of the CLC process at higher temperatures with excess air supply in order to achieve higher power efficiency.

In order to evaluate the effect of air supply on energy output, we consider the baseline case of Table 3 and vary the air flow rate. The results are presented in Figure 2 and Table A1. From Figure 2, it can be seen that with an increase in the air flow rate, the net energy output increases and achieves a maximum for a certain air flow rate. If the air flow rate is further increased from its maximum value (i.e., value corresponding to maximum energy output), the energy output starts decreasing albeit very slowly. This result implies that there exists a certain rate of air supply around 900 kg/h to obtain the maximum energy output for 100 kg/h of coal supply. At this flow rate in the air reactor, 131.06 kW of additional energy is generated, which is 23.6% more than the baseline case given in Table 2 indicating that the reaction in the air reactor is not complete for the baseline case. Excess air supply ensures the 80% conversion of Fe_3O_4 to Fe_2O_3.

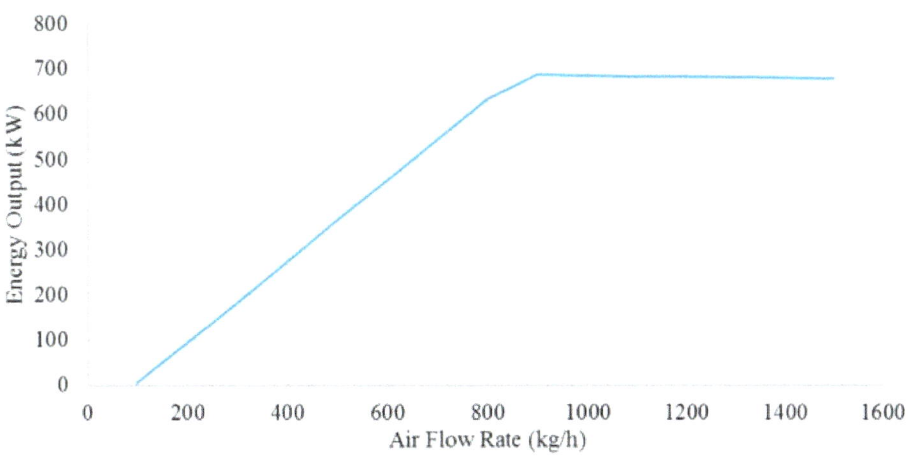

Figure 2: Energy output for various air flow rates for 100 kg/h of coal supply

Effect of varying the oxygen carrier feeding rate on energy output of baseline case with 100 kg/h coal feed rate

The oxygen carrier (OC) plays a vital role in the CLC process; it reacts with the syngas in the fuel reactor and reacts with the air in the air reactor. Both of these reactions contribute a large amount to the net energy output. Figure 3 and Tables A1–A6 present the energy output for different OC feeding rates in the system with varying air flow rates. As expected, Figure 3 shows that for a given air flow rate, a higher OC feeding rate yields more energy output. However, when the OC feeding rate increases above a certain threshold value, the marginal increase in energy output by increasing the OC rate becomes extremely small. The red line in Figure 3 represents the baseline case (Fe_2O_3 at 5921 kg/h), for which the maximum energy output is 685.26 kW with 900 kg/h air flow rate. For the threshold Fe_2O_3 rate of 7000 kg/h, the maximum energy output of 824.33 kW occurs at the 1000 kg/h air flow rate. 138.97 kW of additional energy output is obtained by increasing the OC rate from 5921 kg/h to 7000 kg/h. Therefore, for maximum energy output with a coal feeding rate of 100 kg/h, the optimum rates of air flow and OC feeding are 1000 kg/h and 7000 kg/h respectively. In other words, the optimum ratio of Coal: Air: OC is 1: 10: 70.

Scaled-up simulation

Scaling up is an essential step for the realization and optimization of industrial-scale power plants. Two different scaled-up simulations were conducted to investigate the relationship between the coal feeding rate and energy output. The first scaled-up simulation employed the initial values of the baseline case and the other was based on the optimum values of coal: air supply: oxygen carrier rate. The details of the scaled-up simulations are given in Table A7 and Table A8 respectively. In both cases, the coal feeding rate is scaled up by a factor of up to 12. The OC feeding rate and air supply rate are also scaled-up accordingly to meet the demand of the increased coal feeding. Other modeling parameters such as the reactor efficiency and coal decomposition rate are considered unchanged for both the scaled up simulations. The total thermal power output for the scaled-up simulations is summarized in Figure 4 and Table 4 below. From Figure 4, it can be seen that the total power output increases linearly with increase in coal feeding rate. Considering the principles of energy and mass balance on which the ASPEN Plus modeling is based, linearity in the scaled-up results is expected since the non-linear effects (e.g., the energy loss at multiple locations in the flow sheet) are omitted in the modeling process.

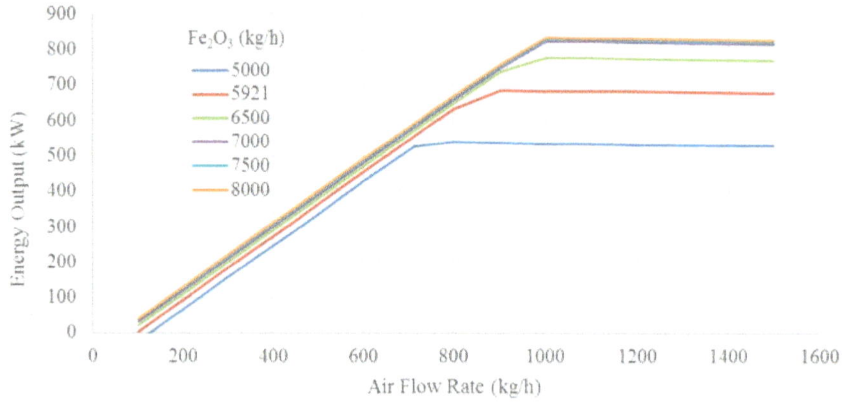

Figure 3: Energy output for various oxygen carrier feeding rates and air flow rates for 100 kg/h of coal supply

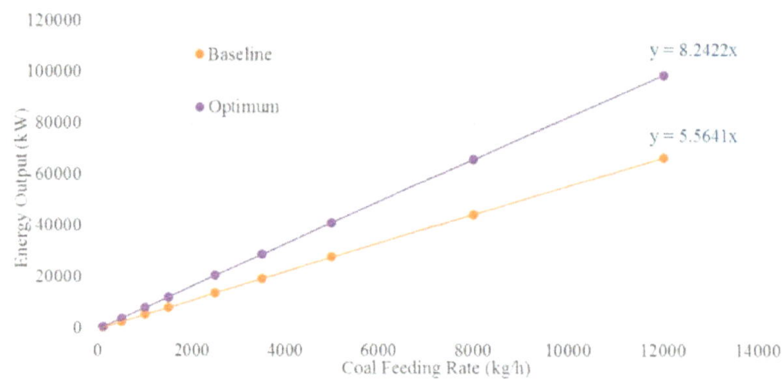

Figure 4: Energy output of two scaled-up simulations for various coal feeding rates

Table 4: Results of two scaled-up simulations for different ratios of Coal: Air: OC

Coal (kg/h)		100	500	1000	1500	2500	3500	5000	8000	12000
Energy output (kW)	Baseline	554.2	2782	5564	8346	13910	19474	27820	44513	66769
	Optimum	824.2	4121	8242	12363	20606	28847	41211	65936	98907

Based on these scaled-up simulations, the energy output for the baseline case is given by the equation

Energy-output=5.5641×Coal-feeding-rate (1)

and the energy output for the optimum case is given by the equation

Energy-output=8.2422×Coal-feeding-rate (2)

Validation of optimum values of air flow rate and oxygen carrier feeding rate for scaled-up simulation

To demonstrate that the optimum values of air flow rate and OC feeding rate for maximum energy output are valid for the scaled-up simulations, four more cases with 12,000 kg/h coal feeding rate and varying rates of air flow and OC were studied, which are presented in Figure 5 and Tables A9-A11. Figure 5 shows that the maximum energy output occurs at 120,000 kg/h of air flow rate, and 840,000 kg/h of Fe_2O_3 feeding rate. This suggests that the optimum ratio of Coal: Air: OC in the system still holds for the scaled-up simulations; it is given by

Coal-feeding-rate:-Air-flow-rate:-OC-feeding-rate=1:10:70 (3)

Equation (3) is an important relationship among these three input parameters for obtaining the maximum energy output from a CLC-based power plant. This relationship can be used for the initial estimates in designing a CLC-based industrial-scale power plant.

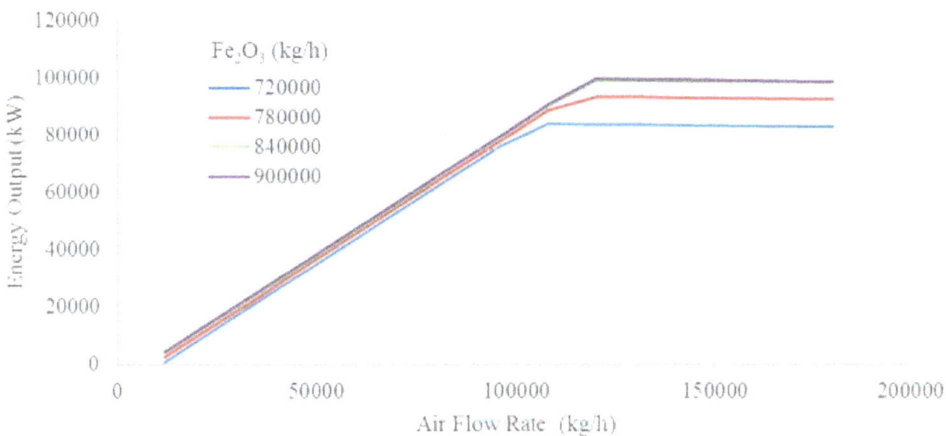

Figure 5: Energy output for different airflow rates and OC rates for a 12000 kg/h coal feeding rate

Energy output of different types of coals

All the results above are dependent on the certain type of coal, the Colombian coal of which the physical and chemical properties are listed in Table 1. Now it is interesting to investigate the performance of different types of coal. Four types of coals are used in this paper which are Colombian, Bituminous, Anthracite and Lignite. The proximate analysis and ultimate analysis of Colombian coal is given in Table 1 and that of other three coals are summarized in Table 5.

Table 5: Physical and chemical properties of bituminous, anthracite and lignite coals

Coal name	Proximate Analysis (wt. %)				Ultimate Analysis (wt. %)						Energy
	Moisture	Volatile matter	Fixed carbon	Ash	C	H	N	S	O	Ash	LHV (kJ/kg)
Bituminous	2.3	33.0	55.9	8.8	65.8	3.3	1.6	0.6	17.6	11.1	21899
Anthracite	1.0	7.5	59.9	31.6	60.7	2.1	0.9	1.3	2.4	32.6	21900
Lignite	12.6	28.6	33.6	25.2	45.4	2.5	0.6	5.2	8.5	37.8	16250

Effect of varying the air flow rate on energy output of four types of coals with 100 kg/h coal feeding rate

Again in order to evaluate the effect of air supply on energy output, we conduct the same process modeling as described in section 3.1 by varying the air flow rate with coal feeding rate of 100 kg/h for four different types of coals. The results are presented in Figure 6 and Table A1 for Colombian coal and in Tables A12-A14 for Bituminous, Anthracite and Lignite coal. From Figure 6, it can be seen that with an increase in air flow rate, all four types of coal show the same trend that net energy output increases and achieves a maximum for a certain air flow rate. Every coal type has a different inflection point which corresponds to the maximum energy output on the y-axis for a certain air flow rate shown on x-axis. It can be seen that the inflection point is different depending upon the type of coal which is expected because of different properties of the coals as given in Table 5. By qualitative analysis, one can infer that higher the concentration of fixed carbon in a coal gives more fuel to burn, and the higher concentration of volatile matter and ash cost less energy to decompose the coal. Next, we determine the optimal ratio of Coal: Air: OC for the other three types of coal.

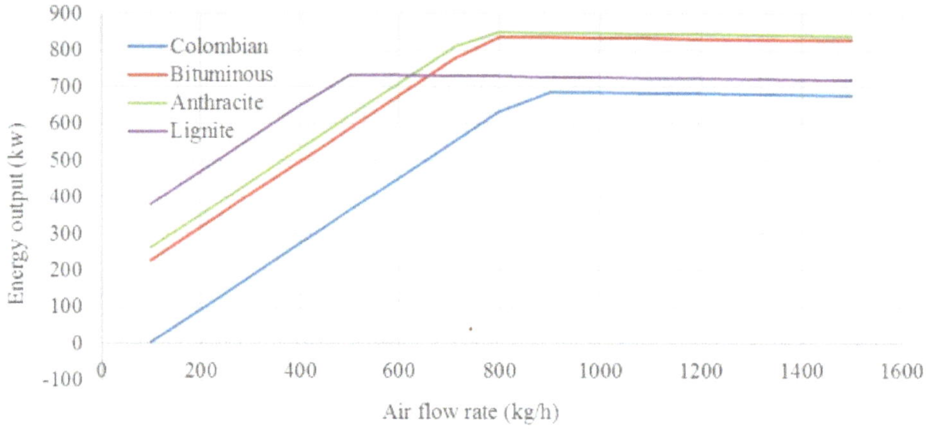

Figure 6: Energy output for various air flow rates for 100 kg/h of four types of coals

Effect of varying the oxygen carrier feeding rate on energy output of four types of coals with 100 kg/h coal feeding rate

The effect of varying the oxygen carrier feeding rate on energy output of Colombian coal was shown in Figure 3 and Tables A1-A6. The results of other three types of coal, Bituminous, Anthracite and Lignite are presented in Figures 7-9 and Tables A12-A24. As expected, as with the Colombian coal, there is the maximum energy output based on optimal coal feeding rate: air flow rate: OC feeding rate for Bituminous, Anthracite and Lignite coal as well. Table 6 summarizes the maximum energy output and optimal ratio of Coal: Air: OC for four types of coal with 100 kg/h coal feed rate.

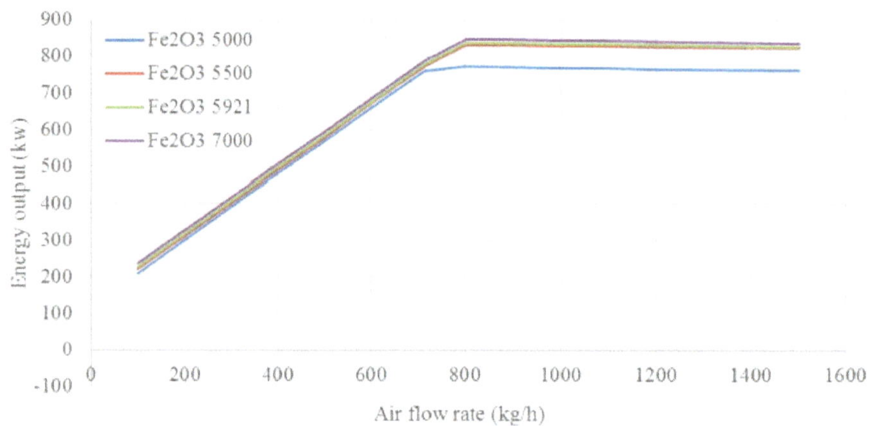

Figure 7: Energy output for various oxygen carrier feeding rates and air flow rates for 100 kg/h of Bituminous coal

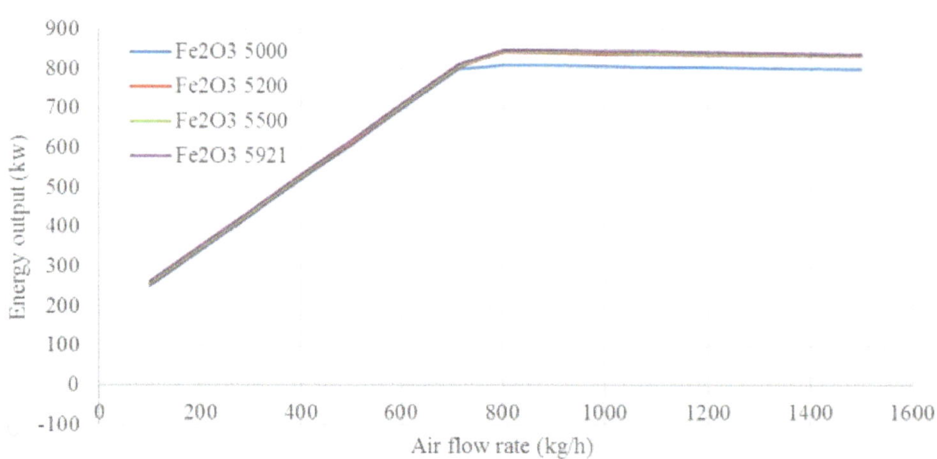

Figure 8: Energy output for various oxygen carrier feeding rates and air flow rates for 100 kg/h of Anthracites coal

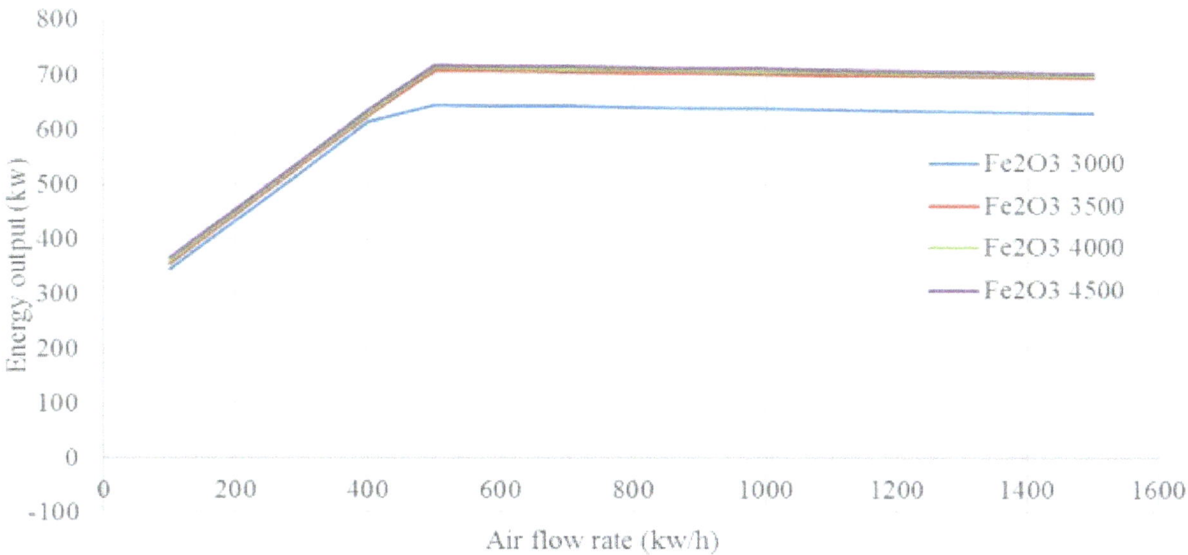

Figure 9: Energy output for various oxygen carrier feeding rates and air flow rates for 100 kg/h of Lignite coal

Table 6: Maximum energy output and optimal ratio of Coal: Air: OC for four types of coal with 100 kg/h coal feed rate

Coal Name	Maximum Energy (kw)	Optimal Ratio of Coal: Air: OC
Colombian	824.229	1: 70: 10
Bituminous	832.373	1: 55: 8
Anthracite	841.258	1: 52: 8
Lignite	707.905	1: 35: 5

CONCLUSIONS

In this paper, ASPEN Plus is employed to model and study the complete CLC process from the coal gasification to the reduction and eventual re-oxidation of the oxygen carrier (OC). The CLC process model is validated against the experiment of Sahir et al [10] and shows good agreement between the experimental data and the simulation results. Based on further studies on the effect of varying air flow rates and OC feeding rates, it is found that the maximization of energy output from CLC can be accomplished by using the optimum ratio of Coal: Air: OC in the system equal to 1: 10: 70 for Colombian coal. Compared to the baseline case based on the experiment of Sahir et al [10], a net increase in power of 48% can be obtained by increasing the air flow rate by 40.25% and the OC feeding rate by 18.22% to attain this optimum ratio for the Colombian coal for the given coal feeding rate of 100kg/h. Scaled-up simulations are also conducted using different coal feeding rates. The results show that the total power output is linear with increase in coal feeding rate. In general, such linearity is not expected for actual industrial-level scale-up since the ASPEN Plus system modeling software neglects miscellaneous energy losses in the system due to changes in the hydrodynamic characteristics of the two fluidized bed reactors. To account for the changes in the hydrodynamics characteristics, detailed hydrodynamic simulations are needed using the computational fluid dynamics software. Three other types of coal (Bituminous, Anthracite, and Lignite) are also investigated, and the optimal ratio of coal: airflow: OC is determined for each of these coal types. There are other parameters that may also influence the energy output such as the temperature and pressure of the reactors, particle size, etc., which are not investigated in this paper.

APPENDIX

Table A1: CLC process simulation results for different air flow rates with Colombian coal at 100 kg/h and Fe_2O_3/Al_2O_3 at 5921/3951 kg/h

	Coal (kg/h)	100	100	100	100	100	100
	Water (kg/h)	140	140	140	140	140	140
	Air Flow Rate (kg/h)	100	300	400	500	600	713
Initial	Temperature of Fuel Reactor (°C)	950	950	950	950	950	950
values	Temperature of Air Reactor (°C)	935	935	935	935	935	935
	Fe_2O_3 flow in the Fuel Reactor (kg/h)	5921	5921	5921	5921	5921	5921
	Al_2O_3 in the System (kg/h)	3951	3951	3951	3951	3951	3951
	Particle Density (kg/m³)	3200	3200	3200	3200	3200	3200
	Fuel Reactor	-161.8	-161.8	-161.8	-161.8	-161.8	-161.8
	Air Reactor	96.498	289.5	386	482.49	578.99	688
	Cool air reactor exhaust	18.996	56.988	75.985	94.981	113.98	135.4
Energy	Cool flue gas	148.3	148.3	148.3	148.3	148.3	148.3
balance	Cool OC for air reactor	40.9	40.9	40.9	40.9	40.9	40.9
(kW)	Reheat OC for fuel reactor	-42.7	-42.7	-42.7	-42.7	-42.7	-42.7
	Heat steam	-69.8	-69.8	-69.8	-69.8	-69.8	-69.8
	Heat air	-25.82	-77.47	-103.3	-129.1	-154.9	-184.1
	Net	4.57	183.92	273.59	363.26	452.93	554.2
	Coal (kg/h)	100	100	100	100	100	100
	Water (kg/h)	140	140	140	140	140	140
	Air Flow Rate (kg/h)	800	900	1000	1100	1200	1500
Initial	Temperature of Fuel Reactor (°C)	950	950	950	950	950	950
values (continued)	Temperature of Air Reactor (°C)	935	935	935	935	935	935
	Fe_2O_3 flow in the Fuel Reactor (kg/h)	5921	5921	5921	5921	5921	5921
	Al_2O_3 in the System (kg/h)	3951	3951	3951	3951	3951	3951
	Particle Density (kg/m³)	3200	3200	3200	3200	3200	3200
	Fuel Reactor	-161.8	-161.8	-161.8	-161.8	-161.8	-161.8
	Air Reactor	771.99	829.67	829.79	829.93	830.07	830.49
	Cool air reactor exhaust	151.97	173.08	197.33	221.58	245.83	318.57
Energy	Cool flue gas	148.33	148.33	148.33	148.33	148.33	148.33
balance	Cool OC for air reactor	40.9	40.9	40.9	40.9	40.9	40.9
(kW) (continued)	Reheat OC for fuel reactor	-42.7	-42.7	-42.7	-42.7	-42.7	-42.7
	Heat steam	-69.8	-69.8	-69.8	-69.8	-68.8	-69.8
	Heat air	-206.6	-232.4	-258.2	-284.1	-309.9	-387.3
	Net	632.3	685.26	683.81	682.37	681.94	676.63

Table A2: CLC process simulation results for different air flow rates with Colombian coal at 100 kg/h and Fe_2O_3/Al_2O3 at 5000/3000 kg/h

	Coal (kg/h)	100	100	100	100	100	100
	Water (kg/h)	140	140	140	140	140	140
	Air Flow Rate (kg/h)	100	300	400	500	600	713
Initial	Temperature of Fuel Reactor (°C)	950	950	950	950	950	950
values	Temperature of Air Reactor (°C)	935	935	935	935	935	935
	Fe_2O_3 flow in the Fuel Reactor (kg/h)	5000	5000	5000	5000	5000	5000
	Al_2O_3 in the System (kg/h)	3000	3000	3000	3000	3000	3000
	Particle Density (kg/m³)	3200	3200	3200	3200	3200	3200
	Fuel Reactor	-183.2	-183.2	-183.2	-183.2	-183.2	-183.2
	Air Reactor	96.498	289.5	386	482.49	578.99	688
	Cool air reactor exhaust	18.996	56.999	75.985	94.981	113.98	135.4
Energy	Cool flue gas	142.63	142.63	142.63	142.63	142.63	142.63
balance	Cool OC for air reactor	32.792	32.792	32.792	32.792	32.792	32.792
(kW)	Reheat OC for fuel reactor	-34.27	-34.27	-34.27	-34.27	-34.27	-34.27
	Heat steam	-69.8	-69.8	-69.8	-69.8	-69.8	-69.8
	Heat air	-25.82	-77.47	-103.3	-129.1	-154.9	-184.1
	Net	-22.21	157.14	246.8	336.47	426.14	527.41
	Coal (kg/h)	100	100	100	100	100	100
	Water (kg/h)	140	140	140	140	140	140
	Air Flow Rate (kg/h)	800	900	1000	1100	1200	1500
Initial	Temperature of Fuel Reactor (°C)	950	950	950	950	950	950
values (continued)	Temperature of Air Reactor (°C)	935	935	935	935	935	935
	Fe_2O_3 flow in the Fuel Reactor (kg/h)	5000	5000	5000	5000	5000	5000
	Al_2O_3 in the System (kg/h)	3000	3000	3000	3000	3000	3000
	Particle Density (kg/m³)	3200	3200	3200	3200	3200	3200
	Fuel Reactor	-183.2	-183.2	-183.2	-183.2	-183.2	-183.2
	Air Reactor	700.66	700.79	700.93	701.11	701.21	701.64
	Cool air reactor exhaust	155.86	180.1	204.35	228.6	252.85	325.59
Energy	Cool flue gas	142.63	142.63	142.63	142.63	142.63	142.63
balance	Cool OC for air reactor	32.792	32.792	32.792	32.792	32.792	32.792
(kW) (continued)	Reheat OC for fuel reactor	-34.27	-34.27	-34.27	-34.27	-34.27	-34.27
	Heat steam	-69.8	-69.8	-69.8	-69.8	-69.8	-69.8
	Heat air	-206.6	-232.4	-258.2	-284.1	-309.9	-387.3
	Net	538.05	536.6	535.16	533.77	532.29	527.99

Table A3: CLC process simulation results for different air flow rates with Colombian coal at 100 kg/h and Fe_2O_3/Al_2O_3 at 6500/4500 kg/h

Initial values	Coal (kg/h)	100	100	100	100	100	100
	Water (kg/h)	140	140	140	140	140	140
	Air Flow Rate (kg/h)	100	300	400	500	600	713
	Temperature of Fuel Reactor (°C)	950	950	950	950	950	950
	Temperature of Air Reactor (°C)	935	935	935	935	935	935
	Fe_2O_3 flow in the Fuel Reactor (kg/h)	6500	6500	6500	6500	6500	6500
	Al_2O_3 in the System (kg/h)	4500	4500	4500	4500	4500	4500
	Particle Density (kg/m³)	3200	3200	3200	3200	3200	3200
Energy balance (kW)	Fuel Reactor	-148.4	-148.4	-148.4	-148.4	-148.4	-148.4
	Air Reactor	96.498	289.5	386	482.49	578.99	688.04
	Cool air reactor exhaust	18.996	56.999	75.985	94.981	113.98	135.4
	Cool flue gas	151.96	151.96	151.96	151.96	151.96	151.96
	Cool OC for air reactor	45.781	45.781	45.781	45.781	45.781	45.781
	Reheat OC for fuel reactor	-47.71	-47.71	-47.71	-47.71	-47.71	-47.71
	Heat steam	-69.8	-69.8	-69.8	-69.8	-69.8	-69.8
	Heat air	-25.82	-77.47	-103.3	-129.1	-154.9	-184.1
	Net	21.473	200.83	290.49	380.16	469.83	571.14

Initial values (continued)	Coal (kg/h)	100	100	100	100	100	100
	Water (kg/h)	140	140	140	140	140	140
	Air Flow Rate (kg/h)	800	900	1000	1100	1200	1500
	Temperature of Fuel Reactor (°C)	950	950	950	950	950	950
	Temperature of Air Reactor (°C)	935	935	935	935	935	935
	Fe_2O_3 flow in the Fuel Reactor (kg/h)	6500	6500	6500	6500	6500	6500
	Al_2O_3 in the System (kg/h)	4500	4500	4500	4500	4500	4500
	Particle Density (kg/m³)	3200	3200	3200	3200	3200	3200
Energy balance (kW) (continued)	Fuel Reactor	-148.4	-148.4	-148.4	-148.4	-148.4	-148.4
	Air Reactor	771.99	868.49	910.81	910.94	911.07	911.49
	Cool air reactor exhaust	151.97	170.97	192.91	217.16	241.41	314.15
	Cool flue gas	151.96	151.96	151.96	151.96	151.96	151.96
	Cool OC for air reactor	45.781	45.781	45.781	45.781	45.781	45.781
	Reheat OC for fuel reactor	-47.71	-47.71	-47.71	-47.71	-47.71	-47.71
	Heat steam	-69.8	-69.8	-69.8	-69.8	-69.8	-69.8
	Heat air	-206.6	-232.4	-258.2	-284.1	-309.9	-387.3
	Net	649.18	738.85	777.29	775.85	774.41	770.1

Table A4: CLC process simulation results for different air flow rates with Colombian coal at 100 kg/h and Fe_2O_3/Al_2O_3 at 7000/5000 kg/h

Initial values	Coal (kg/h)	100	100	100	100	100	100
	Water (kg/h)	140	140	140	140	140	140
	Air Flow Rate (kg/h)	100	300	400	500	600	713
	Temperature of Fuel Reactor (°C)	950	950	950	950	950	950
	Temperature of Air Reactor (°C)	935	935	935	935	935	935
	Fe_2O_3 flow in the Fuel Reactor (kg/h)	7000	7000	7000	7000	7000	7000
	Al_2O_3 in the System (kg/h)	5000	5000	5000	5000	5000	5000
	Particle Density (kg/m³)	3200	3200	3200	3200	3200	3200
Energy balance (kW)	Fuel Reactor	-139.6	-139.6	-139.6	-139.6	-139.6	-139.6
	Air Reactor	96.498	289.5	386	482.49	578.99	688.04
	Cool air reactor exhaust	18.996	56.999	75.985	94.981	113.98	135.4
	Cool flue gas	153.71	153.71	153.71	153.71	153.71	153.71
	Cool OC for air reactor	50.175	50.175	50.175	50.175	50.175	50.175
	Reheat OC for fuel reactor	-52.19	-52.19	-52.19	-52.19	-52.19	-52.19
	Heat steam	-69.8	-69.8	-69.8	-69.8	-69.8	-69.8
	Heat air	-25.82	-77.47	-103.3	-129.1	-154.9	-184.1
	Net	31.917	211.27	300.93	390.6	480.28	581.58
Initial values (continued)	Coal (kg/h)	100	100	100	100	100	100
	Water (kg/h)	140	140	140	140	140	140
	Air Flow Rate (kg/h)	800	900	1000	1100	1200	1500
	Temperature of Fuel Reactor (°C)	950	950	950	950	950	950
	Temperature of Air Reactor (°C)	935	935	935	935	935	935
	Fe_2O_3 flow in the Fuel Reactor (kg/h)	7000	7000	7000	7000	7000	7000
	Al_2O_3 in the System (kg/h)	5000	5000	5000	5000	5000	5000
	Particle Density (kg/m³)	3200	3200	3200	3200	3200	3200
Energy balance (kW) (continued)	Fuel Reactor	-139.6	-139.6	-139.6	-139.6	-139.6	-139.6
	Air Reactor	771.99	868.49	949.41	949.52	949.66	950.07
	Cool air reactor exhaust	151.97	170.97	190.81	215.06	239.31	312.05
	Cool flue gas	153.71	153.71	153.71	153.71	153.71	153.71
	Cool OC for air reactor	50.175	50.175	50.175	50.175	50.175	50.175
	Reheat OC for fuel reactor	-52.19	-52.19	-52.19	-52.19	-52.19	-52.19
	Heat steam	-69.8	-69.8	-69.8	-69.8	-68.8	-69.8
	Heat air	-206.6	-232.4	-258.2	-284.1	-309.9	-387.3
	Net	659.62	749.29	824.23	822.77	822.33	817.02

Table A5: CLC process simulation results for different air flow rates with Colombian coal at 100 kg/h and Fe_2O_3/Al_2O_3 at 7500/5500 kg/h

Initial values	Coal (kg/h)	100	100	100	100	100	100
	Water (kg/h)	140	140	140	140	140	140
	Air Flow Rate (kg/h)	100	300	400	500	600	713
	Temperature of Fuel Reactor (°C)	950	950	950	950	950	950
	Temperature of Air Reactor (°C)	935	935	935	935	935	935
	Fe_2O_3 flow in the Fuel Reactor (kg/h)	7500	7500	7500	7500	7500	7500
	Al_2O_3 in the System (kg/h)	5500	5500	5500	5500	5500	5500
	Particle Density (kg/m³)	3200	3200	3200	3200	3200	3200
Energy balance (kW)	Fuel Reactor	-134.6	-134.6	-134.6	-134.6	-134.6	-134.6
	Air Reactor	96.498	289.5	386	482.49	578.99	688.04
	Cool air reactor exhaust	18.996	56.999	75.985	94.981	113.98	135.4
	Cool flue gas	153.71	153.71	153.71	153.71	153.71	153.71
	Cool OC for air reactor	54.647	54.647	54.647	54.647	54.647	54.647
	Reheat OC for fuel reactor	-56.67	-56.67	-56.67	-56.67	-56.67	-56.67
	Heat steam	-69.8	-69.8	-69.8	-69.8	-69.8	-69.8
	Heat air	-25.82	-77.47	-103.3	-129.1	-154.9	-184.1
	Net	36.942	216.3	305.96	395.63	485.3	586.61
Initial values (continued)	Coal (kg/h)	100	100	100	100	100	100
	Water (kg/h)	140	140	140	140	140	140
	Air Flow Rate (kg/h)	800	900	1000	1100	1200	1500
	Temperature of Fuel Reactor (°C)	950	950	950	950	950	950
	Temperature of Air Reactor (°C)	935	935	935	935	935	935
	Fe_2O_3 flow in the Fuel Reactor (kg/h)	7500	7500	7500	7500	7500	7500
	Al_2O_3 in the System (kg/h)	5500	5500	5500	5500	5500	5500
	Particle Density (kg/m³)	3200	3200	3200	3200	3200	3200
Energy balance (kW) (continued)	Fuel Reactor	-134.6	-134.6	-134.6	-134.6	-134.6	-134.6
	Air Reactor	771.99	868.49	949.41	949.52	949.66	950.07
	Cool air reactor exhaust	151.97	170.97	190.81	215.06	239.31	312.05
	Cool flue gas	153.71	153.71	153.71	153.71	153.71	153.71
	Cool OC for air reactor	54.647	54.647	54.647	54.647	54.647	54.647
	Reheat OC for fuel reactor	-56.67	-56.67	-56.67	-56.67	-56.67	-56.67
	Heat steam	-69.8	-69.8	-69.8	-69.8	-69.8	-69.8
	Heat air	-206.6	-232.4	-258.2	-284.1	-309.9	-387.3
	Net	664.65	754.32	829.25	827.8	826.35	822.04

Table A6: CLC process simulation results for different air flow rates with Colombian coal at 100 kg/h and Fe_2O_3/Al_2O_3 at 8000/6000 kg/h

Initial values	Coal (kg/h)	100	100	100	100	100	100
	Water (kg/h)	140	140	140	140	140	140
	Air Flow Rate (kg/h)	100	300	400	500	600	713
	Temperature of Fuel Reactor (°C)	950	950	950	950	950	950
	Temperature of Air Reactor (°C)	935	935	935	935	935	935
	Fe_2O_3 flow in the Fuel Reactor (kg/h)	8000	8000	8000	8000	8000	8000
	Al_2O_3 in the System (kg/h)	6000	6000	6000	6000	6000	6000
	Particle Density (kg/m³)	3200	3200	3200	3200	3200	3200
Energy balance (kW)	Fuel Reactor	-129.6	-129.6	-129.6	-129.6	-129.6	-129.6
	Air Reactor	96.498	289.5	386	482.49	578.99	688.04
	Cool air reactor exhaust	18.996	56.999	75.985	94.981	113.98	135.4
	Cool flue gas	153.71	153.71	153.71	153.71	153.71	153.71
	Cool OC for air reactor	59.121	59.121	59.121	59.121	59.121	59.121
	Reheat OC for fuel reactor	-61.15	-61.15	-61.15	-61.15	-61.15	-61.15
	Heat steam	-69.8	-69.8	-69.8	-69.8	-69.8	-69.8
	Heat air	-25.82	-77.47	-103.3	-129.1	-154.9	-184.1
	Net	41.968	221.32	310.98	400.66	490.33	591.63
Initial values (continued)	Coal (kg/h)	100	100	100	100	100	100
	Water (kg/h)	140	140	140	140	140	140
	Air Flow Rate (kg/h)	800	900	1000	1100	1200	1500
	Temperature of Fuel Reactor (°C)	950	950	950	950	950	950
	Temperature of Air Reactor (°C)	935	935	935	935	935	935
	Fe_2O_3 flow in the Fuel Reactor (kg/h)	8000	8000	8000	8000	8000	8000
	Al_2O_3 in the System (kg/h)	6000	6000	6000	6000	6000	6000
	Particle Density (kg/m³)	3200	3200	3200	3200	3200	3200
Energy balance (kW) (continued)	Fuel Reactor	-129.6	-129.6	-129.6	-129.6	-129.6	-129.6
	Air Reactor	771.99	868.49	949.41	949.52	949.66	950.07
	Cool air reactor exhaust	151.97	170.97	190.81	215.06	239.31	312.05
	Cool flue gas	153.71	153.71	153.71	153.71	153.71	153.71
	Cool OC for air reactor	59.121	59.121	59.121	59.121	59.121	59.121
	Reheat OC for fuel reactor	-61.15	-61.15	-61.15	-61.15	-61.15	-61.15
	Heat steam	-69.8	-69.8	-69.8	-69.8	-68.8	-69.8
	Heat air	-206.6	-232.4	-258.2	-284.1	-309.9	-387.3
	Net	669.67	759.34	834.28	832.82	832.38	827.07

Table A7: Scaled-up simulation results for different coal feeding rates using the baseline ratios of air flow rate and oxygen carrier flow rate from the experiment of Sahir et al [10]

Initial values	Coal (kg/h)	100	500	1000	1500	2500
	Water (kg/h)	140	700	1400	2100	3500
	Air Flow Rate (kg/h)	713	3565	7130	10695	17825
	Temperature of Fuel Reactor (°C)	950	950	950	950	950
	Temperature of Air Reactor (°C)	935	935	935	935	935
	Fe_2O_3 flow in the Fuel Reactor (kg/h)	5921	30000	60000	90000	150000
	Al_2O_3 in the System (kg/h)	3951	20000	40000	60000	100000
	Particle Density (kg/m³)	3200	3200	3200	3200	3200
Energy balance (KW)	Fuel Reactor	-161.8	-800.8	-1602	-2402	-4.004
	Air Reactor	688	3440.2	6880.4	10321	17.201
	Cool air reactor exhaust	135.4	677.22	1354.4	2031.7	3.3861
	Cool flue gas	148.3	744.09	1488.2	2232.3	3.7205
	Cool OC for air reactor	40.9	207.26	414.52	621.77	1.0363
	Reheat OC for fuel reactor	-42.7	-216.2	-432.3	-648.5	-1.081
	Heat steam	-69.8	-349.1	-698.3	-1047	-1.746
	Heat air	-184.1	-920.6	-1841	-2762	-4.603
	Net	554.2	2782	5564.1	8346.1	13910
Initial values (continued)	Coal (kg/h)	3500	5000	8000	12000	
	Water (kg/h)	4900	7000	11200	16800	
	Air Flow Rate (kg/h)	24955	35650	57040	85560	
	Temperature of Fuel Reactor (°C)	950	950	950	950	
	Temperature of Air Reactor (°C)	935	935	935	935	
	Fe_2O_3 flow in the Fuel Reactor (kg/h)	210000	300000	480000	720000	
	Al_2O_3 in the System (kg/h)	140000	200000	320000	480000	
	Particle Density (kg/m³)	3200	3200	3200	3200	
Energy balance (KW) (continued)	Fuel Reactor	-5.606	-8.008	-12.81	-19.22	
	Air Reactor	24.081	34.402	55.043	82.564	
	Cool air reactor exhaust	4.7405	6.7722	10.836	16.253	
	Cool flue gas	5.2087	7.4409	11.906	17.858	
	Cool OC for air reactor	1.4508	2.0726	3.3161	4.9742	
	Reheat OC for fuel reactor	-1.513	-2.162	-3.459	-5.188	
	Heat steam	-2.444	-3.491	-5.586	-8.379	
	Heat air	-6.444	-9.206	-14.73	-22.09	
	Net	19474	27820	44513	66769	

Table A8: Scaled-up simulation results for different coal feeding rates using the optimum ratios of air flow rate and oxygen carrier flow rate

Initial values	Coal (kg/h)	100	500	1000	1500	2500
	Water (kg/h)	140	700	1400	2100	3500
	Air Flow Rate (kg/h)	1000	5000	10000	15000	25000
	Temperature of Fuel Reactor (°C)	950	950	950	950	950
	Temperature of Air Reactor (°C)	935	935	935	935	935
	Fe_2O_3 flow in the Fuel Reactor (kg/h)	7000	35000	70000	105000	175000
	Al_2O_3 in the System (kg/h)	5000	25000	50000	75000	125000
	Particle Density (kg/m³)	3200	3200	3200	3200	3200
Energy balance (KW)	Fuel Reactor	-139.6	-698.2	-1484	-2226	-3711
	Air Reactor	949.41	4747	9108.1	13662	22770
	Cool air reactor exhaust	190.81	954.06	1929.1	2893.7	4822.9
	Cool flue gas	153.71	768.54	1519.6	2279.4	3799
	Cool OC for air reactor	50.175	250.87	457.81	686.72	1144.5
	Reheat OC for fuel reactor	-52.19	-261	-477.1	-715.7	-1193
	Heat steam	-69.8	-349.1	-698.3	-1047	-1746
	Heat air	-258.2	-1291	-2582	-3873	-6456
	Net	824.23	4121	8242.3	12363	20606
Initial values (continued)	Coal (kg/h)	3500	5000	8000	12000	
	Water (kg/h)	4900	7000	11200	16800	
	Air Flow Rate (kg/h)	35000	50000	80000	120000	
	Temperature of Fuel Reactor (°C)	950	950	950	950	
	Temperature of Air Reactor (°C)	935	935	935	935	
	Fe_2O_3 flow in the Fuel Reactor (kg/h)	245000	350000	560000	840000	
	Al_2O_3 in the System (kg/h)	175000	250000	400000	600000	
	Particle Density (kg/m³)	3200	3200	3200	3200	
Energy balance (KW) (continued)	Fuel Reactor	-5195	-7421	-11874	-17811	
	Air Reactor	31878	45541	72865	109297	
	Cool air reactor exhaust	6752	9645.7	15433	23150	
	Cool flue gas	5318.6	7598.1	12157	18235	
	Cool OC for air reactor	1602.3	2289.1	3662.5	5493.7	
	Reheat OC for fuel reactor	-1670	-2386	-3817	-5725	
	Heat steam	-2444	-3491	-5586	-8379	
	Heat air	-9038	-12912	-20659	-30988	
	Net	28847	41211	65936	98907	

Table A9: CLC process simulation results for different air flow rates with Colombian coal at 12000 kg/h and Fe_2O_3/Al_2O_3 at 780000/540000 kg/h

Initial values	Coal (kg/h)	12000	12000	12000	12000	12000	12000
	Water (kg/h)	16800	16800	16800	16800	16800	16800
	Air Flow Rate (kg/h)	12000	36000	48000	60000	72000	84000
	Temperature of Fuel Reactor (°C)	950	950	950	950	950	950
	Temperature of Air Reactor (°C)	935	935	935	935	935	935
	Fe_2O_3 flow in the Fuel Reactor (kg/h)	780000	780000	780000	780000	780000	780000
	Al_2O_3 in the System (kg/h)	540000	540000	540000	540000	540000	540000
	Particle Density (kg/m³)	3200	3200	3200	3200	3200	3200
Energy balance (KW)	Fuel Reactor	-17811	-17811	-17811	-17811	-17811	-17811
	Air Reactor	11580	34740	46320	57899	69479	81059
	Cool air reactor exhaust	2279.6	6838.7	9118.2	11398	13677	15957
	Cool flue gas	18235	18235	18235	18235	18235	18235
	Cool OC for air reactor	5493.7	5493.7	5493.7	5493.7	5493.7	5493.7
	Reheat OC for fuel reactor	-5725	-5725	-5725	-5725	-5725	-5725
	Heat steam	-8379	-8379	-8379	-8379	-8379	-8379
	Heat air	-3099	-9296	-12395	-15494	-18593	-21692
	Net	2573.3	24095	34855	45616	56376	67137
Initial values (continued)	Coal (kg/h)	12000	12000	12000	12000	12000	12000
	Water (kg/h)	16800	16800	16800	16800	16800	16800
	Air Flow Rate (kg/h)	96000	108000	120000	132000	144000	180000
	Temperature of Fuel Reactor (°C)	950	950	950	950	950	950
	Temperature of Air Reactor (°C)	935	935	935	935	935	935
	Fe_2O_3 flow in the Fuel Reactor (kg/h)	780000	780000	780000	780000	780000	780000
	Al_2O_3 in the System (kg/h)	540000	540000	540000	540000	540000	540000
	Particle Density (kg/m³)	3200	3200	3200	3200	3200	3200
Energy balance (KW) (continued)	Fuel Reactor	-17811	-17811	-17811	-17811	-17811	-17811
	Air Reactor	92639	104219	109297	109313	109329	109379
	Cool air reactor exhaust	18236	20516	23150	26059	28969	37698
	Cool flue gas	18235	18235	18235	18235	18235	18235
	Cool OC for air reactor	5493.7	5493.7	5493.7	5493.7	5493.7	5493.7
	Reheat OC for fuel reactor	-5725	-5725	-5725	-5725	-5725	-5725
	Heat steam	-8379	-8379	-8379	-8379	-8379	-8379
	Heat air	-24790	-27889	-30988	-34087	-37186	-46482
	Net	77898	88658	93271	93098	92925	92408

Table A10: CLC process simulation results for different air flow rates with Colombian coal at 12000 kg/h and Fe_2O_3/Al_2O_3 at 840000/600000 kg/h

Initial values	Coal (kg/h)	12000	12000	12000	12000	12000	12000
	Water (kg/h)	16800	16800	16800	16800	16800	16800
	Air Flow Rate (kg/h)	12000	36000	48000	60000	72000	84000
	Temperature of Fuel Reactor (°C)	950	950	950	950	950	950
	Temperature of Air Reactor (°C)	935	935	935	935	935	935
	Fe_2O_3 flow in the Fuel Reactor (kg/h)	840000	840000	840000	840000	840000	840000
	Al_2O_3 in the System (kg/h)	600000	600000	600000	600000	600000	600000
	Particle Density (kg/m³)	3200	3200	3200	3200	3200	3200
Energy balance (KW)	Fuel Reactor	-16757	-16757	-16757	-16757	-16757	-16757
	Air Reactor	11580	34740	46320	57899	69479	81059
	Cool air reactor exhaust	2279.6	6838.7	9118.2	11398	13677	15957
	Cool flue gas	18445	18445	18445	18445	18445	18445
	Cool OC for air reactor	6021	6021	6021	6021	6021	6021
	Reheat OC for fuel reactor	-6263	-6263	-6263	-6263	-6263	-6263
	Heat steam	-8379	-8379	-8379	-8379	-8379	-8379
	Heat air	-3099	-9296	-12395	-15494	-18593	-21692
	Net	3826.8	25348	36109	46869	57630	68390
Initial values (continued)	Coal (kg/h)	12000	12000	12000	12000	12000	12000
	Water (kg/h)	16800	16800	16800	16800	16800	16800
	Air Flow Rate (kg/h)	96000	108000	120000	132000	144000	180000
	Temperature of Fuel Reactor (°C)	950	950	950	950	950	950
	Temperature of Air Reactor (°C)	935	935	935	935	935	935
	Fe_2O_3 flow in the Fuel Reactor (kg/h)	840000	840000	840000	840000	840000	840000
	Al_2O_3 in the System (kg/h)	600000	600000	600000	600000	600000	600000
	Particle Density (kg/m³)	3200	3200	3200	3200	3200	3200
Energy balance (KW) (continued)	Fuel Reactor	-16757	-16757	-16757	-16757	-16757	-16757
	Air Reactor	92639	104219	113929	113943	113959	114009
	Cool air reactor exhaust	18236	20516	22897	25807	28717	37446
	Cool flue gas	18445	18445	18445	18445	18445	18445
	Cool OC for air reactor	6021	6021	6021	6021	6021	6021
	Reheat OC for fuel reactor	-6263	-6263	-6263	-6263	-6263	-6263
	Heat steam	-8379	-8379	-8379	-8379	-8379	-8379
	Heat air	-24790	-27889	-30988	-34087	-37186	-46482
	Net	79151	89912	98904	98730	98556	98039

Table A11: CLC process simulation results for different air flow rates with Colombian coal at 12000 kg/h and Fe_2O_3/Al_2O_3 at 900000/660000 kg/h

Initial values	Coal (kg/h)	12000	12000	12000	12000	12000	12000
	Water (kg/h)	16800	16800	16800	16800	16800	16800
	Air Flow Rate (kg/h)	12000	36000	48000	60000	72000	84000
	Temperature of Fuel Reactor (°C)	950	950	950	950	950	950
	Temperature of Air Reactor (°C)	935	935	935	935	935	935
	Fe_2O_3 flow in the Fuel Reactor (kg/h)	900000	900000	900000	900000	900000	900000
	Al_2O_3 in the System (kg/h)	660000	660000	660000	660000	660000	660000
	Particle Density (kg/m³)	3200	3200	3200	3200	3200	3200
Energy balance (KW)	Fuel Reactor	-16154	-16154	-16154	-16154	-16154	-16154
	Air Reactor	11580	34740	46320	57899	69479	81059
	Cool air reactor exhaust	2279.6	6838.7	9118.2	11398	13677	15957
	Cool flue gas	18445	18445	18445	18445	18445	18445
	Cool OC for air reactor	6557.7	6557.7	6557.7	6557.7	6557.7	6557.7
	Reheat OC for fuel reactor	-6801	-6801	-6801	-6801	-6801	-6801
	Heat steam	-8379	-8379	-8379	-8379	-8379	-8379
	Heat air	-3099	-9296	-12395	-15494	-18593	-21692
	Net	4429.8	25951	36712	47472	58233	68993
Initial values (continued)	Coal (kg/h)	12000	12000	12000	12000	12000	12000
	Water (kg/h)	16800	16800	16800	16800	16800	16800
	Air Flow Rate (kg/h)	96000	108000	120000	132000	144000	180000
	Temperature of Fuel Reactor (°C)	950	950	950	950	950	950
	Temperature of Air Reactor (°C)	935	935	935	935	935	935
	Fe_2O_3 flow in the Fuel Reactor (kg/h)	900000	900000	900000	900000	900000	900000
	Al_2O_3 in the System (kg/h)	660000	660000	660000	660000	660000	660000
	Particle Density (kg/m³)	3200	3200	3200	3200	3200	3200
Energy balance (KW) (continued)	Fuel Reactor	-16154	-16154	-16154	-16154	-16154	-16154
	Air Reactor	92639	104219	113929	113943	113959	114009
	Cool air reactor exhaust	18236	20516	22897	25807	28717	37446
	Cool flue gas	18445	18445	18445	18445	18445	18445
	Cool OC for air reactor	6557.7	6557.7	6557.7	6557.7	6557.7	6557.7
	Reheat OC for fuel reactor	-6801	-6801	-6801	-6801	-6801	-6801
	Heat steam	-8379	-8379	-8379	-8379	-8379	-8379
	Heat air	-24790	-27889	-30988	-34087	-37186	-46482
	Net	79754	90515	99507	99333	99159	98642

Table A12: CLC process simulation results for different air flow rates with Bituminous coal at 100 kg/h and Fe_2O_3/Al_2O_3 at 5921/3951 kg/h

Initial values	Coal (kg/h)	100	100	100	100	100	100
	Water (kg/h)	140	140	140	140	140	140
	Air Flow Rate (kg/h)	100	300	400	500	600	713
	Temperature of Fuel Reactor (°C)	950	950	950	950	950	950
	Temperature of Air Reactor (°C)	935	935	935	935	935	935
	Fe_2O_3 flow in the Fuel Reactor (kg/h)	5921	5921	5921	5921	5921	5921
	Al_2O_3 in the System (kg/h)	3951	3951	3951	3951	3951	3951
	Particle Density (kg/m³)	3200	3200	3200	3200	3200	3200
Energy balance (kW)	Fuel Reactor	72.119	72.119	72.119	72.119	72.119	72.119
	Air Reactor	96.498	289.497	385.995	482.494	578.993	688
	Cool air reactor exhaust	18.996	56.988	75.985	94.981	113.978	135.4
	Cool flue gas	137.064	137.064	137.064	137.064	137.064	137.064
	Cool OC for air reactor	41.082	41.082	41.082	41.082	41.082	41.082
	Reheat OC for fuel reactor	-42.7	-42.7	-42.7	-42.7	-42.7	-42.7
	Heat steam	-69.8	-69.8	-69.8	-69.8	-69.8	-69.8
	Heat air	-25.823	-77.469	-103.29	-129.12	-154.94	-184.1
	Net	227.436	406.781	496.452	586.124	675.796	777.065
Initial values (continued)	Coal (kg/h)	100	100	100	100	100	100
	Water (kg/h)	140	140	140	140	140	140
	Air Flow Rate (kg/h)	800	900	1000	1100	1200	1500
	Temperature of Fuel Reactor (°C)	950	950	950	950	950	950
	Temperature of Air Reactor (°C)	935	935	935	935	935	935
	Fe_2O_3 flow in the Fuel Reactor (kg/h)	5921	5921	5921	5921	5921	5921
	Al_2O_3 in the System (kg/h)	3951	3951	3951	3951	3951	3951
	Particle Density (kg/m³)	3200	3200	3200	3200	3200	3200
Energy balance (kW) (continued)	Fuel Reactor	72.119	72.119	72.119	72.119	72.119	72.119
	Air Reactor	752.554	752.667	752.812	752.951	753.091	753.515
	Cool air reactor exhaust	153.029	177.277	201.524	225.772	250.02	322.764
	Cool flue gas	137.064	137.064	137.064	137.064	137.064	137.064
	Cool OC for air reactor	41.082	41.082	41.082	41.082	41.082	41.082
	Reheat OC for fuel reactor	-42.7	-42.7	-42.7	-42.7	-42.7	-42.7
	Heat steam	-69.8	-69.8	-69.8	-69.8	-69.8	-69.8
	Heat air	-206.59	-232.41	-258.23	-284.06	-309.88	-387.35
	Net	836.762	835.299	833.868	832.432	830.996	826.695

Table A13: CLC process simulation results for different air flow rates with Anthracite coal at 100 kg/h and Fe_2O_3/Al_2O_3 at 5921/3951 kg/h

Initial values	Coal (kg/h)	100	100	100	100	100	100
	Water (kg/h)	140	140	140	140	140	140
	Air Flow Rate (kg/h)	100	300	400	500	600	713
	Temperature of Fuel Reactor (°C)	950	950	950	950	950	950
	Temperature of Air Reactor (°C)	935	935	935	935	935	935
	Fe_2O_3 flow in the Fuel Reactor (kg/h)	5921	5921	5921	5921	5921	5921
	Al_2O_3 in the System (kg/h)	3951	3951	3951	3951	3951	3951
	Particle Density (kg/m³)	3200	3200	3200	3200	3200	3200
Energy balance (kW)	Fuel Reactor	116.145	116.145	116.145	116.145	116.145	116.145
	Air Reactor	96.498	289.497	385.995	482.494	578.993	688
	Cool air reactor exhaust	18.996	56.988	75.985	94.981	113.978	135.4
	Cool flue gas	127.834	127.834	127.834	127.834	127.834	127.834
	Cool OC for air reactor	41.132	41.132	41.132	41.132	41.132	41.132
	Reheat OC for fuel reactor	-42.7	-42.7	-42.7	-42.7	-42.7	-42.7
	Heat steam	-69.8	-69.8	-69.8	-69.8	-69.8	-69.8
	Heat air	-25.823	-77.469	-103.29	-129.11	-154.94	-184.1
	Net	262.282	441.627	531.298	620.97	710.642	811.911
Initial values (continued)	Coal (kg/h)	100	100	100	100	100	100
	Water (kg/h)	140	140	140	140	140	140
	Air Flow Rate (kg/h)	800	900	1000	1100	1200	1500
	Temperature of Fuel Reactor (°C)	950	950	950	950	950	950
	Temperature of Air Reactor (°C)	935	935	935	935	935	935
	Fe_2O_3 flow in the Fuel Reactor (kg/h)	5921	5921	5921	5921	5921	5921
	Al_2O_3 in the System (kg/h)	3951	3951	3951	3951	3951	3951
	Particle Density (kg/m³)	3200	3200	3200	3200	3200	3200
Energy balance (kW)	Fuel Reactor	116.145	116.145	116.145	116.145	116.145	116.145
	Air Reactor	728.287	728.416	728.553	728.692	728.832	729.257
	Cool air reactor exhaust	154.351	178.599	202.847	227.094	251.342	324.086
(continued)	Cool flue gas	127.834	127.834	127.834	127.834	127.834	127.834
	Cool OC for air reactor	41.132	41.132	41.132	41.132	41.132	41.132
	Reheat OC for fuel reactor	-42.7	-42.7	-42.7	-42.7	-42.7	-42.7
	Heat steam	-69.8	-69.8	-69.8	-69.8	-69.8	-69.8
	Heat air	-206.58	-232.41	-258.23	-284.05	-309.88	-387.34
	Net	848.663	847.216	845.778	844.341	842.905	838.605

Table A14: CLC process simulation results for different air flow rates with Lignite coal at 100 kg/h and Fe_2O_3/Al_2O_3 at 5921/3951 kg/h

Initial values	Coal (kg/h)	100	100	100	100	100	100
	Water (kg/h)	140	140	140	140	140	140
	Air Flow Rate (kg/h)	100	300	400	500	600	713
	Temperature of Fuel Reactor (°C)	950	950	950	950	950	950
	Temperature of Air Reactor (°C)	935	935	935	935	935	935
	Fe_2O_3 flow in the Fuel Reactor (kg/h)	5921	5921	5921	5921	5921	5921
	Al_2O_3 in the System (kg/h)	3951	3951	3951	3951	3951	3951
	Particle Density (kg/m³)	3200	3200	3200	3200	3200	3200
Energy balance (kW)	Fuel Reactor	246.014	246.014	246.014	246.014	246.014	246.014
	Air Reactor	96.498	289.497	385.995	475.591	475.716	475.872
	Cool air reactor exhaust	18.996	56.988	75.985	95.357	119.605	147.005
	Cool flue gas	115.4	115.4	115.4	115.4	115.4	115.4
	Cool OC for air reactor	41.649	41.649	41.649	41.649	41.649	41.649
	Reheat OC for fuel reactor	-42.7	-42.7	-42.7	-42.7	-42.7	-42.7
	Heat steam	-69.8	-69.8	-69.8	-69.8	-69.8	-69.8
	Heat air	-25.823	-77.46	-103.29	-129.11	-154.94	-184.1
	Net	380.234	559.579	649.25	732.395	730.944	729.34
Initial values (continued)	Coal (kg/h)	100	100	100	100	100	100
	Water (kg/h)	140	140	140	140	140	140
	Air Flow Rate (kg/h)	800	900	1000	1100	1200	1500
	Temperature of Fuel Reactor (°C)	950	950	950	950	950	950
	Temperature of Air Reactor (°C)	935	935	935	935	935	935
	Fe_2O_3 flow in the Fuel Reactor (kg/h)	5921	5921	5921	5921	5921	5921
	Al_2O_3 in the System (kg/h)	3951	3951	3951	3951	3951	3951
	Particle Density (kg/m³)	3200	3200	3200	3200	3200	3200
Energy balance (kW) (continued)	Fuel Reactor	246.014	246.014	246.014	246.014	246.014	246.014
	Air Reactor	475.994	476.136	476.278	476.42	476.563	476.992
	Cool air reactor exhaust	168.101	192.349	216.596	240.884	265.092	337.836
	Cool flue gas	115.4	115.4	115.4	115.4	115.4	115.4
	Cool OC for air reactor	41.649	41.649	41.649	41.649	41.649	41.649
	Reheat OC for fuel reactor	-42.7	-42.7	-42.7	-42.7	-42.7	-42.7
	Heat steam	-69.8	-69.8	-69.8	-69.8	-69.8	-69.8
	Heat air	-206.58	-232.41	-258.23	-284.05	-309.88	-387.34
	Net	728.072	726.638	725.204	723.811	722.338	718.042

Table A15: CLC process simulation results for different air flow rates with Bituminous coal at 100 kg/h and Fe_2O_3/Al_2O_3 at 5000/3000 kg/h

Initial values	Coal (kg/h)	100	100	100	100	100	100
	Water (kg/h)	140	140	140	140	140	140
	Air Flow Rate (kg/h)	100	300	400	500	600	713
	Temperature of Fuel Reactor (°C)	950	950	950	950	950	950
	Temperature of Air Reactor (°C)	935	935	935	935	935	935
	Fe_2O_3 flow in the Fuel Reactor (kg/h)	5000	5000	5000	5000	5000	5000
	Al_2O_3 in the System (kg/h)	3000	3000	3000	3000	3000	3000
	Particle Density (kg/m³)	3200	3200	3200	3200	3200	3200
Energy balance (kW)	Fuel Reactor	57.719	57.719	57.719	57.719	57.719	57.719
	Air Reactor	96.498	289.497	385.995	482.494	578.993	688
	Cool air reactor exhaust	18.996	56.988	75.985	94.981	113.978	135.4
	Cool flue gas	134.734	134.734	134.734	134.734	134.734	134.734
	Cool OC for air reactor	32.792	32.792	32.792	32.792	32.792	32.792
	Reheat OC for fuel reactor	-34.273	-34.273	-34.273	-34.273	-34.273	-34.273
	Heat steam	-69.8	-69.8	-69.8	-69.8	-69.8	-69.8
	Heat air	-25.82	-77.46	-103.29	-129.11	-154.94	-184.1
	Net	210.843	390.188	479.859	569.531	659.203	760.472
Initial values (continued)	Coal (kg/h)	100	100	100	100	100	100
	Water (kg/h)	140	140	140	140	140	140
	Air Flow Rate (kg/h)	800	900	1000	1100	1200	1500
	Temperature of Fuel Reactor (°C)	950	950	950	950	950	950
	Temperature of Air Reactor (°C)	935	935	935	935	935	935
	Fe_2O_3 flow in the Fuel Reactor (kg/h)	5000	5000	5000	5000	5000	5000
	Al_2O_3 in the System (kg/h)	3000	3000	3000	3000	3000	3000
	Particle Density (kg/m³)	3200	3200	3200	3200	3200	3200
Energy balance (kW) (continued)	Fuel Reactor	57.719	57.719	57.719	57.719	57.719	57.719
	Air Reactor	700.662	700.794	700.932	701.072	701.213	701.639
	Cool air reactor exhaust	155.856	180.104	204.352	228.6	252.848	325.591
	Cool flue gas	134.734	134.734	134.734	134.734	134.734	134.734
	Cool OC for air reactor	32.792	32.792	32.792	32.792	32.792	32.792
	Reheat OC for fuel reactor	-34.273	-34.273	-34.273	-34.273	-34.273	-34.273
	Heat steam	-69.8	-69.8	-69.8	-69.8	-69.8	-69.8
	Heat air	-206.58	-232.41	-258.23	-284.05	-309.88	-387.34
	Net	771.104	769.66	768.223	766.788	765.353	761.053

Table A16: CLC process simulation results for different air flow rates with Bituminous coal at 100 kg/h and Fe_2O_3/Al_2O_3 at 5500/3500 kg/h

Initial values	Coal (kg/h)	100	100	100	100	100	100
	Water (kg/h)	140	140	140	140	140	140
	Air Flow Rate (kg/h)	100	300	400	500	600	713
	Temperature of Fuel Reactor (°C)	950	950	950	950	950	950
	Temperature of Air Reactor (°C)	935	935	935	935	935	935
	Fe_2O_3 flow in the Fuel Reactor (kg/h)	5500	5500	5500	5500	5500	5500
	Al_2O_3 in the System (kg/h)	3500	3500	3500	3500	3500	3500
	Particle Density (kg/m³)	3200	3200	3200	3200	3200	3200
Energy balance (kW)	Fuel Reactor	67.705	67.705	67.705	67.705	67.705	67.705
	Air Reactor	96.498	289.497	385.995	482.494	578.993	688.037
	Cool air reactor exhaust	18.9963	56.9989	75.985	94.981	113.978	135.4
	Cool flue gas	137.064	137.064	137.064	137.064	137.064	137.064
	Cool OC for air reactor	37.159	37.159	37.159	37.159	37.159	37.159
	Reheat OC for fuel reactor	-38.752	-38.752	-38.752	-38.752	-38.752	-38.752
	Heat steam	-69.8	-69.8	-69.8	-69.8	-69.8	-69.8
	Heat air	-25.82	-77.46	-103.29	-129.11	-154.94	-184.12
	Net	223.047	402.4029	492.063	581.735	671.407	772.693
Initial values (continued)	Coal (kg/h)	100	100	100	100	100	100
	Water (kg/h)	140	140	140	140	140	140
	Air Flow Rate (kg/h)	800	900	1000	1100	1200	1500
	Temperature of Fuel Reactor (°C)	950	950	950	950	950	950
	Temperature of Air Reactor (°C)	935	935	935	935	935	935
	Fe_2O_3 flow in the Fuel Reactor (kg/h)	5500	5500	5500	5500	5500	5500
	Al_2O_3 in the System (kg/h)	3500	3500	3500	3500	3500	3500
	Particle Density (kg/m³)	3200	3200	3200	3200	3200	3200
Energy balance (kW) (continued)	Fuel Reactor	67.705	67.705	67.705	67.705	67.705	67.705
	Air Reactor	752.554	752.677	752.812	752.951	753.091	753.515
	Cool air reactor exhaust	153.029	177.277	201.524	225.773	250.02	322.764
	Cool flue gas	137.064	137.064	137.064	137.064	137.064	137.064
	Cool OC for air reactor	37.159	37.159	37.159	37.159	37.159	37.159
	Reheat OC for fuel reactor	-38.752	-38.752	-38.752	-38.752	-38.752	-38.752
	Heat steam	-69.8	-69.8	-69.8	-69.8	-69.8	-69.8
	Heat air	-206.58	-232.41	-258.23	-284.05	-309.88	-387.34
	Net	832.373	830.92	829.479	828.044	826.607	822.306

Table A17: CLC process simulation results for different air flow rates with Bituminous coal at 100 kg/h and Fe_2O_3/Al_2O_3 at 7000/5000 kg/h

Initial values	Coal (kg/h)	100	100	100	100	100	100
	Water (kg/h)	140	140	140	140	140	140
	Air Flow Rate (kg/h)	100	300	400	500	600	713
	Temperature of Fuel Reactor (°C)	950	950	950	950	950	950
	Temperature of Air Reactor (°C)	935	935	935	935	935	935
	Fe_2O_3 flow in the Fuel Reactor (kg/h)	7000	7000	7000	7000	7000	7000
	Al_2O_3 in the System (kg/h)	5000	5000	5000	5000	5000	5000
	Particle Density (kg/m^3)	3200	3200	3200	3200	3200	3200
Energy balance (kW)	Fuel Reactor	82.8	82.8	82.8	82.8	82.8	82.8
	Air Reactor	96.498	289.497	385.995	482.494	578.993	688.037
	Cool air reactor exhaust	18.9963	56.9989	75.985	94.981	113.978	135.4
	Cool flue gas	137.064	137.064	137.064	137.064	137.064	137.064
	Cool OC for air reactor	50.577	50.577	50.577	50.577	50.577	50.577
	Reheat OC for fuel reactor	-52.192	-52.192	-52.192	-52.192	-52.192	-52.192
	Heat steam	-69.8	-69.8	-69.8	-69.8	-69.8	-69.8
	Heat air	-25.82	-77.46	-103.29	-129.11	-154.94	-184.12
	Net	238.12	417.4759	507.136	596.808	686.48	787.766
Initial values (continued)	Coal (kg/h)	100	100	100	100	100	100
	Water (kg/h)	140	140	140	140	140	140
	Air Flow Rate (kg/h)	800	900	1000	1100	1200	1500
	Temperature of Fuel Reactor (°C)	950	950	950	950	950	950
	Temperature of Air Reactor (°C)	935	935	935	935	935	935
	Fe_2O_3 flow in the Fuel Reactor (kg/h)	7000	7000	7000	7000	7000	7000
	Al_2O_3 in the System (kg/h)	5000	5000	5000	5000	5000	5000
	Particle Density (kg/m^3)	3200	3200	3200	3200	3200	3200
Energy balance (kW)	Fuel Reactor	82.8	82.8	82.8	82.8	82.8	82.8
	Air Reactor	752.554	752.677	752.812	752.951	753.091	753.515
	Cool air reactor exhaust	153.029	177.277	201.524	225.772	250.02	322.764
	Cool flue gas	137.064	137.064	137.064	137.064	137.064	137.064
	Cool OC for air reactor	50.577	50.577	50.577	50.577	50.577	50.577
	Reheat OC for fuel reactor	-52.192	-52.192	-52.192	-52.192	-52.192	-52.192
	Heat steam	-69.8	-69.8	-69.8	-69.8	-69.8	-69.8
	Heat air	-206.58	-232.41	-258.23	-284.05	-309.88	-387.34
	Net	847.446	845.993	844.552	843.116	841.68	837.379

Table A18: CLC process simulation results for different air flow rates with Anthracite coal at 100 kg/h and Fe_2O_3/Al_2O_3 at 5000/3000 kg/h

Initial values	Coal (kg/h)	100	100	100	100	100	100
	Water (kg/h)	140	140	140	140	140	140
	Air Flow Rate (kg/h)	100	300	400	500	600	713
	Temperature of Fuel Reactor (°C)	950	950	950	950	950	950
	Temperature of Air Reactor (°C)	935	935	935	935	935	935
	Fe_2O_3 flow in the Fuel Reactor (kg/h)	5000	5000	5000	5000	5000	5000
	Al_2O_3 in the System (kg/h)	3000	3000	3000	3000	3000	3000
	Particle Density (kg/m^3)	3200	3200	3200	3200	3200	3200
Energy balance (kW)	Fuel Reactor	104.05	104.05	104.05	104.05	104.05	104.05
	Air Reactor	96.50	289.50	386.00	482.49	578.99	688.04
	Cool air reactor exhaust	19.00	56.99	75.99	94.98	113.98	135.44
	Cool flue gas	126.59	126.59	126.59	126.59	126.59	126.59
	Cool OC for air reactor	32.79	32.79	32.79	32.79	32.79	32.79
	Reheat OC for fuel reactor	-34.27	-34.27	-34.27	-34.27	-34.27	-34.27
	Heat steam	-69.80	-69.80	-69.80	-69.80	-69.80	-69.80
	Heat air	-25.82	-77.47	-103.29	-129.12	-154.94	-184.12
	Net	249.03	428.37	518.05	607.72	697.39	798.72
Initial values (continued)	Coal (kg/h)	100	100	100	100	100	100
	Water (kg/h)	140	140	140	140	140	140
	Air Flow Rate (kg/h)	800	900	1000	1100	1200	1500
	Temperature of Fuel Reactor (°C)	950	950	950	950	950	950
	Temperature of Air Reactor (°C)	935	935	935	935	935	935
	Fe_2O_3 flow in the Fuel Reactor (kg/h)	5000	5000	5000	5000	5000	5000
	Al_2O_3 in the System (kg/h)	3000	3000	3000	3000	3000	3000
	Particle Density (kg/m^3)	3200	3200	3200	3200	3200	3200
Energy balance (kW) (continued)	Fuel Reactor	104.05	104.05	104.05	104.05	104.05	104.05
	Air Reactor	700.66	700.79	700.93	701.07	701.21	701.64
	Cool air reactor exhaust	155.86	180.10	204.35	228.60	252.85	325.59
	Cool flue gas	126.59	126.59	126.59	126.59	126.59	126.59
	Cool OC for air reactor	32.79	32.79	32.79	32.79	32.79	32.79
	Reheat OC for fuel reactor	-34.27	-34.27	-34.27	-34.27	-34.27	-34.27
	Heat steam	-69.80	-69.80	-69.80	-69.80	-69.80	-69.80
	Heat air	-206.59	-232.41	-258.23	-284.06	-309.88	-387.35
	Net	809.29	807.85	806.41	804.97	803.54	799.24

Table A19: CLC process simulation results for different air flow rates with Anthracite coal at 100 kg/h and Fe_2O_3/Al_2O_3 at 5200/3200 kg/h

Initial values	Coal (kg/h)	100	100	100	100	100	100
	Water (kg/h)	140	140	140	140	140	140
	Air Flow Rate (kg/h)	100	300	400	500	600	713
	Temperature of Fuel Reactor (°C)	950	950	950	950	950	950
	Temperature of Air Reactor (°C)	935	935	935	935	935	935
	Fe_2O_3 flow in the Fuel Reactor (kg/h)	5200	5200	5200	5200	5200	5200
	Al_2O_3 in the System (kg/h)	3200	3200	3200	3200	3200	3200
	Particle Density (kg/m³)	3200	3200	3200	3200	3200	3200
Energy balance (kW)	Fuel Reactor	108.71	108.71	108.71	108.71	108.71	108.71
	Air Reactor	96.50	289.50	386.00	482.49	578.99	688.04
	Cool air reactor exhaust	19.00	56.99	75.99	94.98	113.98	135.44
	Cool flue gas	127.83	127.83	127.83	127.83	127.83	127.83
	Cool OC for air reactor	34.53	34.53	34.53	34.53	34.53	34.53
	Reheat OC for fuel reactor	-36.07	-36.07	-36.07	-36.07	-36.07	-36.07
	Heat steam	-69.80	-69.80	-69.80	-69.80	-69.80	-69.80
	Heat air	-25.82	-77.47	-103.29	-129.12	-154.94	-184.12
	Net	254.88	434.22	523.89	613.57	703.24	804.57
Initial values (continued)	Coal (kg/h)	100	100	100	100	100	100
	Water (kg/h)	140	140	140	140	140	140
	Air Flow Rate (kg/h)	800	900	1000	1100	1200	1500
	Temperature of Fuel Reactor (°C)	950	950	950	950	950	950
	Temperature of Air Reactor (°C)	935	935	935	935	935	935
	Fe_2O_3 flow in the Fuel Reactor (kg/h)	5200	5200	5200	5200	5200	5200
	Al_2O_3 in the System (kg/h)	3200	3200	3200	3200	3200	3200
	Particle Density (kg/m³)	3200	3200	3200	3200	3200	3200
Energy balance (kW) (continued)	Fuel Reactor	108.71	108.71	108.71	108.71	108.71	108.71
	Air Reactor	728.29	728.42	728.55	728.69	728.83	729.26
	Cool air reactor exhaust	154.35	178.60	202.85	227.09	251.34	324.09
	Cool flue gas	127.83	127.83	127.83	127.83	127.83	127.83
	Cool OC for air reactor	34.53	34.53	34.53	34.53	34.53	34.53
	Reheat OC for fuel reactor	-36.07	-36.07	-36.07	-36.07	-36.07	-36.07
	Heat steam	-69.80	-69.80	-69.80	-69.80	-69.80	-69.80
	Heat air	-206.59	-232.41	-258.23	-284.06	-309.88	-387.35
	Net	841.26	839.81	838.37	836.94	835.50	831.20

Table A20: CLC process simulation results for different air flow rates with Anthracite coal at 100 kg/h and Fe_2O_3/Al_2O_3 at 5500/3500 kg/h

Initial values	Coal (kg/h)	100	100	100	100	100	100
	Water (kg/h)	140	140	140	140	140	140
	Air Flow Rate (kg/h)	100	300	400	500	600	713
	Temperature of Fuel Reactor (°C)	950	950	950	950	950	950
	Temperature of Air Reactor (°C)	935	935	935	935	935	935
	Fe_2O_3 flow in the Fuel Reactor (kg/h)	5500	5500	5500	5500	5500	5500
	Al_2O_3 in the System (kg/h)	3500	3500	3500	3500	3500	3500
	Particle Density (kg/m³)	3200	3200	3200	3200	3200	3200
Energy balance (kW)	Fuel Reactor	111.73	111.73	111.73	111.73	111.73	111.73
	Air Reactor	96.50	289.50	386.00	482.49	578.99	688.04
	Cool air reactor exhaust	19.00	56.99	75.99	94.98	113.98	135.40
	Cool flue gas	127.83	127.83	127.83	127.83	127.83	127.83
	Cool OC for air reactor	37.03	37.03	37.03	37.03	37.03	37.03
	Reheat OC for fuel reactor	-38.75	-38.75	-38.75	-38.75	-38.75	-38.75
	Heat steam	-69.80	-69.80	-69.80	-69.80	-69.80	-69.80
	Heat air	-25.82	-77.47	-103.29	-129.12	-154.94	-184.12
	Net	257.71	437.06	526.73	616.40	706.07	807.36
Initial values (continued)	Coal (kg/h)	100	100	100	100	100	100
	Water (kg/h)	140	140	140	140	140	140
	Air Flow Rate (kg/h)	800	900	1000	1100	1200	1500
	Temperature of Fuel Reactor (°C)	950	950	950	950	950	950
	Temperature of Air Reactor (°C)	935	935	935	935	935	935
	Fe_2O_3 flow in the Fuel Reactor (kg/h)	5500	5500	5500	5500	5500	5500
	Al_2O_3 in the System (kg/h)	3500	3500	3500	3500	3500	3500
	Particle Density (kg/m³)	3200	3200	3200	3200	3200	3200
Energy balance (kW) (continued)	Fuel Reactor	111.73	111.73	111.73	111.73	111.73	111.73
	Air Reactor	728.29	728.42	728.55	728.69	728.83	729.26
	Cool air reactor exhaust	154.35	178.60	202.85	227.09	251.34	324.09
	Cool flue gas	127.83	127.83	127.83	127.83	127.83	127.83
	Cool OC for air reactor	37.03	37.03	37.03	37.03	37.03	37.03
	Reheat OC for fuel reactor	-38.75	-38.75	-38.75	-38.75	-38.75	-38.75
	Heat steam	-69.80	-69.80	-69.80	-69.80	-69.80	-69.80
	Heat air	-206.59	-232.41	-258.23	-284.06	-309.88	-387.35
	Net	844.09	842.65	841.21	839.77	838.34	834.04

Table A21: CLC process simulation results for different air flow rates with Lignite coal at 100 kg/h and Fe_2O_3/Al_2O_3 at 3000/1000 kg/h

Initial values	Coal (kg/h)	100	100	100	100	100	100
	Water (kg/h)	140	140	140	140	140	140
	Air Flow Rate (kg/h)	100	300	400	500	600	713
	Temperature of Fuel Reactor (°C)	950	950	950	950	950	950
	Temperature of Air Reactor (°C)	935	935	935	935	935	935
	Fe_2O_3 flow in the Fuel Reactor (kg/h)	3000	3000	3000	3000	3000	3000
	Al_2O_3 in the System (kg/h)	1000	1000	1000	1000	1000	1000
	Particle Density (kg/m³)	3200	3200	3200	3200	3200	3200
Energy balance (kW)	Fuel Reactor	212.04	212.04	212.04	212.04	212.04	212.04
	Air Reactor	96.50	289.50	386.00	420.42	420.60	420.72
	Cool air reactor exhaust	19.00	56.99	75.99	98.36	122.61	150.01
	Cool flue gas	113.10	113.10	113.10	113.10	113.10	113.10
	Cool OC for air reactor	15.47	15.47	15.47	15.47	15.47	15.47
	Reheat OC for fuel reactor	-16.35	-16.35	-16.35	-16.35	-16.35	-16.35
	Heat steam	-69.80	-69.80	-69.80	-69.80	-69.80	-69.80
	Heat air	-25.82	-77.47	-103.29	-129.12	-154.94	-184.10
	Net	344.13	523.48	613.15	644.13	642.73	641.09
Initial values (continued)	Coal (kg/h)	100	100	100	100	100	100
	Water (kg/h)	140	140	140	140	140	140
	Air Flow Rate (kg/h)	800	900	1000	1100	1200	1500
	Temperature of Fuel Reactor (°C)	950	950	950	950	950	950
	Temperature of Air Reactor (°C)	935	935	935	935	935	935
	Fe_2O_3 flow in the Fuel Reactor (kg/h)	3000	3000	3000	3000	3000	3000
	Al_2O_3 in the System (kg/h)	1000	1000	1000	1000	1000	1000
	Particle Density (kg/m³)	3200	3200	3200	3200	3200	3200
Energy balance (kW)	Fuel Reactor	212.04	212.04	212.04	212.04	212.04	212.04
	Air Reactor	420.84	420.98	421.13	421.27	421.41	421.84
	Cool air reactor exhaust	171.11	195.36	219.60	243.85	268.10	340.84
(continued)	Cool flue gas	113.10	113.10	113.10	113.10	113.10	113.10
	Cool OC for air reactor	15.47	15.47	15.47	15.47	15.47	15.47
	Reheat OC for fuel reactor	-16.35	-16.35	-16.35	-16.35	-16.35	-16.35
	Heat steam	-69.80	-69.80	-69.80	-69.80	-69.80	-69.80
	Heat air	-206.59	-232.41	-258.23	-284.06	-309.88	-387.35
	Net	639.82	638.39	636.96	635.52	634.09	629.79

Table A22: CLC process simulation results for different air flow rates with Lignite coal at 100 kg/h and Fe_2O_3/Al_2O_3 at 3500/1500 kg/h

Initial values	Coal (kg/h)	100	100	100	100	100	100
	Water (kg/h)	140	140	140	140	140	140
	Air Flow Rate (kg/h)	100	300	400	500	600	713
	Temperature of Fuel Reactor (°C)	950	950	950	950	950	950
	Temperature of Air Reactor (°C)	935	935	935	935	935	935
	Fe_2O_3 flow in the Fuel Reactor (kg/h)	3500	3500	3500	3500	3500	3500
	Al_2O_3 in the System (kg/h)	1500	1500	1500	1500	1500	1500
	Particle Density (kg/m³)	3200	3200	3200	3200	3200	3200
Energy balance (kW)	Fuel Reactor	221.47	221.47	221.47	221.47	221.47	221.47
	Air Reactor	96.50	289.50	386.00	475.59	475.72	475.87
	Cool air reactor exhaust	19.00	56.99	75.99	95.36	119.61	147.01
	Cool flue gas	115.40	115.40	115.40	115.40	115.40	115.40
	Cool OC for air reactor	19.83	19.83	19.83	19.83	19.83	19.83
	Reheat OC for fuel reactor	-20.83	-20.83	-20.83	-20.83	-20.83	-20.83
	Heat steam	-69.80	-69.80	-69.80	-69.80	-69.80	-69.80
	Heat air	-25.82	-77.47	-103.29	-129.12	-154.94	-184.10
	Net	355.74	535.09	624.76	707.91	706.45	704.85
Initial values (continued)	Coal (kg/h)	100	100	100	100	100	100
	Water (kg/h)	140	140	140	140	140	140
	Air Flow Rate (kg/h)	800	900	1000	1100	1200	1500
	Temperature of Fuel Reactor (°C)	950	950	950	950	950	950
	Temperature of Air Reactor (°C)	935	935	935	935	935	935
	Fe_2O_3 flow in the Fuel Reactor (kg/h)	3500	3500	3500	3500	3500	3500
	Al_2O_3 in the System (kg/h)	1500	1500	1500	1500	1500	1500
	Particle Density (kg/m³)	3200	3200	3200	3200	3200	3200
Energy balance (kW) (continued)	Fuel Reactor	221.47	221.47	221.47	221.47	221.47	221.47
	Air Reactor	475.99	476.14	476.28	476.42	476.56	476.99
	Cool air reactor exhaust	168.10	192.35	216.60	240.88	265.09	337.84
	Cool flue gas	115.40	115.40	115.40	115.40	115.40	115.40
	Cool OC for air reactor	19.83	19.83	19.83	19.83	19.83	19.83
	Reheat OC for fuel reactor	-20.83	-20.83	-20.83	-20.83	-20.83	-20.83
	Heat steam	-69.80	-69.80	-69.80	-69.80	-69.80	-69.80
	Heat air	-206.59	-232.41	-258.23	-284.06	-309.88	-387.35
	Net	703.58	702.15	700.71	699.32	697.85	693.55

Table A23: CLC process simulation results for different air flow rates with Lignite coal at 100 kg/h and Fe_2O_3/Al_2O3 at 4000/2000 kg/h

Initial values	Coal (kg/h)	100	100	100	100	100	100
	Water (kg/h)	140	140	140	140	140	140
	Air Flow Rate (kg/h)	100	300	400	500	600	713
	Temperature of Fuel Reactor (°C)	950	950	950	950	950	950
	Temperature of Air Reactor (°C)	935	935	935	935	935	935
	Fe_2O_3 flow in the Fuel Reactor (kg/h)	4000	4000	4000	4000	4000	4000
	Al_2O_3 in the System (kg/h)	2000	2000	2000	2000	2000	2000
	Particle Density (kg/m³)	3200	3200	3200	3200	3200	3200
Energy balance (kW)	Fuel Reactor	226.51	226.51	226.51	226.51	226.51	226.51
	Air Reactor	96.50	289.50	386.00	475.59	475.72	475.87
	Cool air reactor exhaust	19.00	56.99	75.99	95.36	119.61	147.01
	Cool flue gas	115.40	115.40	115.40	115.40	115.40	115.40
	Cool OC for air reactor	24.31	24.31	24.31	24.31	24.31	24.31
	Reheat OC for fuel reactor	-25.31	-25.31	-25.31	-25.31	-25.31	-25.31
	Heat steam	-69.80	-69.80	-69.80	-69.80	-69.80	-69.80
	Heat air	-25.82	-77.47	-103.29	-129.12	-154.94	-184.10
	Net	360.77	540.11	629.79	712.93	711.48	709.88
Initial values (continued)	Coal (kg/h)	100	100	100	100	100	100
	Water (kg/h)	140	140	140	140	140	140
	Air Flow Rate (kg/h)	800	900	1000	1100	1200	1500
	Temperature of Fuel Reactor (°C)	950	950	950	950	950	950
	Temperature of Air Reactor (°C)	935	935	935	935	935	935
	Fe_2O_3 flow in the Fuel Reactor (kg/h)	4000	4000	4000	4000	4000	4000
	Al_2O_3 in the System (kg/h)	2000	2000	2000	2000	2000	2000
	Particle Density (kg/m³)	3200	3200	3200	3200	3200	3200
Energy balance (kW) (continued)	Fuel Reactor	226.51	226.51	226.51	226.51	226.51	226.51
	Air Reactor	475.99	476.14	476.28	476.42	476.56	476.99
	Cool air reactor exhaust	168.10	192.35	216.60	240.88	265.09	337.84
	Cool flue gas	115.40	115.40	115.40	115.40	115.40	115.40
	Cool OC for air reactor	24.31	24.31	24.31	24.31	24.31	24.31
	Reheat OC for fuel reactor	-25.31	-25.31	-25.31	-25.31	-25.31	-25.31
	Heat steam	-69.80	-69.80	-69.80	-69.80	-69.80	-69.80
	Heat air	-206.59	-232.41	-258.23	-284.06	-309.88	-387.35
	Net	708.61	707.17	705.74	704.35	702.87	698.58

Table A24: CLC process simulation results for different air flow rates with Lignite coal at 100 kg/h and Fe_2O_3/Al_2O_3 at 4500/2500 kg/h

Initial values	Coal (kg/h)	100	100	100	100	100	100
	Water (kg/h)	140	140	140	140	140	140
	Air Flow Rate (kg/h)	100	300	400	500	600	713
	Temperature of Fuel Reactor (°C)	950	950	950	950	950	950
	Temperature of Air Reactor (°C)	935	935	935	935	935	935
	Fe_2O_3 flow in the Fuel Reactor (kg/h)	4500	4500	4500	4500	4500	4500
	Al_2O_3 in the System (kg/h)	2500	2500	2500	2500	2500	2500
	Particle Density (kg/m³)	3200	3200	3200	3200	3200	3200
Energy balance (kW)	Fuel Reactor	231.54	231.54	231.54	231.54	231.54	231.54
	Air Reactor	96.50	289.50	386.00	475.59	475.72	475.87
	Cool air reactor exhaust	19.00	56.99	75.99	95.36	119.61	147.01
	Cool flue gas	115.40	115.40	115.40	115.40	115.40	115.40
	Cool OC for air reactor	28.78	28.78	28.78	28.78	28.78	28.78
	Reheat OC for fuel reactor	-29.79	-29.79	-29.79	-29.79	-29.79	-29.79
	Heat steam	-69.80	-69.80	-69.80	-69.80	-69.80	-69.80
	Heat air	-25.82	-77.47	-103.29	-129.12	-154.94	-184.10
	Net	365.79	545.14	634.81	717.96	716.50	714.90
Initial values (continued)	Coal (kg/h)	100	100	100	100	100	100
	Water (kg/h)	140	140	140	140	140	140
	Air Flow Rate (kg/h)	800	900	1000	1100	1200	1500
	Temperature of Fuel Reactor (°C)	950	950	950	950	950	950
	Temperature of Air Reactor (°C)	935	935	935	935	935	935
	Fe_2O_3 flow in the Fuel Reactor (kg/h)	4500	4500	4500	4500	4500	4500
	Al_2O_3 in the System (kg/h)	2500	2500	2500	2500	2500	2500
	Particle Density (kg/m³)	3200	3200	3200	3200	3200	3200
Energy balance (kW) (continued)	Fuel Reactor	231.54	231.54	231.54	231.54	231.54	231.54
	Air Reactor	475.99	476.14	476.28	476.42	476.56	476.99
	Cool air reactor exhaust	168.10	192.35	216.60	240.88	265.09	337.84
	Cool flue gas	115.40	115.40	115.40	115.40	115.40	115.40
	Cool OC for air reactor	28.78	28.78	28.78	28.78	28.78	28.78
	Reheat OC for fuel reactor	-29.79	-29.79	-29.79	-29.79	-29.79	-29.79
	Heat steam	-69.80	-69.80	-69.80	-69.80	-69.80	-69.80
	Heat air	-206.59	-232.41	-258.23	-284.06	-309.88	-387.35
	Net	713.63	712.20	710.76	709.37	707.90	703.60

REFERENCE

1. Cuadrat A., Abad A., de Diego L.F. (2012) Prompt considerations on the design of chemicallooping combustion of coal from experimental tests, Fuel 97, pp. 219-232.
2. Wang J., Anthony E.J. (2008) Clean combustion of solid fuels, Applied Energy 85 (2), pp. 73-79.
3. Gnanapragasam N.V., Reddy B.V., Rosen M.A. (2009) Hydrogen production from coal using coal direct chemical looping and syngas chemical looping combustion systems: assessment of system operation and resource requirements, Int. J. Hydrogen Energy 34 (6), pp. 2606-2615.
4. Adánez J., Gayán P., Celaya J. (2006) Chemical looping combustion in a 10 kWth prototype using a CuO/Al2O3 oxygen carrier: Effect of operating conditions on methane combustion, Industrial & Engineering Chemistry Research 45 (17), pp. 6075-6080.
5. Arjmand M., Azad A.M., Leion H., Lyngfelt A., Mattisson T. (2011) Prospects of Al2O3 and MgAl2O4-supported CuO oxygen carriers in chemical-looping combustion (CLC) and chemicallooping with oxygen uncoupling (CLOU), Energy & Fuels 25 (11), pp. 5493-5502.
6. Leion H., Lyngfelt A., Johansson M., Jerndal E., Mattisson T. (2008) The use of ilmenite as an oxygen carrier in chemical-looping combustion, Chemical Engineering Research and Design 86 (9), pp. 1017-1026.
7. Leion H., Mattisson T., Lyngfelt A. (2009) Using chemical-looping with oxygen uncoupling (CLOU) for combustion of six different solid fuels, Energy Procedia 1 (1), pp. 447-453.
8. Mattisson T., Lyngfelt A., Leion H. (2009) Chemical-looping with oxygen uncoupling for combustion of solid fuels, Int. J. Greenhouse Gas Control 3 (1), pp. 11-19.
9. [9] Cuadrat A., Abad A., Adánez J., de Diego L.F. (2012) Behavior of ilmenite as oxygen carrier in chemical-looping combustion, Fuel Processing Technology 94 (1), pp. 101-112.
10. Sahir A.H., Cadore A.L., Dansie J.K. (2012) Process analysis of chemical looping with oxygen uncoupling (CLOU) and chemical looping combustion (CLC) for solid fuels, 2nd International Conference on Chemical looping, Darmstadt, Germany.
11. Mukherjee S., Kumar P., Hosseini A. (2014) Comparative assessment of gasification based coal power plants with various CO2 capture technologies producing electricity and hydrogen Energy & Fuels 28 (2), pp. 1028-1040.

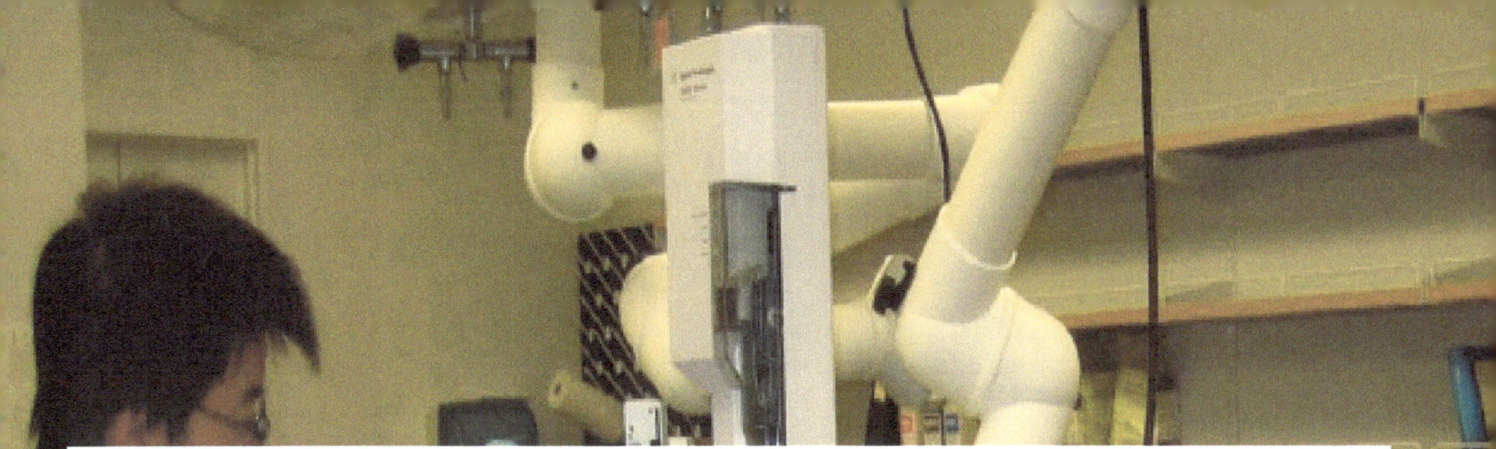

CHAPTER 14

VOLATILE CONSTITUENTS IN RADIX FLEMINGIAE MACROPHYLLAE AND A SUBSTITUTE BY GAS CHROMATOGRAPHY-MASS SPECTROMETRY AND CHEMOMETRIC METHODS

Shen-Yu Cheng, Yan Xie, Xiao-Liang Feng and Lan-Fang Huang

Keywords: Radix Flemingiae Macrophyllae; chemometrics; gas chromatography-mass spectrometry; subwindow factor analysis; spectral correlative chromatography; correlative constituent

INTRODUCTION

Radix Flemingiae Macrophyllae, belongs to the leguminous plant family. As a traditional Chinese medicine (TCM) listed in the Chinese Pharmacopoeia, it has been widely used to cure rheumatism, lumbar muscle strains, myasthenia of limbs, injuries from falls, swelling, sore throats, etc. [1–3]. Although Radix Flemingiae Macrophyllae contains tens or even hundreds of compounds, only a limited number of compounds, such as flavonoids and the volatile constituents, might be responsible for the pharmaceutical or toxic effects [4–6]. For example, the volatile constituents are known to exhibit pharmacological and biological activity, and are used for the therapy of various liver and gall bladder diseases [7]. In recent years, because Radix Flemingiae Macrophyllae has been widely harvested for a long time, its supplies have become scarce and many fakes or substitutes, such as Radix Flemingiae Latifolia Benth, appeared on the market. To ensure the reliability and repeatability of pharmacological and clinical research and understand their bioactivities and possible side effects of active compounds, it is necessary to study all of the phytochemical constituents of botanical extracts and develop a method for quality control of Radix Flemingiae Macrophyllae.

Analysis of the volatile compounds in Radix Flemingiae Macrophyllae is usually performed with gas chromatography (GC) and gas chromatography–mass spectrometry (GC-MS) [8,9]. Because the composition of Radix Flemingiae Macrophyllae is very complicated and the concentrations of many important volatile components in Radix Flemingiae Macrophyllae are very low, suitable sample-preparation methods are necessary before detection by GC-MS, such as steam distillation [8,9]. However, although sample preparation methods are used for the complicated Radix Flemingiae Macrophyllae samples, it is still difficult to achieve a complete separation unless rigorous conditions are imposed on the chromatographic separation process. In these reports, the qualitative and quantitative analysis of the volatile components determined by gas chromatography–mass spectrometry (GC-MS) is based on retention index of gas chromatography and mass spectra. However, it is difficult to assess the purity of chromatographic peaks by general GC and the peaks inspected as one component may be mixtures of several components, so the results obtained by the methods mentioned above would be questionable. Therefore, development of a simple and reliable method for the determination of the volatile constituents in Radix Flemingiae Macrophyllae and its substitutes is

Citation: Shen-Yu Cheng, Yan Xie, Xiao-Liang Feng and Lan-Fang Huang, Volatile Constituents in Radix Flemingiae Macrophyllae and a Substitute by Gas Chromatography-Mass Spectrometry and Chemometric Methods, doi:10.3390/molecules171214111

necessary. Fortunately, many associated chemometric methods [10–16] have been developed to provide more information for chemical analysis both in chromatographic separation and in spectral identification, which makes it possible to interpret these complex systems [17,18].

In this work, combined chemometrics methods, subwindow factor analysis (SFA) [14] and spectral correlative chromatography (SCC) [15], are used for analysis of the volatile constituents in Radix Flemingiae Macrophyllae and Radix Flemingiae Latifolia Benth., one of its substitutes. Then, a simple, reliable and reproducible procedure to reveal the sameness and differences between Radix Flemingiae Macrophyllae and its substitutes is developed. After extraction by the water distillation method, the volatile components in Radix Flemingiae Macrophyllae and Radix Flemingiae Latifolia Benth. were detected by GC-MS under appropriate conditions. The qualitative and quantitative analysis of volatile components in Radix Flemingiae Macrophyllae was completed when subwindow factor analysis was used as an auxiliary means. Then, spectral correlative chromatography (SCC) was used to extract the correlative constituent from Radix Flemingiae Latifolia Benth. The obtained results proved the power of the developed approach for the analysis of complex samples.

RESULTS AND DISCUSSION

Resolution of Overlapped Peaks with SFA

The total ion chromatogram (TIC) of the volatile components in Radix Flemingiae Macrophyllae is shown in Figure 1. Apparently, it is indeed a very complicated system, and the majority of the peaks are grouped with retention times within the 12.50–25.00 min. range. If we searched in the NIST mass database directly without further data processing, incorrect results or impossible identification would result obtained. The main reason is that although the chromatographic separation conditions were optimized, some of the eluting components overlap with one another, as indicated by the fact that different mass spectra were obtained at different positions in the peaks when we searched in the NIST database, and the furthermore, the concentrations of many volatile components are very low. In order to obtain reliable qualitative and qualitative results of the volatile components in Radix Flemingiae Macrophyllae, it was necessary for the overlapped peaks and the components with low content to be resolved into pure spectra and chromatograms using combined chemometrics resolution methods. Subwindow factor analysis (SFA) [14], a common used chemometric method for solving these problem, was used for this purpose.

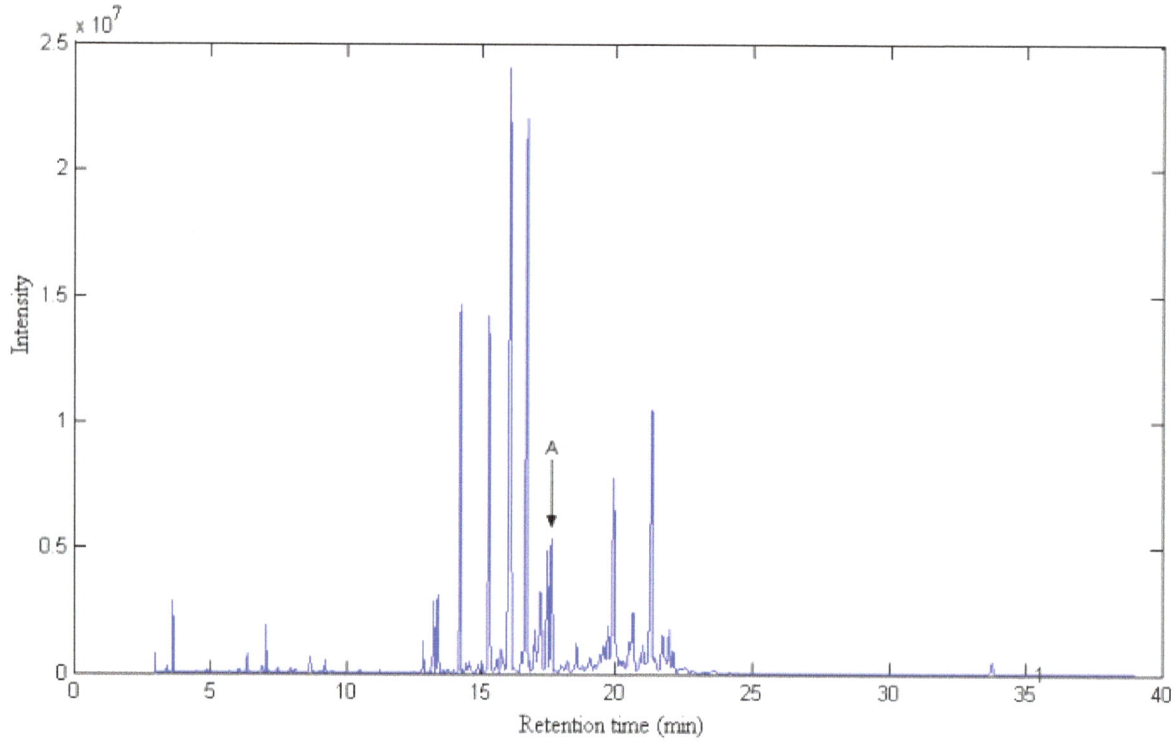

Figure 1: The total ion chromatogram (TIC) of the volatile components in Radix Flemingiae Macrophyllae.

The TIC of peak cluster A (16.90–17.60 min) in Figure 1 is taken as an example to demonstrate the whole procedure of this approach. Figure 2 shows the total ion chromatogram (TIC) of this peak cluster, which seems to be a three component overlapped peak, but the diverse spectra at different parts of the cluster indicates there are actually more than three components in the cluster. A satisfactory result can't be obtained by the usual direct search in the mass spectrum library. Of course, the quantitative analysis of this cluster can't be completed because the area of each component can't be determined. Thus, sub-window factor analysis is used to extract the pure spectra and pure chromatogram.

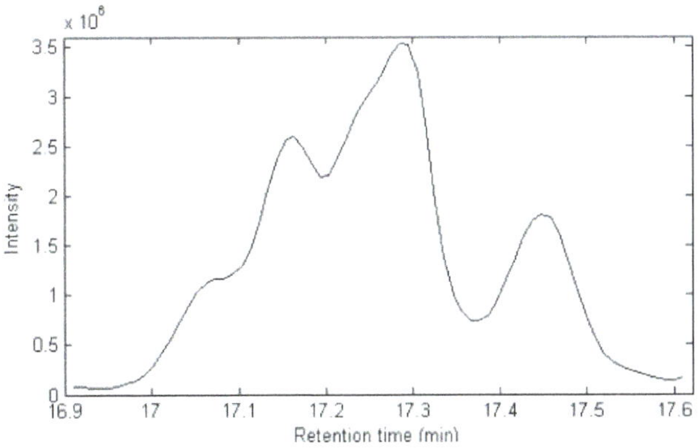

Figure 2: The total ion Chromatogram(TIC) of the peak cluster A.

After background and baseline shift in the raw dimensional data are preprocessed by the appropriate method [13], the elution sequences of individual components (rank map of peak A) can be estimated by FSMWEFA [16], which is shown in Figure 3. It can be seen there are seven components in this peak cluster according to the rank estimation method, which are marked as components 1–7 according to elution sequence. By analyzing the correlation of two subwindows, the pure spectrum of each component can be extracted by SFA directly without previous resolution of their concentration profiles. The resolved mass spectrum of component 1 is shown in Figure 4.

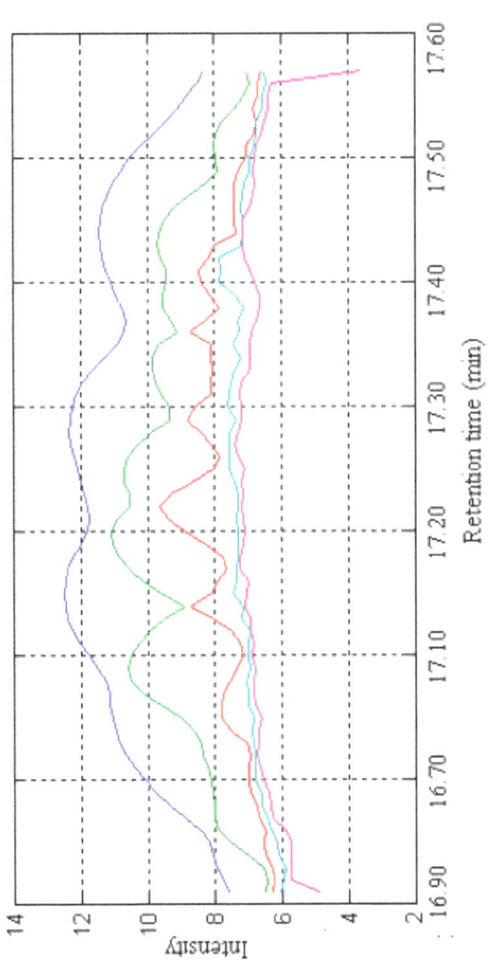

Figure 3: The evolving eigenvalues of peak A obtained using FSMWEFA with a window size of 5.

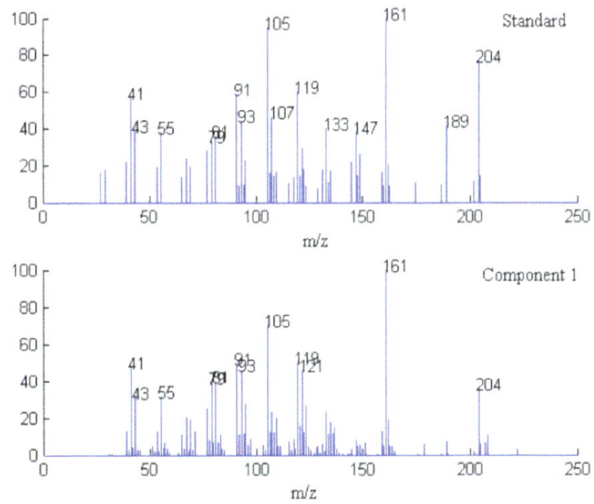

Figure 4: Resolved mass spectrum of component 1 by SFA and standard mass spectrum of β-guaiene.

In the same way, the spectra of the other components were obtained, and the corresponding resolved mass spectra of each component are shown in Figures 5–8. After all the pure spectra were obtained, the concentration profiles could be achieved by using the prior information of the spectra and the linear regression $C = XS(S^T S)^{-1}$, as shown in Figure 9.

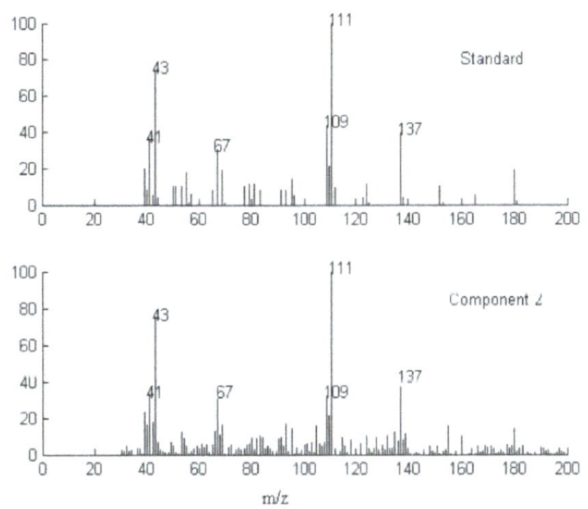

Figure 5: Resolved mass spectrum of component 2 by SFA and standard mass spectrum of dihydroactinidiolide.

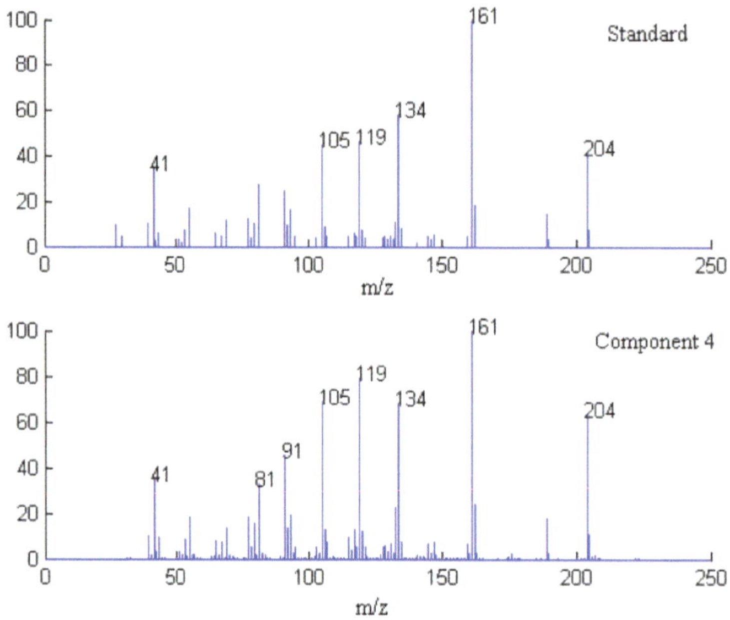

Figure 6: Resolved mass spectrum of component 4 by SFA and standard mass spectrum of δ-cadinene.

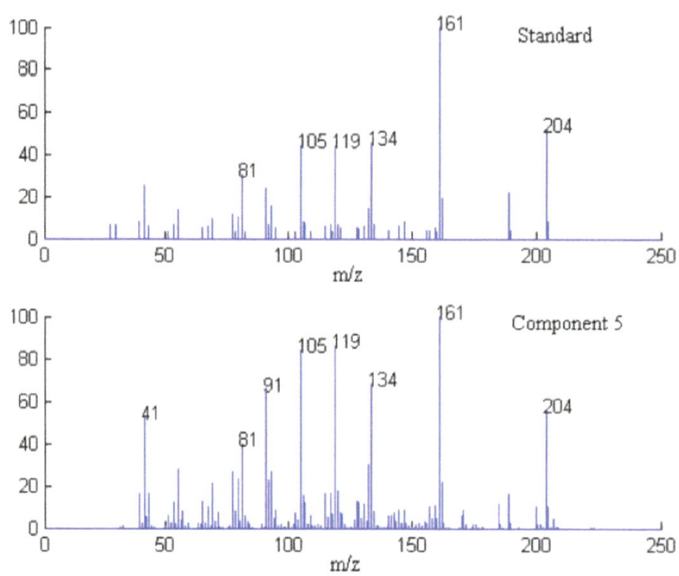

Figure 7: Resolved mass spectrum of component 5 by SFA and standard mass spectrum of β-cadinece.

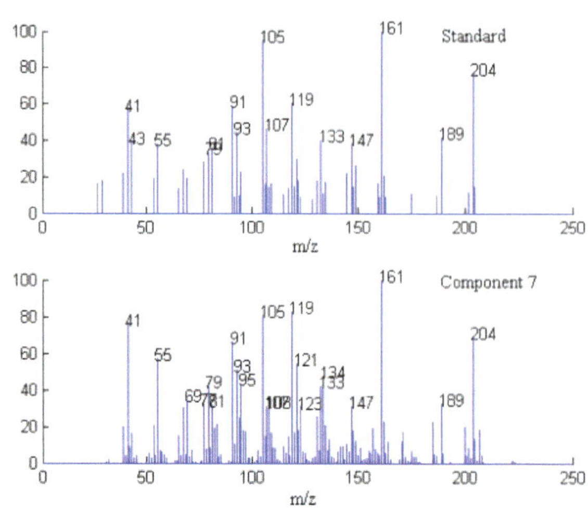

Figure 8: Resolved mass spectrum of component 7 by SFA and standard mass spectrum of valencene.

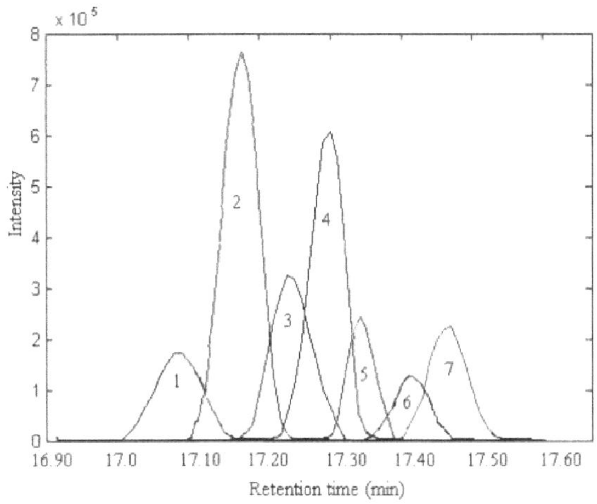

Figure 9: Resolved chromatographic profiles of peak cluster A.

Finally, when each pure spectrum was extracted and the resolved chromatographic profiles of these seven components were obtained, their identification can be done by similarity searches in the NIST mass spectral database. Components 1, 2, 4, 5, and 7 in this cluster may be β-guaiene, dihydroactinidiolide, δ-cadinene, β-cadinece and valencene, based on their similarity matching results of 0.93, 0.99, 0.97, 0.93, and 0.95, respectively. The accuracy and reliability of the results were thus increased greatly. The corresponding standard spectra of each component in cluster A are shown in Figures 4–8. Components 3 and 6 were not identified due to their very low match values. Likewise, the spectra of components in other segments can be obtained. The qualitative results are listed in Table 1. Eighty two constituents were resolved, and among them 65 components were identified. Unfortunately, 27 components remain unidentified, either because of the low signal-to-noise ratio or the absence of the compound from the mass spectral database, and some of the component identifications should be viewed as tentative.

Table 1: Identification and quantification of compounds in the essential oil from Flemingiae Macrophyllae and Radix Flemingiae Latifolia Benth.

Series No.	Retention time (min) X_1 [a]	X_2 [b]	Compound name	Molecular formula	Relative content (%)
1	3.439	-	Prenal	C_5H_8O	0.07
2	3.653	3.658	Hexanal	$C_6H_{12}O$	0.35
3	6.368	6.372	2-Amylfuran	$C_9H_{14}O$	0.03
4	7.065	-	D-Limonene	C_9H_{20}	0.27
5	7.979	7.984	Linalool	$C_{10}H_{18}O$	0.98
6	8.682	8.691	L-Camphor	$C_{10}H_{16}O$	1.91
7	9.104	-	l-2-Bornanol	$C_{10}H_{18}O$	0.06
8	9.267	9.275	*p*-Menth-1-en-4-ol	$C_{10}H_{18}O$	0.66
9	9.419	9.428	*p*-Menth-1-en-8-ol	$C_{10}H_{18}O$	0.87
10	10.529	10.536	Nonanoic acid	$C_9H_{18}O_2$	0.46
11	10.671	-	Perillyl aldehyde	$C_{10}H_{14}O$	0.04
12	10.914	10.925	Anethole	$C_{10}H_{12}O$	1.62
13	11.065	11.081	Bornyl acetate	$C_{12}H_{20}O_2$	0.72
14	11.294	11.308	Tricyclo[3.2.1.02,7]oct-3-ene, 2,3,4,5-tetramethyl-	$C_{12}H_{18}$	0.57
15	11.504	-	Furan, 2,5-dibutyl-	$C_{12}H_{20}O$	0.03
16	12.465	12.478	Decanoic acid	$C_{10}H_{20}O_2$	0.06
17	12.720	12.731	α-Cubebene	$C_{15}H_{24}$	0.56
18	12.859	12.875	α-Longipinene	$C_{15}H_{24}$	1.47
19	13.325	13.331	Ylangene	$C_{15}H_{24}$	1.69
20	13.476	13.493	Longicyclene	$C_{15}H_{24}$	2.84
21	13.771	-	Copaene	$C_{15}H_{24}$	0.64
22	14.305	14.312	Longofolene	$C_{15}H_{24}$	3.21
23	14.437	-	Di-*epi*-α-cedrene	$C_{15}H_{26}$	0.08
24	14530	14.542	β-Caryophyllene	$C_{15}H_{24}$	0.82
25	14.575	14.583	β-Cedrene	$C_{15}H_{24}$	5.92
26	14.744	14.762	Germacrene D	$C_{15}H_{24}$	0.65
27	15.057	15.071	(*Z*)-β-Farnesene	$C_{15}H_{24}$	0.61
28	15.108	-	Himachala-2,4-diene	$C_{15}H_{24}$	0.07
29	15.385	15.435	α-Himachalene	$C_{15}H_{24}$	4.13
30	15.572	15.590	α-Caryophyllene	$C_{15}H_{24}$	0.44
31	15.593	-	Acoradiene	$C_{15}H_{24}$	0.33
32	15.673	15.685	Dihydrocurcumene	$C_{15}H_{24}$	0.03
33	15.858	15.869	Eremophilene	$C_{15}H_{24}$	1.30
34	16.006	16.021	(+)-Cycloisosativene	$C_{15}H_{24}$	0.05
35	16.108	16.117	Humulen-(v1)	$C_{15}H_{24}$	0.92
36	16.210	16.215	Longifolene-(V4)	$C_{15}H_{24}$	6.82
37	16.321	16.327	α-Guaiene	$C_{15}H_{24}$	0.49

Series No.	Retention time (min) X_1 [a]	X_2 [b]	Compound name	Molecular Formula	Relative content (%)
38	16.419	-	Patchoulene	$C_{15}H_{24}$	0.88
39	16.721	16.730	β-Himachalene	$C_{15}H_{24}$	5.26
40	17.077	17.086	β-Guaiene	$C_{15}H_{24}$	0.72
41	17.162	17.184	Dihydroactinidiolide	$C_{11}H_{16}O_2$	2.14
42	17.275	17.292	δ-Cadinene	$C_{15}H_{24}$	1.53
43	17.321	17.334	β-Cadinece	$C_{15}H_{24}$	0.56
44	17.510	-	Valencene	$C_{15}H_{24}$	0.67
45	18.710	18.722	Caryophyllenyl alcohol	$C_{15}H_{26}O$	0.31
46	19.076	19.085	Caryophyllene oxide	$C_{15}H_{24}O$	0.85
47	19.142	19.155	Drimenol	$C_{15}H_{26}O$	1.27
48	19.470	-	3-Isobutyl-4,5-dimethyl-3*H*-isobenzofuran-1-one	$C_{14}H_{18}O_2$	0.06
49	19.732	19.740	Cedrol	$C_{15}H_{26}O$	0.83
50	19.918	19.932	Bulnesol	$C_{15}H_{26}O$	4.17
51	20.235	-	Mansonone C	$C15H_{26}O_2$	0.07
52	20.342	20.360	Epiglobulol	$C15H_{26}O$	0.78
53	20.621	20.647	Cubenol	$C_{15}H_{26}O$	1.74
54	20.967	20.982	ι-Cadinol	$C_{15}H_{26}O$	0.56
55	21.151	21.165	δ-Cadinol	$C_{15}H_{26}O$	1.82
56	21.494	21.522	Torreyol	$C_{15}H_{26}O$	0.38
57	21.662	21.684	β-Selinenol	$C_{15}H_{26}O$	7.12
58	21.867	21.887	α-Eudesmol	$C_{15}H_{26}O$	5.16
59	22.061	22.077	Cadalene	$C_{15}H_{18}$	2.22
60	23.167	23.185	Hedycaryol	$C_{15}H_{26}O$	0.51
61	23.286	23.310	Peruviol	$C_{15}H_{26}O$	1.26
62	26.172	26.190	Farnesol isomer a	$C_{15}H_{26}O$	4.87
63	26.362	26.384	*cis*-Farnesal	$C_{15}H_{24}O$	0.46
64	26.471	-	Tetradecanoic acid	$C_{14}H_{28}O_2$	0.24
65	33.571	33.590	Hexadecanoic acid	$C_{16}H_{32}O_2$	0.58

[a] Representing the sample from Flemingiae Macrophyllae. [b] Representing the sample from Flemingiae Latifolia Benth. - Correlative component is not found in X_2.

Quantitative Analysis

After the pure chromatographic profile and mass spectrum of each component were resolved, the total two-way response of each component can be obtained from the outer product of the concentration vector and the spectrum vector for each component, namely $C_iS_i^T$. Similar to the general chromatographic quantitative method with peak area or height, the concentration of each component is proportional to the overall volume of its two-way response$^{(C_iS_i^T)}$. The final relative quantitative results were also listed in Table 1. The 65 components which have been identified in Radix Flemingiae Macrophyllae accounted for 88.79% of the volatile components.

Identification of Common Components

Curve 1 and curve 2 in Figure 10 are the TIC chromatograms of response X_1 from Radix Flemingiae Macrophyllae and X_2 from Radix Flemingiae Latifolia Benth obtained from GC-MS, respectively. It can be seen from Figure 10 that X_2 was consistent in eluting components with X_1, but the concentration distribution of some individual components were different. To compare and distinguish Radix Flemingiae Macrophyllae and its substitutes, such as Radix Flemingiae Latifolia Benth, one may extract pure component spectra from response X_2 and identify them once again. However, the a priori information of X_1 may be used to reduce some unnecessary work when we compare the quality of Radix Flemingiae Macrophyllae and its substitutes, such as Radix Flemingiae Latifolia Benth. Here, spectral correlative chromatography (SCC) [16] was used to identify each common component directly from the known information of X_1 instead of resolving each sample data one by one.

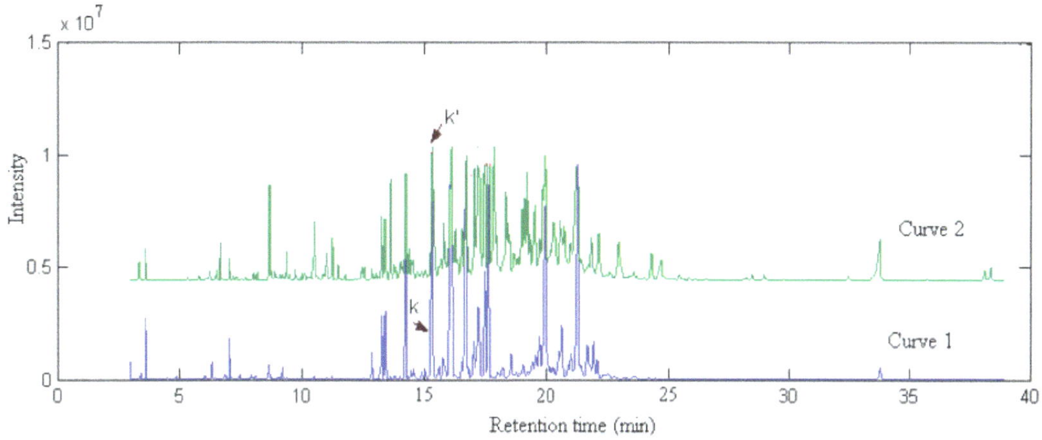

Figure 10: The total ionic current chromatograms of the volatile components in Radix Flemingiae Macrophyllae (X_1: Curve1) and of the volatile components in Radix Flemingiae Latifolia Benth (X_2: Curve 2) obtained from GC-MS, respectively. k is target component in Radix Flemingiae Macrophyllae and and k' is the correlative component in Flemingiae Latifolia Benth.

A brief depiction of spectral correlative chromatography was given as follows. Generally one substance has only a spectrum, no matter whether it exists in different samples, of course, it is best if they are comparable experiments. Suppose that X1 and X2 are two response matrices from two different samples of Chinese herbs, where s_i is the spectrum of the ith component of X_1 and s_j the spectrum of the jth component of X_2, respectively. The correlation coefficient of these two components can be written as the product of two spectral vectors as follows:

$$r(i,j) = \frac{(s_i - \bar{s}_i)^T \cdot (s_j - \bar{s}_j)}{\|(s_i - \bar{s}_i)\| \cdot \|(s_j - \bar{s}_j)\|} \quad (1)$$

where $\bar{\bullet}$ denotes the mean of the spectral vector, and $\|\bullet\|$ is Frobenius norm. The values r(i, j) of spectral correlation coefficient are in the range $-1 < r < 1$, and the larger the value of r is the more correlative are the components between the ith and the jth peak. When r equals to 1, these two components are identical. On account of errors and interference from noise and background, etc., in actual systems, the maximum of r is not equal to 1 but close to 1. In order to check if the component existing in X_1 really also exists in X_2, we may first extract s_i (the component to be investigated) from X1 and then calculate its corresponding correlation coefficient with every row vector, say $x_j^T (j = 1, \ldots, m)$, in X_2 as shown in Equation (2)?:

$$r(i,j) = \frac{(\mathbf{s}_i - \bar{\mathbf{s}}_i)^T \cdot (\mathbf{x}_j - \bar{\mathbf{x}}_j)}{\|(\mathbf{s}_i - \bar{\mathbf{s}}_i)\| \cdot \|(\mathbf{x}_j - \bar{\mathbf{x}}_j)\|}, (j = 1, 2, \ldots, m) \quad (2)$$

Accordingly, a spectral correlation coefficient curve is obtained by calculating r(i, j) of the spectral vector s_i against every row of X_2 in sequence. The curve with the same length of the eluting chromatogram contains therefore the spectral correlation information in X_2 corresponding to the component spectrum, say s_i, in X_1. With the help of the correlation coefficient curve thus obtained, one may easily pick up the correlation information between X_1 and X_2. Consequently, the curve was named spectral correlative chromatography (SCC) in this work. The detailed procedure is illustrated by an example, pure component k (α-himachalene) of X_1. Its spectrum has been extracted with the SFA method. How to extract the common components from X_2 with the SCC method? A series of spectral correlation coefficients (r) were obtained by calculating the spectrum sk of component k against every row of X_2 in sequence (see Figure 11) according to Equation (2). As shown in Figure 11, the correlation coefficients r(k, k') corresponding to components k' of X_2 were larger than the others. As shown in middle top part of Figure 11, the value of r(k, k') is 0.9981. If the errors and interference from noise and background, etc. in actual systems was taken into account, the correlation coefficient was quite close to 1. Then it was concluded that components k' and k were correlative based on spectral-dependent principle of substance identification. In this way, other correlative components in X_2 could be obtained. There were 51 common components between Radix Flemingiae Macrophyllae and Flemingiae Latifolia Benth. The results were also listed in Table 1.

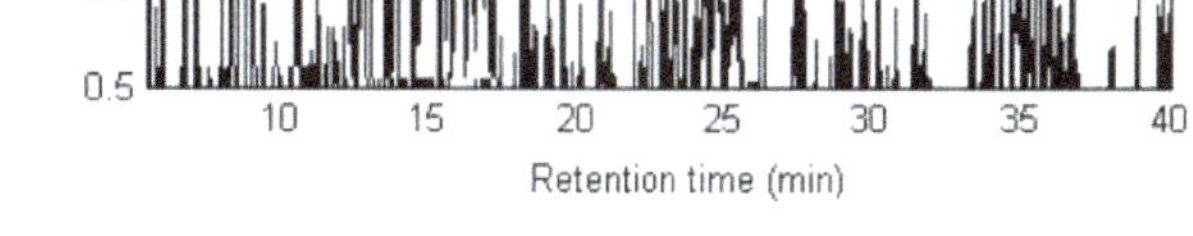

Figure 11: Correlative chromatogram of component k of sample X^1 against those of X^2. Right top part is the amplified segment where component k′ from Flemingiae Latifolia Benth exists and r(k, k′) is the correlation coefficient of component k and k′.

Comparison of Samples

The procedure performed for Flemingiae Macrophyllae was used for those peaks in Flemingiae Latifolia Benth, which were not identified by the SCC method. Thirty components in the essential oil of Flemingiae Latifolia Benth, which were not found in Flemingiae Macrophyllae, were identified with the SFA method. The results are listed in Table 2. Among the

79 separated constituents in the essential oil of Flemingiae Latifolia Benth, a total of 64 components were identified. Fifty one identified components existed in both Flemingiae Macrophyllae and Flemingiae Latifolia Benth. Alkenes, such as longifolene-V4, β-cedrene and β-himachalene, and alcohols, such as β-selinenol, α-eudesmol and farnesol isomer, were the main volatile constituents in both Flemingiae Macrophyllae and Flemingiae Latifolia Benth, but the contents of each component were different. The obtained results can provide useful information for further development of Flemingiae Macrophyllae.

Table 2: Qualitative results of some other constituents in the essential oil from Flemingiae Latifolia Benth.

Series No.	Retention time (min)	Compound name	Molecule structure
1	6.914	*m*-Cymene	$C_{10}H_{16}$
2	9.128	Octanoic acid	$C_8H_{16}O_2$
3	10.719	4-Hydroxy-3-methylacetophenone	$C_9H_{10}O_2$
4	11.526	Di-epi-α-cedrene	$C_{10}H_{18}O$
5	14.430	α-Cedrene	$C_{15}H_{32}$
6	15.112	1*H*-Benzocycloheptene, 2,4a,5,6,7,8,9,9a-octahydro-3,5,5-trimethyl-9-methylene-	$C_{15}H_{24}$
7	15.672	Dihydrocurcumene	$C_{15}H_{24}$
8	16.541	α-Muurolene	$C_{15}H_{24}$
9	17.512	Eudesma-3,7(11)-diene	$C_{15}H_{24}$
10	19.513	Caryophyllenyl alcohol	$C_{15}H_{26}O$
11	20.247	1-Cyclohexen-1-ol, 2,6-dimethyl-, acetate	$C_{10}H_{16}O_2$
12	26.490	(*E*)-10-Pentadecenol	$C_{15}H_{30}O$
13	34.172	*cis*-7-Tetradecen-1-ol	$C_{14}H_{28}O$

EXPERIMENTAL

Instrumentation and Materials

Radix Flemingiae Macrophyllae was purchased from a Zhejiang pharmaceutical store and was identified by a researcher from the Institute of Materia Medica, Hunan Academy of Traditional Chinese Medicine and Materia Medica. Radix Flemingiae Latifolia Benth, a substitute of Radix Flemingiae Macrophyllae was also purchased from a local pharmaceutical store in Zhejiang Province. Anhydrous sodium sulfate and ether were purchased from Shanghai Chemical Reagent Research Institute (Shanghai, P.R. China) and were of analytical grade. GC-MS was performed with a Shimadzu GC-2010 gas chromatography instrument coupled with a Shimadzu 2010 mass spectrometer. The volatile constituents in Radix Flemingiae Macrophyllae and Radix Flemingiae Latifolia Benth were separated on a 30 m × 0.25 mm I.D. fused silica capillary column coated with 0.25 μm OV-1 film.

Extraction of the Essential Oil

The samples were dried at constant temperature (40 °C) for 2 h. Then, 500 g distilled water and 200 g Radix Flemingiae Macrophyllae or Radix Flemingiae Latifolia Benth were added into a standard apparatus, and the volatile constituents were extracted by water distillation for 8 h according to the procedure described in the Chinese Pharmacopoeia [19]. Effluent was extracted with diethyl ether and the diethyl ether was removed under low temperature. The obtained essential oils were dried with anhydrous sodium sulfate and stored in the refrigerator at 4 °C prior to analysis.

Gas Chromatography-Mass Spectrometry

Analytical conditions were as follows: the oven was held at 50 °C for 1 min during injection, then programmed at 10 °C·min−1 to a temperature of 150 °C, finally ramped at 2 °C·min−1 to a temperature of 200 °C and held for 5 min. Inlet temperature was kept at 260°C all the time. A 1.0 μL volume of essential oil was injected into the GC. Helium carrier gas at a constant flow-rate of 1.0 mL·min−1 and a 10:1 split ratio were used simultaneously. Mass spectrometer was operated in full scan and electron impact (EI+) modes with an electron energy of 70 eV. Interface temperature was 250 °C. MS source temperature was 200 °C. In the range of m/z 30 to 350, mass spectra were recorded with 0.2 s·scan^{-1} velocity.

Data Analysis

Data analysis was performed on a Pentium based IBM compatible personal computer. All programs for the chemometrical resolution methods were coded in MATLAB 6.5 for Windows. The library searches and spectral matching of the resolved pure components were conducted on the National Institute of Standards and Technology (NIST) MS database containing about 107,000 compounds.

CONCLUSIONS

In this work, a combined chemometrics method, subwindow factor analysis (SFA) and spectral correlative chromatography (SCC), were used for analysis of the volatile constituents in Radix Flemingiae Macrophyllae and Radix Flemingiae Latifolia Benth, one of its common substitutes. This study showed that the application of this combined approach is a powerful tool, which comprehensively reveals the quality and quantity of chemical constituents of traditional medicines for effective evaluation of similarity or differences between complex analytical samples.

ACKNOWLEDGMENTS

This research work was financially supported by Province Natural Science Foundation of Zhejiang of China (Grant No. Y4110075 and Y2100908) and Zhejiang Qianjiang Talent Project (Grant No. 2010R10054).

REFERENCES

1. Chen, Q.; Li, X.Y.; Cai, Y.; Zhu, Y.L.; Chen, S.Q. Pharmacognostical identification of Flemingia fruticulosa. West China J. Pharm. Sci. 2012, 27, 83–85.
2. Zeng, C.L.; Zhong, Z.X.; Lu, W.J.; Wang, J.; Li, S.B.; Li, X.J. Study on the Pharmacological Action of Flemingia macrophylla(Willd.) Prain. Guiding J. Tradit. Chin. Med. Pharm. 2011, 17, 79–81.
3. Shiao, Y.J.; Wang, C.N.; Wang, W.Y. Neuroprotective flavonoids from Flemingia macrophylla. Planta Med. 2005, 71, 835–840.
4. Li, D.Y.; Fu, Y.B.; Hua, H.M.; Li, Z.L. Chemical constituents from roots of Flemingia macrophylla. Chi. Tradit. Herbal Drugs 2012, 43, 1259–1262.
5. Li, L.; Liu, Z.H.; Qin, M.J. Simultaneous Determination of Seven Flavonoids in the Flemingia macrophylla by HPLC. Chin. Wild Plant Resour. 2011, 30, 54–58.
6. Ahn, E.M.; Nakamura, N.; Akao, T. Estrogenic and Antiestrogenic Activities of the Roots of Moghania

philippinensis and Their Constituents. Biol. Pharm. Bull. 2004, 27, 548–553.

7. Sun, L.; Li, Z.L.; Han, G.H. Chemical constituents of the roots of Moghania philippinensis. Chin. J. Med. Chem. 2009, 19, 364–371.
8. Wang, X.Q.; Yang, S.D.; Yang, Z.Y. Study on the chemical constituents of volatile components from Radix Flemingiae Macrophyllae. J. Yunnan Univ. Tradit. Chin. Med. 2008, 31, 12–15.
9. Liu, J.H.; Gao, L.X.; Gao, Y.Q. Study on the chemical constituents of volatile oil from Radix Flemingiae Macrophyllae. Chin. Tradit. Pat. Med. 2003, 25, 485–487.
10. Wei, D.; Edmund, R. Malinowski.Investigation of copper (II)- Ethylenediaminetetraacetate complexation by window factor analysis of ultraviolet spectra. J. Chemom. 1993, 7, 89–98.
11. Shen, H.L.; Manne, R.; Xu, Q.S.; Chen, D.Z.; Liang, YZ. Local resolution of hyphenated chromatographic data. Chemom. Intell. Lab. Syst. 1999, 45, 323–330.
12. Xu, C.J.; Jiang, J.H.; Liang, Y.Z. Evolving window orthogonal projections method for two-way Data Resolution. Analyst 1999, 124, 1471–1476.
13. Sanchez, F.C.; Rutan, S.C.; Gil Garcia, N.D. Resolution of multicomponent overlapped peaks by the orthogonal projection approach, evolving factor analysis and window factor analysis. Chemom. Intell. Lab. Syst. 1997, 36, 153–164.
14. Manne, R.; Shen, H.L.; Liang, Y.Z. Subwindow factor analysis. Chemom. Intell. Lab. Syst. 1999, 45, 171–176.
15. Li, B.Y.; Hu, Y.; Liang, Y.Z.; Huang, L.F.; Xu, C.J.; Xie, P.S. Spectral correlative chromatography and its application to analysis of chromatographic fingerprints of herbal medicines. J. Sep. Sci. 2003, 27, 581–588.
16. Liang, Y.Z. White, Grey and Black Multicomponent Systems and Their Chemometric Algorithms; Hunan Publishing House of Science and Technology: Changsha, China, 1996.
17. Guo, F.Q.; Liang, Y.Z.; Xu, C.J.; Huang, L.F. Determination of the volatile chemical constituents of Notoptergium Incium by gas chromatography- mass spectrometry and iterative or non-iterative chemometrics resolution methods. J. Chromatogr. A 2003, 1016, 99–110.
18. Wang, Y.M.; Yi, L.Z.; Liang, Y.Z.; Li, H.D.; Yuan, D.L.; Gao, H.Y.; Zeng, M.M. Comparative analysis of essential oil components in Pericarpium Citri Reticulatae Viride and Pericarpium Citri Reticulatae by GC-MS combined with chemometric resolution method. J. Pharm. Biomed. Anal. 2008, 46, 66–74.
19. Chinese Pharmacopoeia Committee. Chinese Pharmacopoeia; Publishing House of People's Health: Beijing, China, 2000.

CHAPTER 15 RAPID QUANTITATIVE ANALYSIS OF ETHANOL AND PREDICTION OF METHANOL CONTENT IN TRADITIONAL FRUIT BRANDIES USING FTIR SPECTROSCOPY AND CHEMOMETRICS

Teodora Emilia Coldea, Carmen Socaciu, Florinela Fetea, Floricuța Ranga, Raluca Maria Pop and Mira Florea

Keywords: ethanol, fruit brandies, FTIR, methanol, PCA, PLS, Romania

INTRODUCTION

The Central and East European countries have an old tradition in producing different types of fruit brandies. In Romania, there is a great interest to produce home-made traditional fruit brandies obtained by a double distillation of fermented fruits. Țuica and pălinca are the two most popular, traditional distilled fruit beverages made from plums or apples, pears, other fruits respectively. Besides their volatile composition, the principal differences of these two types of fruit brandies are the ethanol content which reaches values between 24 up to 86% v/v, for țuica, and 40 up to 70% v/v, for pălinca. The determination of methanol in these types of fruit brandies it is an important approach, related to its safety for consumers. Depending of the type of fruit brandy, the maximum accepted methanol content was established to be 1350 mg/100 ml anhydrous alcohol and in the case of țuica and pălinca, the methanol content should not exceed 1200 mg/100 ml anh. alc. (REG. 110/2008).

Citation: Teodora Emilia COLDEA , Carmen SOCACIU , Florinela FETEA , Floricuța RANGA , Raluca Maria POP , Mira FLOREA, Rapid Quantitative Analysis of Ethanol and Prediction of Methanol Content in Traditional Fruit Brandies using FTIR Spectroscopy and Chemometrics, ISSN 0255-965X;

In the context of quality control of alcoholic beverages a range of different analytical methods are used, such as distillation and picnometry (Tešević et al., 2009), electronic densimetry (Coldea et al., 2011), colorimety (Joshi and Sandhu, 2000), photometry (Borges Sivanildo et al., 2006), by fluorescent chemical sensor (Bozkurt et al., 2010). The most advanced techniques are the chromatographic ones, either GC-FID, GC-MS or HPLC (Diban et al., 2009; García-Llobodanin et al., 2008; HernándezGómez et al., 2005; López-Vázquez et al., 2010; Wang et al., 2003). The last years, the expensive reference methods are replaced by simpler ones, non-destructive, easy to handle. The FTIR (Fourier Transform Infrared Spectroscopy) technique, in combination with chemometrics is such a fast and reproducible way to identify the authenticity and adulteration of different food and beverage products (Da Costa et al., 2004; Mansor et al., 2011; Rohman et al., 2010). The FTIR is increasenly used for the determination of methanol in alcoholic beverages (Arzberger and Lachenmeier, 2008; Gallignani et al., 2005; Lachenmeier, 2007; Mehrotra et al., 2005; Nagarajan et al., 2006; PérezPonce et al., 1998) or to verify the authenticity of different fruit juices (Leopold et al., 2011). The Near-Infrared (NIR) spectroscopy was also used to estimate the ethanol content in alcoholic beverages or to verify the adulteration of different beverages (Ashok et al., 2011; Pontes et al., 2006).

In our previous studies we determined the ethanol and methanol content of brandies by electronic densimetry and by GC-FID techniques, respectively (Coldea et al., 2011). In this study we realized the FTIR fingerprint of distilled fruit brandies in order to verify their authenticity, to identify the specific signals for methanol and ethanol, to apply the non-targeted PCA chemometric analysis for the evidence of three types of fruit brandies, to verify the plum brandy (ţuica) authenticity from different regions and to predict methanol concentration by applying the PLS (Partial Least Squares) regression. We also compared the results obtained by FTIR method and by GC-FID technique or densimetic method.

MATERIALS AND METHODS

Reagents

Highly pure methanol (Merck, Germany, 99.8% purity) and ethanol (Merck, Germany, 99.9% purity) were used.

Sampling

A total number of 26 samples of fruit brandies (plum - ţuica, apple and pear pălinca) originating from different Counties of Transylvania - Maramureş (MM), Cluj (CJ), Bistriţa Năsăud (BN), Alba (AB), Bihor (BH) were collected from producers. All brandies were produced by traditional method: fermentation in wood barrels and double distilled in copper alembic, kept in wood or glass barrels for maturation. They were kept in glass recipients until analysis. Tab. 1 includes the sample denominations and their region of origin.

Fourier Transform Infrared Spectroscopy

To evaluate the content of methanol, we spiked one sample (P10) with methanol in different proportions, in order to identify the spectra modifications. There were used individually pure ethanol, pure methanol, sample P10 and different concentrations of methanol added to sample P10 at ratios P10:methanol of 1:1, 10:1, 10:0.5, 10:0.25, 9:1, 8:2, 8:4, 8:6. The FTIR spectra were registered in the range 1000 to 1200 cm-1, using Attenuated Transmission (ATR) and an internal reflection accessory made of Composite Zinc Selenide (ZnSe) and Diamond crystals (Shimadzu IR Prestige-21 equipment). For each sample, spectrum was registered from 3500-750 cm-1, and based on the ethanol and methanol specific wavenumbers, in all samples, the methanol and ethanol content was determined.

Table 1: Numbering of fruit brandy samples (plum – P, apple – M, pear – PE brandies) and the provenience regions (CJ-Cluj; MM- Maramureş; BH-Bihor; BN- Bistriţa Năsăud)

Sample no.	Region	Sample no.	Region	Sample no.	Region	Sample no.	Region	Sample no.	Region
P1	CJ	P5	MM	P3	BN	M19	BN	PE23	CJ
P2		P7		P12		M20	AB	PE24	MM
P4		P8		P13		M21	CJ	PE25	BN
P6		P9		P14		M22	BH	PE26	CJ
P17		P10		P15					
P18		P11		P16					

Statistic and chemometric analysis

Based on FTIR spectra and the frequency intensities recorded in the domain 1170 -1000 cm^{-1}, the chemometric analysis was performed. By Principal Component Analysis (PCA), using the Unscrambler 10.1 Software, version 10.1 (Camo Software AS, Oslo, Norway) the variability and discrimination between samples was made. Based on peak areas, from 1170 to 1000 cm^{-1}, the concentration of methanol in samples was predicted by PLS regression. For each type of fruit brandy and each region for plum brandy a correlation analysis it was developed using Origin software (OriginLab, version 8.0).

Reference Procedure

Both, ethanol and methanol content of the same samples were previously determined. Ethanol content was analyzed in all samples by densimetric method. Methanol was evaluated by gas chromatography coupled with flame ionization detector (Coldea et al., 2011).

RESULTS AND DISCUSSION

Identification of methanol and ethanol in the FTIR spectra of pure standards and a spiked sample

The identification of methanol and ethanol was done, firstly, by applying the calibration curve using pure standards, as well mixtures of methanol:ethanol (1:1; 1:9) (Fig. 1). Thus, pure methanol solution has the characteristic vibration frequency at 1020 cm-1 (major signal) and 1112 cm-1 (minor signal), while ethanol had the characteristic frequency at 1047 cm-1 (major signal) and 1087 cm-1 (minor signal). We selected the absorbances of major and minor signals specific for pure methanol and ethanol (Tab. 2). The approximate ratio of the absorbances of major/minor signals in methanol was 5.2 and for ethanol 1.5. These frequencies are specific for stretching vibrations of C-O bonds in these molecules.

Using methanol and ethanol in ratios of 1:9, or 1:1 respectively (Fig. 1) we noticed that the increase of methanol in the mixture, determine decreases of absorbance at 1047 and 1087 cm-1 and increase of absorbance at 1030- 1020 cm^{-1} and 1112 cm^{-1}. Based on these observations, we consider the parameters which can identify and quantify the concentration of methanol in samples, are the ratios A_{1020}/A_{1047} and A_{1112}/A_{1087}. Therefore were selected the areas corresponding to major and minor signals specific for pure methanol and ethanol (Tab. 3).

Table 2: The values and ratios of the absorbance intensities (A) and areas (B) of the minor and major signals specific for methanol (1020 and 1112 cm^{-1}) and ethanol (1047 and 1087 cm^{-1})

Mixture/wavenumber (cm^{-1})	1020	1112	1047	1087	1020 / 1047	1112 / 1087
			A			
Methanol	2.1470	0.4125	-	-	-	-
Ethanol	-	-	1.9539	1.3448	1.1	0.3
Methanol:Ethanol 1:9	-	-	1.9484	1.2658	-	-
Methanol:Ethanol 1:1	1.9286	-	1.6314	0.9471	1.2	-
			B			
Methanol	78.45	27.99	-	-	-	-
Ethanol	-	-	73.76	53.21	1.1	0.5
Methanol:Ethanol 1:9	-	-	-	-	-	-
Methanol:Ethanol 1:1	76.46	-	30.26	44.41	2.5	-

To illustrate the effect of the presence of methanol and its limit of detection in samples of brandies, it was chosen sample P10 (Fig. 2). By a controlled adulteration of the sample, were added various concentrations of pure methanol 2.5%; 5%; 9%; 20%; 33%; 43% and 50%, and noticed the gradual increase of intensity frequencies at 1112 and 1020 cm^{-1}. The gradual evolution of methanol and ethanol signals of absorbtion at 1047, 1087, 1020 and 1112 cm^{-1} is represented in Tab. 3

Therefore, the FTIR technique was useful not only to identify and quantify the ethanol, but also to recognize the presence of methanol and to determine the ratio between ethanol and methanol.

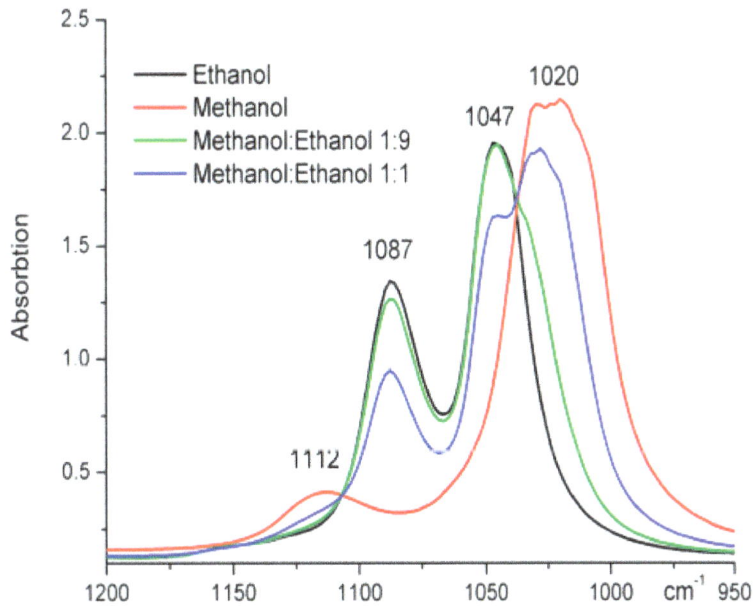

Figure 1: The specific IR spectra for methanol and ethanol, as pure substances and in different concentration in mixture methanol:ethanol

Identification of ethanol by FTIR fingerprinting of brandy samples

FTIR spectroscopy can be used as potential means for quantitative analysis of fruit brandies, considered as a "fingerprinting tool", which means that there are no two fruit brandies samples with the same FTIR spectra, either in the number of peaks or in the absorbance of the maximum peak. Fig. 3 (up) exhibits the general FTIR spectra of 3 representative brandy samples (M21, PE25 and P5) in a region range of 3500-750 cm-1, showing quite similar fingerprints of these three samples in the region 1700-900 cm^{-1}.

Fig. 3 (down) presents also the FTIR spectra in the region 1120-1000 cm-1 and the specific peaks and absorbances for ethanol at 1087 and 1047 cm^{-1}.

Using detailed investigation, the FTIR spectra of plum brandy samples exhibited some differences. Since ethanol has a specific frequency of vibration at 1087 and 1047 cm^{-1}, these signals were compared for samples P5, M21 and PE25 (Fig. 4). The sample P5 has higher absorbance values at these frequencies followed by sample M21 and PE25, in agreement with the concentration of ethanol reported by densimetric method (Coldea et al., 2011). While pear brandies had the lowest ethanol content, plum brandy sample had the highest, the peak area is a better parameter to evaluate the ethanol content.

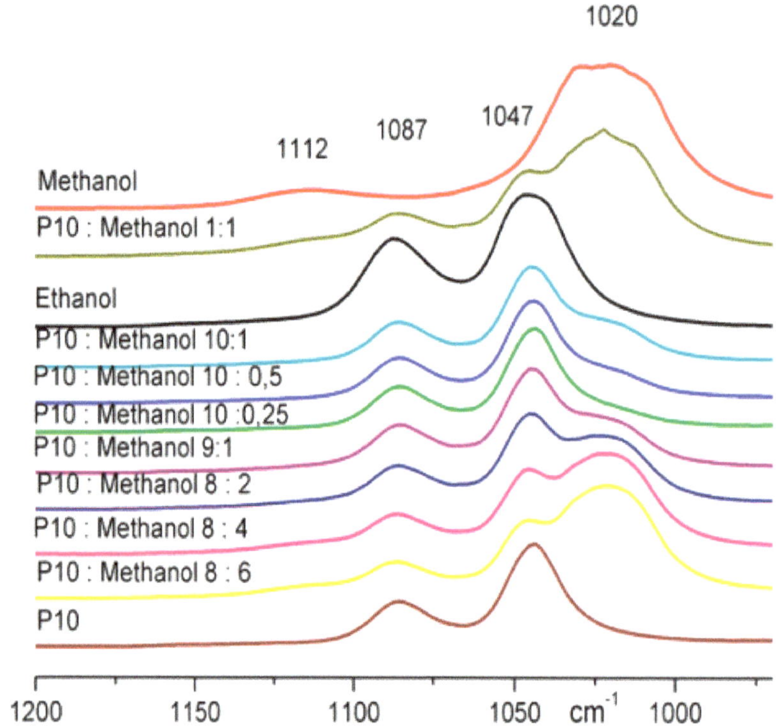

Figure 2: Overlapped IR Spectra for methanol and ethanol in different proportion from sample P10

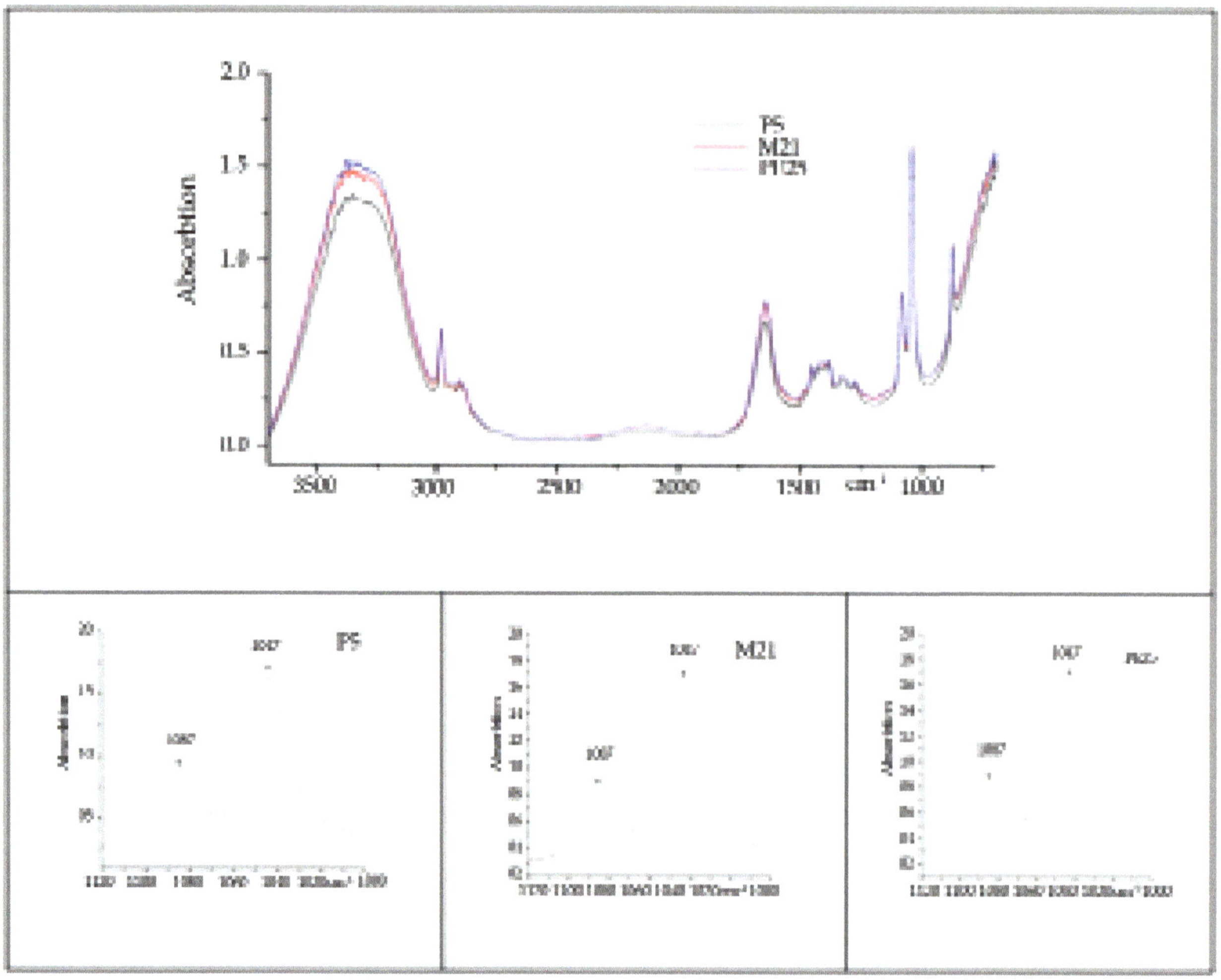

Figure 3: Comparative IR Spectra, overlapped on domain 3500-750 cm-1, for samples: P5, M21 and PE25. Specific ethanol signals (1087 and 1047 cm-1) evidentiated for each sample

Principal component analysis (PCA) based on FTIR data

The multivariate analysis of principal component analysis (PCA) based on the fingerprint of FTIR spectra which includes alcohols (methanol, ethanol) and some unfermented glucides within 1170-1000 cm-1 was carried out in order to differentiate the samples according to the biological origin and the provenience region of the traditional brandies. There were considered absorbances which corresponded to the peak areas within the region 1170- 1000 cm^{-1} for each sample when building the PCA score plot. Fig. 5 shows the score plot of PCA for classification of the fruit brandies using the FTIR spectra values. The PCA representation was made without normalization. As shown in Fig. 6, the first two principal components (PCs) described 92% (PC-1) and 7% (PC-2) of the total variation, respectively. Therefore, it can be pointed that the variation of 99% could be described by the first two PCs. PC^{-1} is associated to methanol with the samples containing the highest methanol level having PC-1 score. PC-2, is associated with ethanol. Although almost all samples have similiar composition in alcohols, we still can distinguish six groups based on their methanol or ethanol content.

Table 3: The gradual evolution of methanol and ethanol signals of absorbtion at 1047, 1087, 1020 and 1112 cm^{-1}

Mixture	1047	1087	1020	1112
Ethanol	1.9539	1.3448	-	-
P10	1.6468	0.8509	-	-
P10:Methanol 10:0.25	1.6674	0.8603	-	-
P10:Methanol 10:0.5	1.6411	0.8585	-	-
P10:Methanol 10:1	1.6136	0.8487	-	-
P10:Methanol 9:1	1.6595	0.8813	-	-
P10:Methanol 8:2	1.5435	0.8232	1.2428	-
P10:Methanol 8:4	1.382	0.7526	1.5956	-
P10:Methanol 8:6	1.2761	0.6986	1.7524	-
P10:Methanol 1:1	1.6674	0.8603	-	-
Methanol	-	-	2.1470	0.4125

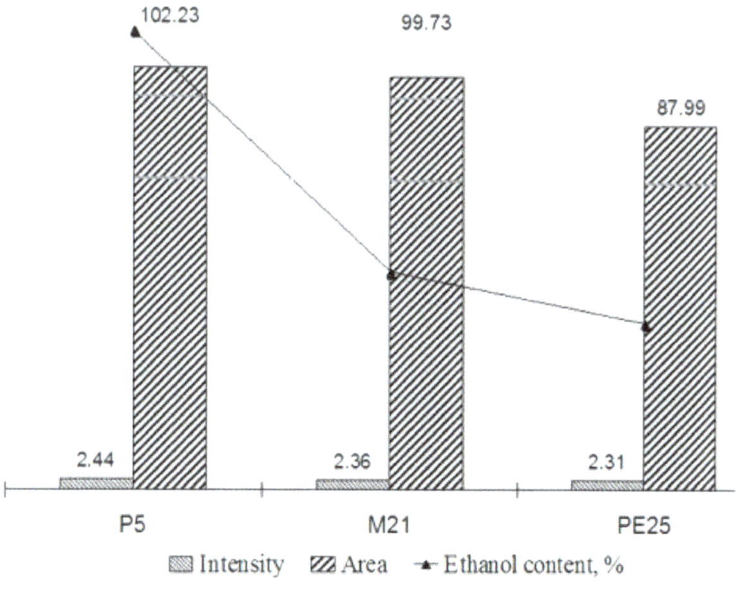

Figure 4: Comparative values of FTIR absorption intensities and areas for signals 1047 and 1087 cm-1, as obtained by FTIR and the ethanol concentration (line- %) as obtained by densimetric method (for sample P5, M21 and PE25)

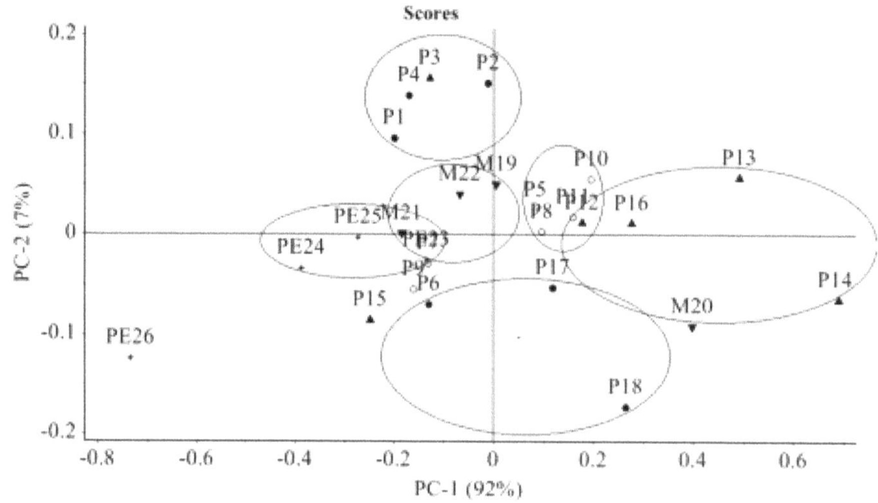

Figure 5: Evidence of differences (scores) from analyzed beverages, depending on FTIR analysis (fingerprint zone area 1170-1000 cm^{-1}) by PCA method

Regarding the methanol content registered by GCFID method, sample P2 had the highest content of all samples from Cluj County. In contrast, samples M19 and P10 had the lowest values. The most evident variability have the plum brandies from BN County. Tab. 4 includes the comparison between the FTIR area signals corresponding to ethanol and the values obtained by densimetric method. There was a signifficant correlation of the results obtained for ethanol by the two methods (Fig. 6). Correlation analysis was made between results obtained by both FTIR and densimetric methods used to determine the concentration of ethanol for all the brandies. It was used two-tailed test of significance. These results have a strong correlation (R = 0.841) for the all samples.

Calibration and validation of PLS procedure for methanol

By applying the PLS regression for methanol calibration curve (Fig. 7 and Tab. 2) we were able to predict its content in brandies. The predicted concentrations of methanol were obtained based on FTIR absorbance at 1020 cm^{-1} (g/100 g fruit brandy) and transformed into g/100 mg anh. alc.

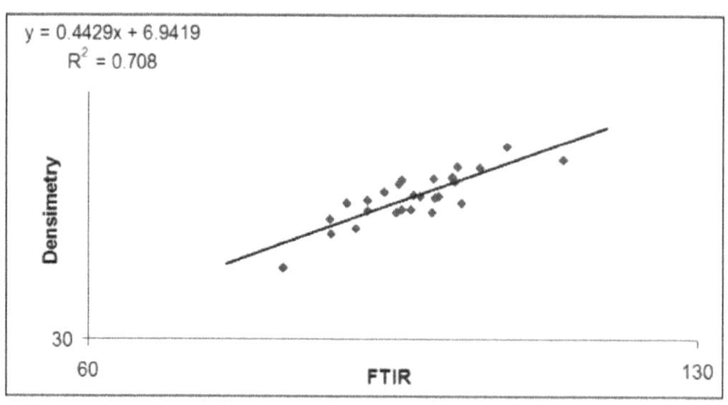

Figure 6: Correlation curve established for the etahnol concentrations obtained by two methods (densimetry and FTIR). Correlation coefficient= 0.708 ($p<0.05$)

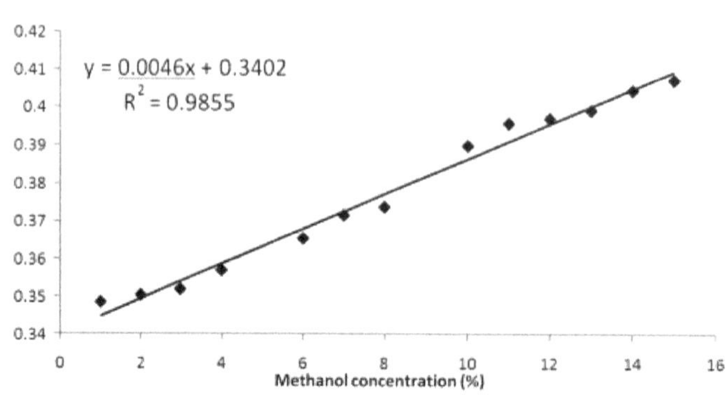

Figure 7: Prediction of methanol concentration in fruit brandy samples, by appling the Partial Least Squares (PLS) Regresion: Calibration curve with methanol at 1020 cm^{-1} (on methanol concentration interval 1-15%)

Tab. 5 exhibits the correlations between the values for methanol content, for the set of all 26 samples, as obtained by the two methods applied, FTIR (predicted values) and GC-FID (experimental values) (Coldea et al., 2011). A mean ratio of 1.1176 ± 0.2402 was obtained between the values obtained by GC-FID and FTIR, demonstrating the good prediction (p

Table 4: Correlations between the values of methanol concentrations for obtained from the two methods FTIR and GC-FID

Sample	FTIR (predicted values)	GC-FID (Actual values)	Ratio GC-FID/FTIR
P1	851.91	1077.59	1.2649
P2	818.39	1266.97	1.5481
P4	850.40	954.31	1.1221
P6	868.18	1244.35	1.4332
P17	861.54	1100.76	1.2776
P18	906.28	953.69	1.0523
P3	843.26	964.96	1.1443
P12	815.72	901.39	1.1050
P13	758.35	1073.70	1.4158
P14	819.42	913.01	1.1142
P15	926.54	765.86	0.8265
P16	795.93	953.73	1.1982
P5	803.50	857.71	1.0674
P7	841.05	504.18	0.5994
P8	804.21	999.81	1.2432
P9	875.00	1256.41	1.4358
P10	755.55	450.50	0.5962
P11	825.07	954.12	1.1564
M19	822.32	597.64	0.7267
M20	866.61	1189.29	1.3723
M21	878.57	1020.89	1.1619
M22	830.68	810.97	0.9762
PE23	881.89	973.61	1.1040
PE24	923.37	818.34	0.8862
PE25	883.67	880.69	0.9966
PE26	1045.89	1290.06	1.2334
	Average ratio		1.1176
	SD		0.2402

By comparison with GC-FID analysis, the obtained values for methanol, based on the forecast calibration and FTIR spectra, were of the same order, from 0 to 1.2% anh. alc. Among all samples, the minimum concentration of methanol was determined in sample P10, and the maximum in PE26, by both FTIR and GC-FID methods.

We noticed also differences among the type of brandy: the methanol from apple brandies was better correlated (R = 0.874), followed by the pear brandies (R = 0.860) and no correlations for the plum brandies (R = 0.213). A high correlation was in the case of MM plum brandies (R = 0.621). Unsignificant correlations were registered in the case of CJ plum brandies (R = -0.565) and BN plum brandies (R = -0.911). The most appropiate results for methanol obtained by the two methods were those for plum distillate from Maramureş County. The standard deviation of values obtained by FTIR method were much lower than those by GC-FID. Finally we conclude that FTIR technique applied for methanol quantification can predict reliable results and in good correlation with GCMS technique.

CONCLUSIONS

FTIR method can be useful not only for determing the methanol and ethanol content in extract-free samples, but also to evaluate the ratio between these two volatile components in brandies. FTIR/PLS offers valuable advantages when choosing versus conventional methods.Their importance it is justified by being a rapid, efficient and non-destructive tool for screening alcoholic beverages. Besides the quantitative PLS analysis, the PCA classification of data obtained by FTIR spectra becomes important for authenticity control of the provenience region and the type of the fruit brandy.

ACKNOWLEDGEMENTS

This work has benefited from financial support through the 2010 POSDRU/ 89/ 1.5 /S 52432 project, „Organizing the National Interest Postdoctoral School of Applied Biotechnologies with Impact on Romanian Bioeconomy", project co-financed by the European Social Fund through the Sectoral Operational Programme Human Resources Development 2007-2013.

REFERENCES

1. Arzberger U, Lachenmeier DW (2008). Fourier transform infrared spectroscopy with multivariate analysis as a novel method for characterizing alcoholic strength, density, and total dry extract in spirits and liqueurs. Food Anal Methods 1(1):18-22.
2. Ashok PC, Praveen BB Dholakia K (2011). Near infrared spectroscopic analysis of single malt Scotch whisky on an optofluidic chip. Opt Express 19(23):22982-22992.
3. Borges Sivanildo S, Rejane M Frizzarin, Boaventura F Reis (2006). An automatic flow injection analysis procedure for photometric determination of ethanol in red wine without using a chromogenic reagent. Anal Bioanal Chem 385(1):197-202.
4. Bozkurt SS, Merdivan E, Benibol Y (2010). A fluorescent chemical sensor for ethanol determination in alcoholic beverages. Microchim Acta 168(1-2):141-145.
5. Coldea TE, Socaciu C, Pârv M, Vodnar D (2011). Gas-chromatographic analysis of major volatile compounds found in traditional fruit brandies from Transylvania, Romania. Not Bot Horti Agrobo 39(2):109-116.
6. Da Costa RS, Santos SRB, Almeida LF, Nascimento ECL, Pontes MJC, Lima RAC, Simões SS, Araújo MCU (2004). A novel strategy to verification of adulteration in alcoholic beverages based on Schlieren effect measurements and chemometric techniques. Microchem J 78(1):27-33.
7. Diban N, Voinea OC, Urtiaga A, Ortiz I (2009). Vacuum membrane distillation of the main pear aroma compound: experimental study and mass transfer modeling. J Membrane Sci 326:64-75.
8. Gallignani M, Ayala C, Brunetto MD, Burguera JL, Burguera M (2005). A simple strategy for determining ethanol in all types of alcoholic beverages based on its on-line liquid-liquid extraction with chloroform, using a flow injection system and Fourier transform infrared spectrometric detection in the mid-IR. Talanta 68(2):470-479.

9. García-Llobodanin L, Ferrando M, Güell C, López F (2008). Pear distillates: influence of the raw material used on final quality. Eur Food Res Technol 228(1):75-82.

10. Hernández-Gómez LF, Úbeda-Iranzo J, García-Romero E, Briones-Pérez A (2005). Comparative production of different melon distillates: chemical and sensory analyses. Food Chem 90:115-125.

11. Joshi VK, Sandhu DK (2000). Influence of ethanol concentration, addition of spices extract, and level of sweetness on physico-chemical characteristics and sensory quality of apple vermouth. Braz Arch Biol Techn 43(5):537-545.

12. Lachenmeier DW (2007). Rapid quality control of spirit drinks and beer using multivariate data analysis of Fourier transform infrared spectra. Food Chem 101(2):825-832.

13. Leopold LF, Leopold N, Diehl HA, Socaciu C (2011). Quantification of carbohydrates in fruit juices using FTIR spectroscopy and multivariate analysis. Spectrosc Int J 26(2):93- 104

14. López-Vázquez C, Bollaín MH, Berstsch K, Orriols I (2010). Fast determination of principal volatile compounds in distilled spirits. Food Control 21:1436-1441.

15. Mansor TST, Man YBC, Rohman A (2011). Application of fast gas chromatography and fourier transform infrared spectroscopy for analysis of lard adulteration in virgin coconut oil. Food Anal Method 4(3):365-372.

16. Mehrotra R, Gupta A, Nagarajan R (2005). NIR spectroscopy and fiber optic probe for determination of alcohol, sugar and tartaric acid in alcoholic beverages. J Sci Ind Res India 64(2):134-137

17. Nagarajan R, Mehrotra R, Bajaj MM (2006). Quantitative analysis of methanol, an adulterant in alcoholic beverages, using attenuated total reflectance spectroscopy. J Sci Ind Res India 65(5):416-419

18. Pérez-Ponce A, Rambla FJ, Garrigues JM, Garrigues S, de la Guardia M (1998). Partial least-squares fourier transform infrared spectrometric determination of methanol and ethanol by vapour-phase generation. Analyst 123(6):1253- 1258.

19. Pontes MJC, Santos SRB, Araújo MCU, Almeida LF, Lima RAC, Gaião EN, Souto U (2006). Classification of distilled alcoholic beverages and verification of adulteration by near infrared spectrometry. Food Res Int 39(2):182-189.

20. Regulation (EC) No 110/2008 of the European Parliament and of the Council of 15 January 2008 on the definition, description, presentation, labelling and the protection of geographical indications of spirit drinks and repealing Council Regulation (EEC) No 1576/89. Official Journal of the European Union, 51, L39, 16-54.

21. Rohman A, Man YBC, Riyanto S (2011). Authentication analysis of red fruit (Pandanus Conoideus Lam) oil using FTIR spectroscopy in combination with chemometrics. Phytochem Analysis 22(5):462-467.

22. Tešević V, Nikićević N, Milosavljević S, Bajić D, Vajs V, Vučković I, Vujisić L, Đorđević I, Stanković M, Veličković M (2009). Characterization of volatile compounds of "Drenja", an alcoholic beverage obtained from the fruits of cornelian cherry. J Serb Chem Soc 74(2):117-128.

23. Wang ML, Choong YM, Su NW, Lee MH (2003). A rapid method for determination of ethanol in alcoholic beverages using capillary gas chromatography. J Food Drug Anal 11(2):133-140.

CHAPTER 16 SIMULTANEOUS SPECTROPHOTOMETRIC DETERMINATION OF RABEPRAZOLE SODIUM AND DOMPERIDONE MALEATE IN CAPSULES BY CHEMOMETRIC METHODS

Marc Ohmenhaeuser, Yulia B. Monakhova, Thomas Kuballa, and Dirk W. Lachenmeier

INTRODUCTION

Honey is a natural, sweet, and syrupy fluid collected by bees from nectar of flowers [1]. The taste and aroma of this liquid vary according to its floral origin, geographical and seasonal conditions [1]. The large number of melliferous sources gives therefore the opportunity to produce many characteristical unifloral and a high number of polyfloral nectar honeys.

Each honey is unique on the basis of chemistry, amount, and combination of the various components that give each honey a unique and individual organoleptic character. The control and characterization of quality and botanical origin of unifloral honeys are of great importance and interest in apiculture. Today the most important techniques to determine or certify the unifloral origin of honeys are the melissopalynological analysis and the evaluation of organoleptic characteristics [2]. Current quality assessment of honey by these methods are time-consuming and often operator dependent. Moreover, some types of adulterations (e.g., the addition of sugar concentrate to honey) can hardly be detected with such methods [3].

Various novel, fast, and accurate chromatographic methods such as high-performance liquid chromatography (HPLC) [4–7], gas chromatography (GC) [8–10], liquid chromatography with electrochemical detector [11], and matrix-assisted-laser-desorption/ionization-time-of-flight-mass-spectrometry (MALDI TOF MS) [12, 13] have been used to obtain the chemical composition and detect possible adulteration of honey. Vibration spectroscopic methods such as FT-Raman [14, 15], NIR [16–18], and FT-IR [19–21] could be additionally used as a screening technique for checking the honey authenticity and for quantifying its major compounds.

Apart from these analytical methods, the application of multivariate data analysis and, in particular, principal component analysis (PCA) [9, 22], canonical variate analysis (CLA) [8, 23], partial least squares (PLS) regression [17, 24, 25], principal component regression (PCR) [17], linear discriminant analysis (LDA) [22], and soft independent

Citation: Marc Ohmenhaeuser, Yulia B. Monakhova, Thomas Kuballa, and Dirk W. Lachenmeier, "Simultaneous spectrophotometric determination of rabeprazole sodium and domperidone maleate in capsules by chemometric methods," doi:10.1155/2013/825318

modeling class analogy (SIMCA) [25] proved to be extremely useful for grouping and detecting honey from different origins. Besides these multivariate methods, modern sensor techniques such as electronic nose (e-nose) and electronic tongue (e-tongue) were successfully applied to classify honey samples according to their floral origin [3].

Nuclear magnetic resonance (NMR) spectroscopy has been also used to assess the botanical origin of honey and quantify some major compounds in it [26–30]. It was shown that NMR has a good potential to become a useful quality control tool in the analysis of honey samples. However, the number of floral honey types and the total number of investigated samples have been insufficient to construct a good discrimination model for routine analysis. Targeted quantitative NMR analysis was limited to major carbohydrates and amino acids [27,31]. Therefore, this paper further advances the investigation of a combined NMR spectroscopy (^{1}H and ^{13}C NMR) and chemometric data analysis approach to distinguish the botanical origin of honey. We also explored the potential of high-resolution ^{1}H NMR to allow the identification and the quantification of 13 selected components in honeys.

EXPERIMENTAL

Samples and Chemicals

A total of 328 samples from different botanical origins were analyzed using NMR. The samples were randomly selected by governmental food inspectors from Baden-Württemberg, Germany, from honey bottling plants, supermarkets and directly by bee keepers. The following reference standards were used in proanalysis quality: hydroxymethylfurfural (HMF), fumaric acid, citric acid, malic acid, erlose, melibiose, xylitol, oxalic acid (anhydrous), D-glucuronic acid, DL-lactic acid (Sigma Aldrich, Steinheim, Germany); formic acid, phthalic acid, and glucose, L(+)-tartaric acid, fructose, D(+)-galactose, maltose, and saccharose, barbituric acid (Merck, Darmstadt, Germany); L(+)-rhamnose, arabinose, maltotriose, D(+)-turanose, D(+)-mannose, D(+)-xylose, D(+)-trehalosedihydrate, D(+)-melezitose monohydrate, D(+)-raffinosepentahydrate, malonic acid, pyruvic acid, and DL-proline (Fluka, Buchs, Switzerland); gluconic acid (calcium salt), and succinic acid (Carl Roth, Karlsruhe, Germany). The NMR buffer was prepared by dissolving 10.21 g of KH_2PO_4 and 9.75 mg of sodium azide in 50 mL of pure water and then by adjusting the pH to 4.5 with H_3PO_4 or KOH.

Sample Preparation and Calibration

The water content was obtained for each honey before NMR measurement using the German reference refractometric method [32]. The equivalent of 200 mg water-free honey (about 240 mg) was weighted and combined with 300 μL of NMR buffer (see above), 700 μL of distilled water and 100 μL of an internal standard (D_2O containing 0.1% of TSP (sodium salt of 3(trimethylsilyl)-propionate acid-d4)). Stock solutions were prepared by mixing of about 20 mg of a pure substance in 300 μL of NMR buffer, 700 μL of distilled water, and 100 μL of an internal standard. For neutralization of organic acids, 1-2 μL of 1 M NaOH were added to solutions because the buffer capacity of our NMR buffer was not otherwise sufficient to maintain a constant pH of 4.5 for these standard solutions. By diluting the stock solutions, several calibration standards were further prepared. 600 μL of the final solution were poured into an NMR tube for direct measurement. For quantification, linear calibration curves were constructed from the standards by integrating the specific resonances for each compound against TSP as an intensity reference.

^{1}H and ^{13}C NMR Measurements at 400 MHz

All NMR measurements were performed on a Bruker Avance 400 Ultrashield spectrometer (Bruker BioSpin, Rheinstetten, Germany) equipped with a 5 mm SEI probe with Z-gradient coils, using a Bruker Automatic Sample Changer (B-ACS 120). ^{1}H NMR spectra were acquired at 300.0 K without sample rotation. 64 scans and 4 prior dummy scans of 65 k points were acquired with a spectral width of 19.9914 ppm, a receiver gain of 22.6, and an acquisition time of 4.096 s. Water suppression was achieved using the NOESY-presaturation pulse sequence (Bruker 1D noesygppr1d pulse sequence) with irradiation at the water frequency (1890.60 Hz) during the recycle and mixing time delays. ^{13}C NMR spectra were acquired using a Bruker zgpg30 pulse sequence with 1024 scans and 4 prior dummy scans. The sweep width was 238.9 ppm, the time domain of the FID was 66 k, receiver gain of 2050, and an acquisition time of 1.38 s. The data were acquired automatically under the control of ICON-NMR (Bruker BioSpin, Rheinstetten, Germany), requiring about 91 min per sample (for both ^{1}H and ^{13}C NMR). All NMR spectra were phased, baseline-corrected, and calibrated by the TSP signal at 0.0 ppm.

NMR Spectra Preprocessing and Chemometrics

Multivariate data analysis was performed using Unscrambler X version 10.0.1 (CAMO Software AS, Oslo, Norway) and Amix version 3.9.4 (Bruker BioSpin, Rheinstetten, Germany). First, to cope with small variations in pH or other sample conditions such as ionic strength or temperature, simple rectangular bucket tables were obtained from the complete sets of ^{1}H and ^{13}C NMR spectra. In both cases, scaling to total intensity was used. Further details on the bucketing process of NMR spectra for multivariate data analysis were previously described [33]. Before multivariate analysis, all data were mean centered. In the context of this study, principal component analysis (PCA) was used for visualization and as a tool for a differentiation between different honey types. During PCA, several new axes instead of old variables (buckets) called principal components (PC) are calculated and each NMR spectrum is projected on the selected PCs resulting in the scatter plot. We tested several spectral regions for calculation: δ 0–3 ppm, δ 3–6 ppm, δ 6–10 ppm, and δ 0–10 ppm for ^{1}H NMR and δ0–45 ppm, δ 45–135 ppm, δ 135–200 ppm, and δ 0–200 ppm for ^{13}C NMR. In cases when the whole spectral range was used, two preprocessing methods (scaling to unit variance and Pareto scaling) [34] as well as no scaling were tested for each data set in order to eliminate the magnitude effect of intensity variations in the δ 0–3 ppm and δ 6–10 ppm regions. The bucket width was 0.01 ppm in all cases. The technique of cross-validation was applied to determine the optimal number of principal components (PCs) required to obtain robust models. Kruskal-Wallis one-way analysis of variance, Shapiro-Wilk test, and Welch's t-test methods were used to analyze loadings plots in order to find out the most important buckets for differentiation. After the construction of the models to evaluate the classification performance, soft independent modeling of class analogy (SIMCA) and partial least squares-discriminant analysis (PLS-DA) classification methods were tested on randomly chosen test-set samples that were not included in the classification models.

RESULTS AND DISCUSSION

Nontargeted Multivariate Analysis

Figure 1 shows the complete ^{1}H NMR spectra of tilia (or linden), Robinia pseudoacacia (or acacia), and fir honeys. It can be seen that the mid-low-frequency region between δ 4.2 and 3.0 ppm is dominated with very intensive signals of the major monosaccharides (glucose and fructose) and disaccharides (maltose and sucrose). Other less intensive resonances are also observed in the δ 9.0–6.0 ppm and δ 2.5–1.0 ppm regions in ^{1}H NMR spectra of honey. The ^{13}C NMR spectra of honey investigated in our study were similar to those obtained previously in D_2O [35]. Most of the ^{13}C NMR signals were related to anomeric carbons of reducing and nonreducing sugars and were present in the δ 105–60 ppm region in the majority of samples. Due to the high spectral complexity, differences between honey types cannot be obtained without multivariate techniques. In general, the NMR spectra of our honey samples could be classified into two major groups: polyfloral samples (with floral and honeydew (forest flower honeys) as subgroups) and unifloral honeys (such as rape, tilia, chestnut, and others).

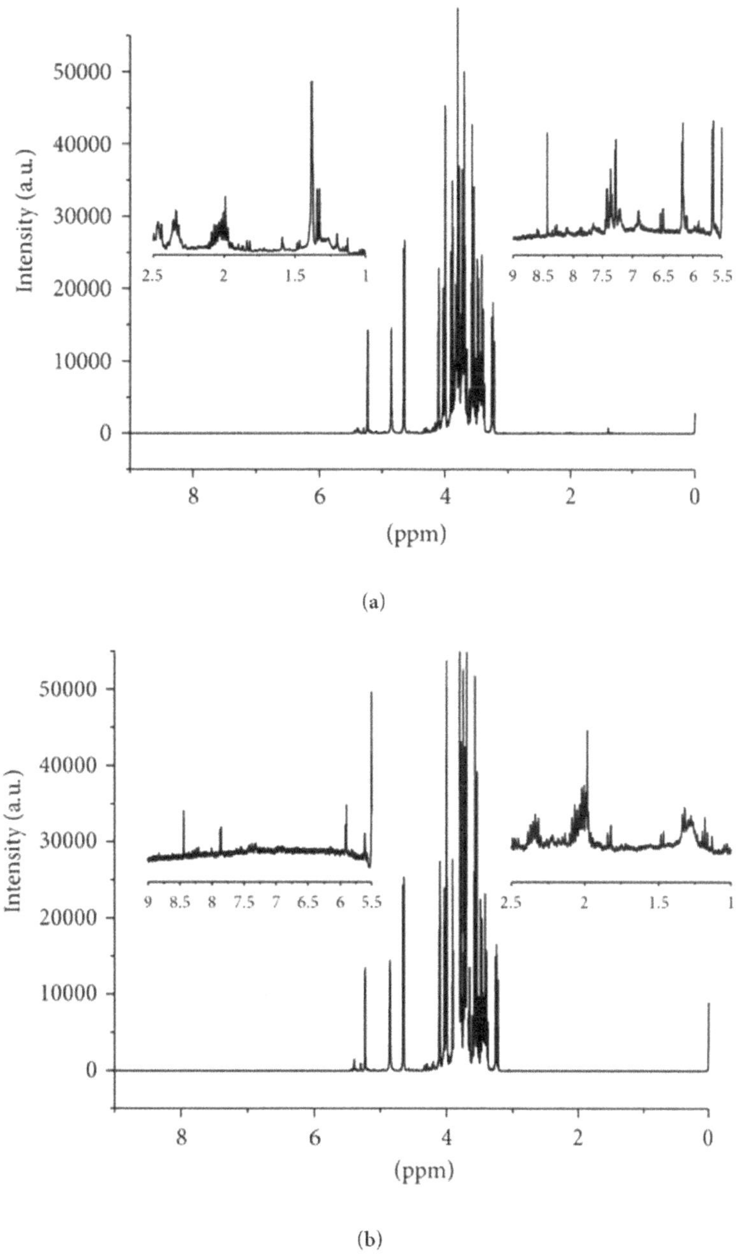

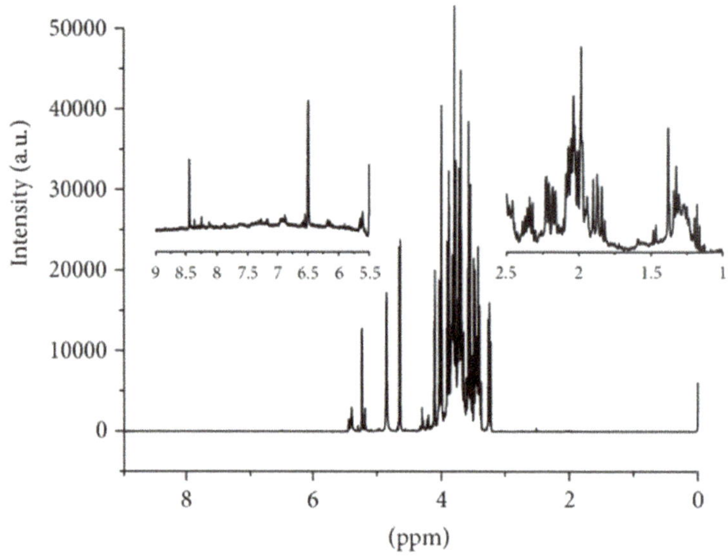

(c)

Figure 1: ^{1}H NMR spectra of Tilia (a), Robinia pseudoacacia (b), and fir (c) honeys. The inserts show expansions of the spectral regions δ 2.5–1.0 ppm and δ 9.0–5.5 ppm.

The PCA score plots generated using PC3-PC4 (^{1}H NMR) and PC1-PC3 (^{13}C NMR) to visualize the separation of the polyfloral honeys are shown in Figure 2, which clearly suggests that the samples can be separated into two groups: honeydew honeys clusters are in the region of positive PC3 (^{1}H NMR) and negative PC1 (^{13}C NMR), respectively; floral honey samples are located in the negative values of PC3 (^{1}H NMR) and positive PC1 (^{13}C NMR) values. Furthermore, Figure 3 suggests that we could not only differentiate the two main polyfloral classes of honey but clusters from several unifloral honeys were also clearly separated from each other. It should be noted that the ^{13}C NMR spectra provided inferior discrimination power as ^{1}H NMR spectra. For example, the PCA scores of rape and sunflower honeys or tilia, sunflower and Robinia pseudoacacia honeys were mixed in the same cluster (Figure 3(b)). With ^{1}H NMR spectra even minor differences in botanical composition can be traced (e.g., rape and rape/clover honeys are occurring in two separate clusters). On both scatter plots, honey samples from coniferous (spruce, fir, and pine trees) were clearly distinguished from the other honey types.

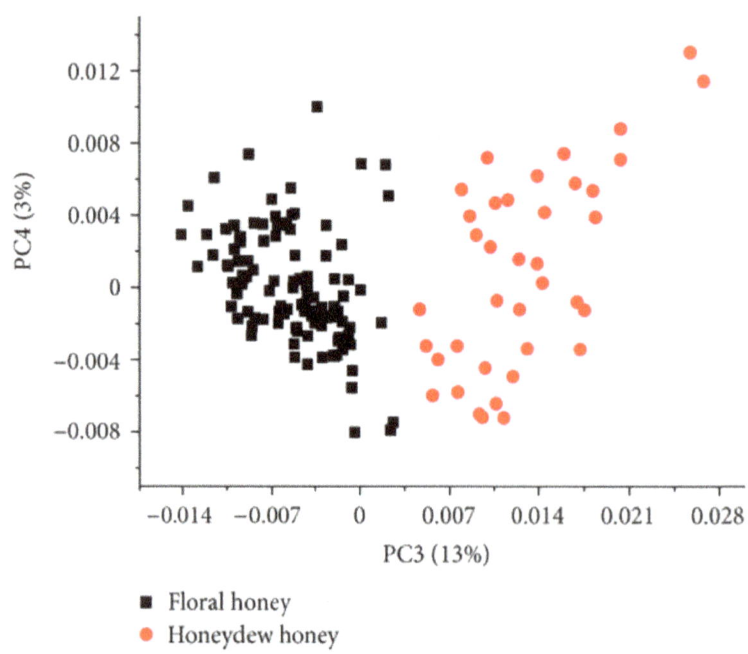

(a)

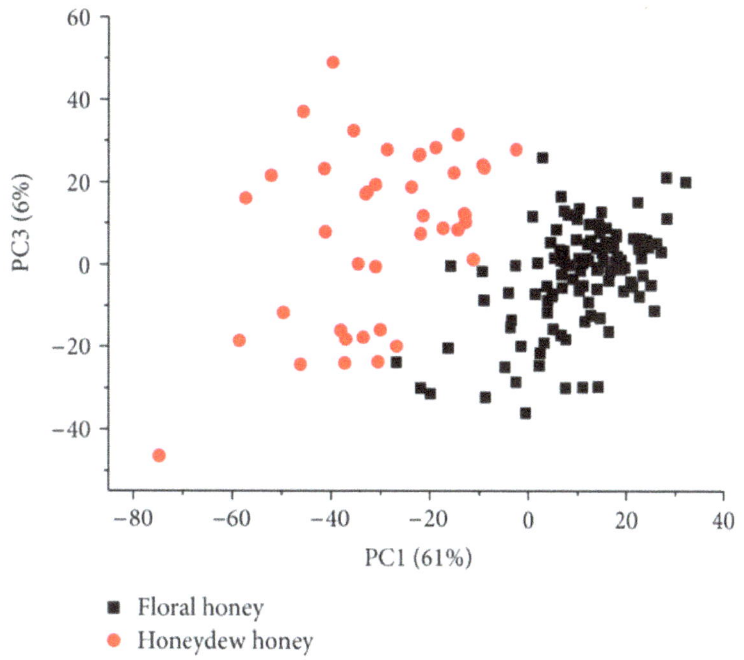

(b)

Figure 2: Scatter plot of the PCA scores of floral and honeydew honeys obtained from ^{1}H NMR (9–0.25 ppm; no scaling (a)) and ^{13}C NMR (200–0.25 ppm; scaling to unit variance (b)).

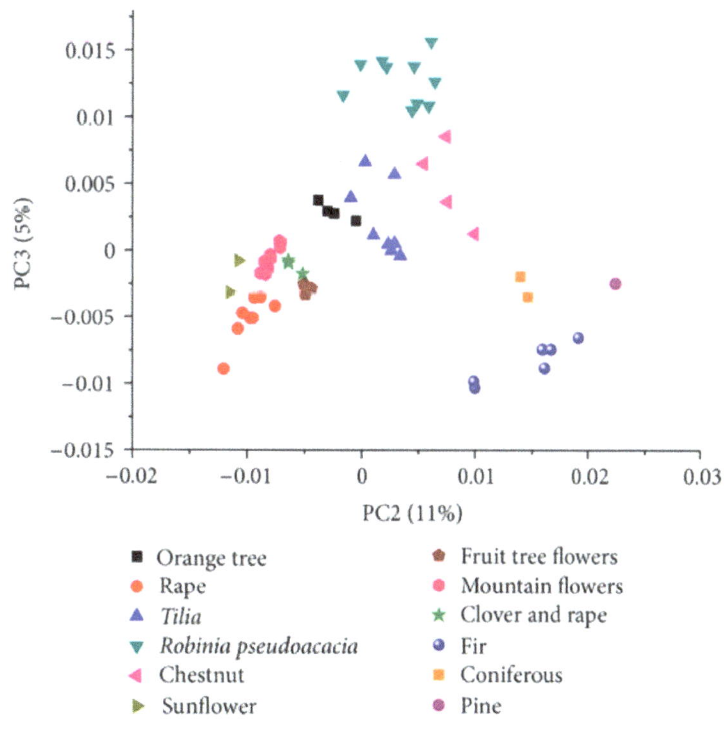

(a)

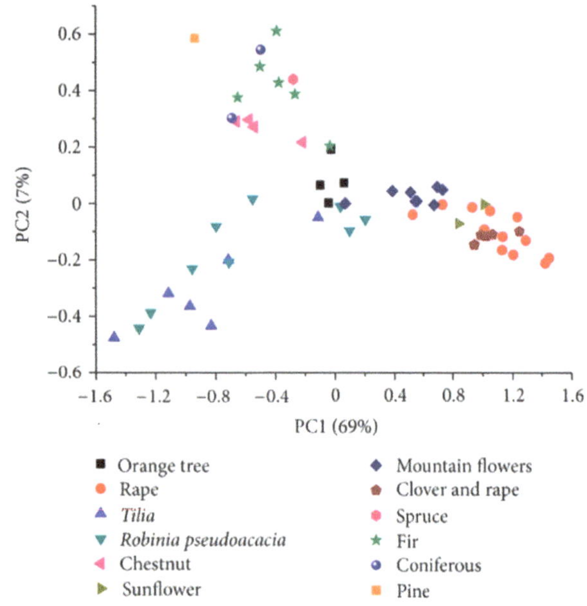

(b)

Figure 3: Scatter plot of the PCA scores of different types of unifloral honeys obtained from ^{1}H NMR (9–0.25 ppm; Pareto scaling (a)) and ^{13}C NMR (200–0.25 ppm; no scaling (b)).

Loadings plots allow to specify the variables (chemical shifts), which are responsible for the observed clustering for both data sets (^{1}H and ^{13}C NMR). Table 1 lists the most important buckets (signals) for different honey types obtained from the loadings plots. It was found that the signals of glucose and fructose play the key factor for differentiation, and this finding is in accordance with another NMR study of honeys [35]. However, resonances of minor compounds also play a certain role such as quinoline alkaloids and kynurenic acid for chestnut honey [36, 37] or unsaturated carboxylic acids for tilia honey [37]. Therefore, the ^{1}H NMR honey profile can be used for the identification of chemical markers of different botanical origin.

Next, it is interesting to show the predictive power of the chemometric methods by classifying new samples. To do this, two data analysis methods (SIMCA and PLS-DA) were evaluated for predicting class membership of honey samples from the ^{1}H NMR spectra. The independent test set for the floral/honeydew honey model (honeydew) consisted of 20 randomly selected objects (10 floral, 10 honeydew honeys). For the unifloral honey model, mountain n (n=2), rape (n=3), coniferous (n=2), Robinia pseudoacacia (n=3), and chestnut (n=1) honeys were selected for the test data set. The rest of the available ^{1}H NMR spectra were included in the calibration data set. All samples from both test sets were correctly recognized by SIMCA method at the 10% significance level. A prediction ability of 95% was obtained by PLS-DA for the honeydew/flower honey model. Thus, our results have shown that ^{1}H NMR coupled with multivariate statistics is an efficient tool for the classification of the different botanical origins of honey samples.

Table 1: The most important variables (buckets) for the differentiation of different botanical origins of honey (chemical shifts are given in ppm).

Botanical origin of honey	^{1}H NMR	^{13}C NMR
Honeydew/floral	1.17–1.20 3.26 3.90–4.12 3.43–3.50 3.57–3.68	101.49 95.85 69.19–76.11 82.01
Coniferous	3.84–3.87 3.61–3.67 3.76–3.77 4.05–4.07 4.20–4.22 5.24	62.84 69.44–69.53 71.49–71.47
Rape	4.64–4.66 3.40–3.46 3.36–3.38 4.05–4.07	75.81–75.88 74.22
Clover and rape	7.65–7.95 5.89 5.59	69.64–69.74 95.92
Sunflower	3.25–3.27	
Robinia pseudoacacia	3.69 3.75 3.43–3.44	69.24–69.75
Tilia	1.38–1.41 1.58–1.60 2.34–2.35 6.10–6.20 7.65–7.66 7.21–7.28	63.96–63.97 60.81 67.64 75.89
Chestnut	4.32–4.34 5.09 7.68–7.71 7.45–7.49	71.45–71.49 60.58
Mountain honey	3.88–3.91	95.90 60.79–60.80 69.68
Orange tree	8.13 3.88–3.91	62.75 98.08 75.49
Fruit tree flowers	3.44–3.45	—[a]

Quantification Studies

Besides the classification of botanical origin of honey samples, it would be advantageous to establish a NMR method for the quantification of main constituents in the honey matrix. As first evaluation, if a quantitative approach is at all possible from the NMR spectra, we measured 34 commercially available compounds that may be present in honey. Then, the spectra of standards were compared to the spectra of honey samples.

For most of the substances studied, direct quantification with integration is not possible due to extensive spectra overlap. As an example, the spectra of four carbohydrates are shown in Figure 4. Clearly, a large number of overlapped signals for all isomeric forms of sugars exist. Thus, for such compounds more advanced techniques, such as multivariate regression or curve deconvolution, are required for quantification. Moreover, the two main carbohydrates—glucose and fructose—have much higher peak intensities than other compounds and, therefore, obscure the rest of the signals.

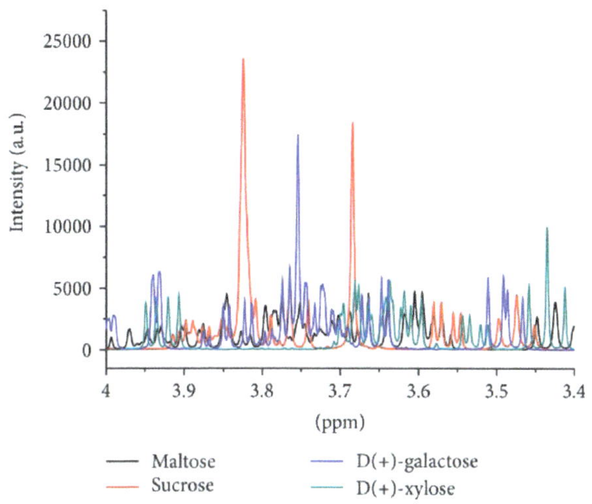

Figure 4: NMR spectra of maltose, sucrose, D(+)-galactose, and D(+)-xylose standards in the mid-field region.

However, we were able to find 13 metabolites for which at least one resolved unambiguous resonance could be identified. Selected ^{1}H NMR peaks (i.e., signals not overlapped or interfered by matrix) corresponding to each substance are shown in Table 2.

Table 2: NMR integration regions and investigated linear concentration ranges.

Compound	NMR range	Working range (mg/kg)
HMF	9.43–9.47 ppm (singlet)	20–600
Formic acid	8.44–8.47 ppm (singlet)	40–1400
Phthalic acid	7.53–7.48 ppm (multiplet)	30–900
Fumaric acid	6.53–6.55 ppm (singlet)	20–670
Pyruvic acid	6.42–6.45 ppm (singlet)	180–5000
L(+)-rhamnose	5.13–5.09 ppm (doublet)	160–2500
Glucose	4.63–4.65 ppm (singlet)	13–43000
Arabinose	4.52–4.54 ppm (singlet)	40–1200
L(+)-tartaric acid	4.32–4.35 (singlet)	90–2600
Fructose	4.14–4.08 (doublet)	11–46000
Malic acid	2.73–2.70 (two singlets)	30–1000
Citric acid	2.69–2.68 (singlet)	40–1300
Succinic acid	2.50–2.52 (singlet)	10–800

The high correlation coefficients ($R > 0.99$) obtained for each calibration graph indicate a good linear response within the concentration range studied for each compound. As an example, Figure 5 shows the NMR peaks of the main carbohydrate fructose and formic acid in authentic honey samples in comparison with two exemplary reference spectra. We applied the aforementioned procedure to the identification and direct quantification of the selected substances in authentic honeys of different floral types ($n = 20$) (Table 3). Only in two cases direct quantification of malic acid was not possible due to spectral interferences.

Table 3: Results of the quantitative determination of substances by NMR (values are given in g/kg honey).

Sample	HMF	Formic acid	Phthalic acid	Fumaric acid	Pyruvic acid	L(+)-rhamnose	Glucose	Arabinose	L(+)-tartaric acid	Fructose	Malic acid	Citric acid	Succinic acid
Manuka honey	0.03	0.15	n.d.	n.d.	n.d.	n.d.	218	0.41	0.31	284	n.d.	0.09	0.03
Flower honey	n.d.[a]	0.05	n.d.	n.d.	n.d.	n.d.	350	0.18	0.31	357	0.2	n.d.	0.01
Sunflower honey	0.05	0.06	n.d.	0.03	n.d.	n.d.	348	0.26	0.38	379	0.4	0.49	0.01
Honeydew honey	n.d.	0.08	n.d.	0.02	n.d.	n.d.	219	0.41	0.58	314	n.d.	0.25	0.17
Chestnut honey	0.06	0.71	n.d.	0.05	n.d.	n.d.	215	0.56	0.64	350	n.d.	n.d.	0.08
Flower honey	n.d.	0.15	n.d.	0.04	n.d.	n.d.	271	0.49	0.59	312	0.9	n.d.	0.1
Robinia pseudoacacia honey	0.06	0.04	n.d.	n.d.	n.d.	0.23	245	0.29	0.48	381	n.d.	0.17	0.01
Orange honey	0.09	0.06	n.d.	n.d.	n.d.	n.d.	281	0.47	0.61	390	0.25	0.26	0.02
Flower honey	n.d.	0.07	n.d.	n.d.	n.d.	2.6	337	0.26	0.35	358	—[b]	n.d.	0.02
Flower honey (mountain)	0.02	0.04	n.d.	n.d.	n.d.	n.d.	318	0.17	0.38	366	n.d.	0.1	n.d.
Rape honey	0.01	0.07	n.d.	n.d.	n.d.	n.d.	349	0.27	0.39	379	n.d.	n.d.	0.03
Honey from fruit trees	n.d.	0.12	n.d.	0.03	n.d.	n.d.	296	0.88	0.44	338	n.d.	n.d.	0.04
Flower honey	0.06	0.08	n.d.	0.03	n.d.	n.d.	297	0.24	0.46	366	n.d.	n.d.	0.03
Flower honey	0.06	0.09	n.d.	0.03	n.d.	n.d.	316	0.32	0.47	382	n.d.	n.d.	0.07
Chestnut honey	n.d.	0.8	n.d.	0.04	n.d.	n.d.	241	0.65	0.74	395	n.d.	0.19	0.18
Honeydew honey	0.03	0.11	n.d.	0.04	n.d.	n.d.	259	0.52	0.66	352	n.d.	0.28	0.28
Flower honey	0.03	0.11	n.d.	n.d.	n.d.	n.d.	260	0.34	0.54	406	n.d.	n.d.	0.03
Eucalyptus honey	0.06	0.1	n.d.	n.d.	n.d.	n.d.	302	1.72	0.43	346	0.28	n.d.	0.06
Flower honey	n.d.	0.63	n.d.	n.d.	1.45	n.d.	215	0.65	0.59	380	n.d.	n.d.	0.05
Flower honey with jelly Royal	0.03	0.05	n.d.	n.d.	n.d.	n.d.	315	0.4	0.46	368	—[b]	0.17	0.02

[a]Not detectable (0.03 g/kg (phthalic acid and malic acid), 0.02 g/kg (HMF and fumaric acid), 0.18 g/kg (pyruvic acid), 0.16 g/kg (L(+)-rhamnose), 0.04 (citric acid), and 0.01 g/kg (succinic acid)).

[b]Overlapped signal, direct quantification is not possible.

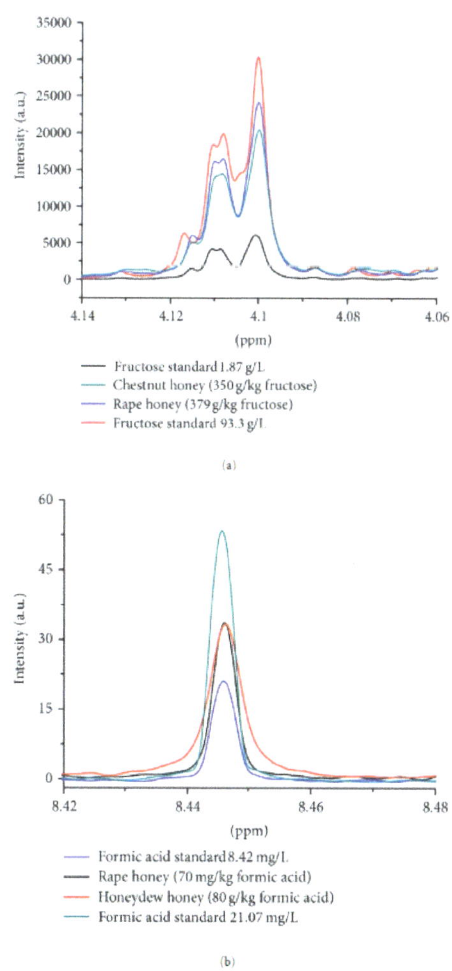

Figure 5: 400 MHz ^{1}H NMR spectra of fructose (a) and formic acid (b) in standard solutions and honey samples.

CONCLUSIONS

NMR spectroscopy has already been used in honey analysis to determine its botanical and geographical origin. In the paper of Lolli et al., 71 Italian honey samples (Robinia, chestnut, citrus, eucalyptus, and polyfloral) were analyzed by ^{1}H NMR and heteronuclear multiple bond correlation (HMBC) spectroscopy [35]. PCA and general discriminant analysis (GDA) were not able to group samples according to their botanical origin by using ^{1}H NMR data. Only with the use of 2D 1H-13C HMBC acceptable clustering occurred [35]. In another article by this research group, HMBC spectroscopy in combination with GDA was used to detect 10%, 20%, and 40% adulteration of authentic honey by commercial sugar syrups [38]. ^{1}H NMR spectroscopy and multivariate analysis techniques have also been used to classify honey into two geographical groups (non-Corsican and Corsican samples) [30]. 96.2% correct classification obtained by cross-validation was obtained for partial least squares-genetic programming (PLS-GP) algorithm. It should be also noted that the site-specific natural isotopic fraction NMR (SNIF-NMR) was not found to be successful for the characterization of geographical and botanical origins of honey [26]. However, to be used in practice, it would be necessary to extend the domain of application of the method for other unifloral honeys and to expand the database. Our study, which is the largest evaluation of honey samples by NMR so far, provides such an opportunity. We can conclude that our models can be used as a method to determine and monitor the botanical origin of honey samples.

With regard to quantification, NMR was only used for determining several saccharides with ^{13}C NMR [27] or methylglyoxal and amino acids with ^{1}H NMR [39] in honey matrices. Clearly, we expanded the range of substances that can be analyzed with NMR spectroscopy without preceding separation; ^{1}H NMR is also suitable for quantification of several aliphatic and aromatic acids as well as aldehydes.

In conclusion, it should be noted that honey is a very complex matrix endowed with very specific physicochemical properties. This complexity makes the analysis of honey difficult in terms of its different properties. Often the determination of botanical origin is complicated because of the incomplete correlation between analytical parameters: sensory properties and botanical identity.

Our investigation has shown that ^{1}H NMR spectra of honeys in combination with appropriate multivariate statistics can provide qualitative information about the botanical origin and represent a good basis for the identification of marker compounds for the specific honey types. Quantitative information about a number of major components is also available from the same spectra without need for chromatographic separation. In combination with multivariate data analysis, NMR spectroscopy possesses the speed, simplicity, and low cost per analysis required for a screening technique.

CONFLICT OF INTERESTS

The authors declare that there is no conflict of interests.

ACKNOWLEDGMENTS

The authors are grateful to Margit Böhm, Bernd Siebler, Jürgen Geisser, Antje Theiner, Beate Wagner, Karin Wolff, and Klaus Klusch for their excellent technical assistance. The views expressed in this paper do not necessarily reflect those of the Ministry of Rural Affairs and Consumer Protection.

REFERENCES

1. A. R. Aparna and D. Rajalakshmi, "Honey—its characteristics, sensory aspects, and applications," Food Reviews International, vol. 15, no. 4, pp. 455–471, 1999.
2. J. Louveaux, A. Maurizio, and G. Vorwohl, "Methods of Melissopalynology," Bee World, vol. 59, no. 4, pp. 139–157, 1978.
3. A. Zakaria, A. Y. Shakaff, M. J. Masnan et al., "A biomimetic sensor for the classification of honeys of different floral origin and the detection of adulteration," Sensors, vol. 11, no. 8, pp. 7799–7822, 2011.
4. P. Arquillue and A. H. Marteache, "Analysis of protein amino acids in some honeys from Los Monegros Spain," Alimentaria, vol. 24, no. 183, pp. 67–71, 1987.
5. J. Prodolliet and C. Hischenhuber, "Food authentication by carbohydrate chromatography," Zeitschrift fur Lebensmittel -Untersuchung und -Forschung, vol. 207, no. 1, pp. 1–12, 1998.
6. L. A. Marghitas, D. S. Dezmirean, C. B. Pocol, M. Ilea, O. Bobis, and I. Gergen, "The development of a biochemical profile of acacia honey by identifying biochemical determinants of its quality," Notulae Botanicae Horti Agrobotanici Cluj-Napoca, vol. 38, no. 2, pp. 84–90, 2010.
7. G. Beretta, R. Artali, E. Caneva, S. Orlandini, M. Centini, and R. M. Facino, "Quinoline alkaloids in honey: further analytical (HPLC-DAD-ESI-MS, multidimensional diffusion-ordered NMR spectroscopy), theoretical and chemometric studies," Journal of Pharmaceutical and Biomedical Analysis, vol. 50, no. 3, pp. 432–439, 2009.
8. J. Gilbert, M. J. Shepherd, M. A. Wallwork, and R. G. Harris, "Determination of the geographical origin of honeys by multivariate-analysis of gas-chromatographic data on their free amino-acid content,"Journal of Apicultural Research, vol. 20, no. 2, pp. 125–135, 1981.
9. A. Terrab, B. V. Castrillón, and M. J. D. Dapena, "Pollen analysis of honeys from the Gharb region (NW Morocco)," Grana, vol. 40, no. 4-5, pp. 210–216, 2001.
10. A. I. Ruiz-Matute, M. Brokl, A. C. Soria, M. L. Sanz, and I. Martínez-Castro, "Gas chromatographic-mass spectrometric characterisation of tri- and tetrasaccharides in honey," Food Chemistry, vol. 120, no. 2, pp. 637–642, 2010.
11. L. Nagy, R. Bátai, G. Nagy, and G. Nagy, "Application of copper electrode based amperometric detector cell for LC analysis of main sugar component of honey and nectar," Analytical Letters, vol. 43, no. 7, pp. 1411–1426, 2010.
12. S. R. Won, D. C. Lee, S. H. Ko, J. W. Kim, and H. I. Rhee, "Honey major protein characterization and its application to adulteration detection," Food Research International, vol. 41, no. 10, pp. 952–956, 2008.
13. J. Wang, M. M. Kliks, W. Qu, S. Jun, G. Shi, and Q.

X. Li, "Rapid determination of the geographical origin of honey based on protein fingerprinting and barcoding using MALDI TOF MS," Journal of Agricultural and Food Chemistry, vol. 57, no. 21, pp. 10081–10088, 2009.

14. J. A. F. Pierna, O. Abbas, P. Dardenne, and V. Baeten, "Discrimination of Corsican honey by FT-Raman spectroscopy and chemometrics," Biotechnology, Agronomy and Society and Environment, vol. 15, no. 1, pp. 75–84, 2011.

15. A. N. Batsoulis, N. G. Siatis, A. C. Kimbaris et al., "FT-Raman spectroscopic simultaneous determination of fructose and glucose in honey," Journal of Agricultural and Food Chemistry, vol. 53, no. 2, pp. 207–210, 2005.

16. M. García-Alvarez, J. F. Huidobro, M. Hermida, and J. L. Rodríguez-Otero, "Major components of honey analysis by near-infrared transflectance spectroscopy," Journal of Agricultural and Food Chemistry, vol. 48, no. 11, pp. 5154–5158, 2000.

17. L. Dvash, O. Afik, S. Shafir et al., "Determination by near-infrared spectroscopy of perseitol used as a marker for the botanical origin of avocado (Persea americana Mill.) honey," Journal of Agricultural and Food Chemistry, vol. 50, no. 19, pp. 5283–5287, 2002.

18. X. Zhu, S. Li, Y. Shan et al., "Detection of adulterants such as sweeteners materials in honey using near-infrared spectroscopy and chemometrics," Journal of Food Engineering, vol. 101, no. 1, pp. 92–97, 2010.

19. D. Cozzolino, E. Corbella, and H. Smyth, "Quality control of honey using infrared spectroscopy: a review," Applied Spectroscopy Reviews, vol. 46, no. 7, pp. 523–538, 2011.

20. L. Svecnjak, N. Biliskov, D. Bubalo, and D. Barisic, "Application of infrared spectroscopy in honey analysis," Agriculturae Conspectus Scientificus, vol. 76, no. 3, pp. 191–195, 2011.

21. S. Hennessy, G. Downey, and C. P. O'Donnell, "Attempted confirmation of the provenance of corsican PDO honey using FT-IR spectroscopy and multivariate data analysis," Journal of Agricultural and Food Chemistry, vol. 58, no. 17, pp. 9401–9406, 2010.

22. S. Sivakesava and J. Irudayaraj, "Classification of simple and complex sugar adulterants in honey by mid-infrared spectroscopy," International Journal of Food Science and Technology, vol. 37, no. 4, pp. 351–360, 2002.

23. M. M. Paradkar, J. Irudayaraj, and S. Sakhamuri, "Discrimination and classification of beet and cane sugars and their inverts in maple syrup by FT-Raman," Applied Engineering in Agriculture, vol. 18, no. 3, pp. 379–383, 2002.

24. H. Wang, Y. Q. Liu, H. M. Yang, Q. L. Guo, H. L. Shi, and L. B. Yan, "Determination of glucose, fructose, sucrose, maltose and lactose in sugar-free products by liquid chromatography-tandem mass spectrometry," Chinese Journal of Analytical Chemistry, vol. 38, no. 6, pp. 873–876, 2010.

25. T. Gallardo-Velázquez, G. Osorio-Revilla, M. Z. D. Loa, and Y. Rivera-Espinoza, "Application of FTIR-HATR spectroscopy and multivariate analysis to the quantification of adulterants in Mexican honeys,"Food Research International, vol. 42, no. 3, pp. 313–318, 2009.

26. J. F. Cotte, H. Casabianca, J. Lhéritier et al., "Study and validity of 13C stable carbon isotopic ratio analysis by mass spectrometry and 2H site-specific natural isotopic fractionation by nuclear magnetic resonance isotopic measurements to characterize and control the authenticity of honey," Analytica Chimica Acta, vol. 582, no. 1, pp. 125–136, 2007.

27. V. Mazzoni, P. Bradesi, F. Tomi, and J. Casanova, "Direct qualitative and quantitative analysis of carbohydrate mixtures using C-13 NMR spectroscopy: application to honey," Magnetic Resonance in Chemistry, vol. 35, pp. S81–S90, 1997.

28. R. Consonni and L. R. Cagliani, "Geographical characterization of polyfloral and acacia honeys by nuclear magnetic resonance and chemometrics," Journal of Agricultural and Food Chemistry, vol. 56, no. 16, pp. 6873–6880, 2008.

29. E. Schievano, E. Peggion, and S. Mammi, "H-1 nuclear magnetic resonance spectra of chloroform extracts of honey for chemometric determination of its botanical origin," Journal of Agricultural and Food Chemistry, vol. 58, no. 1, pp. 57–65, 2010.

30. J. A. Donarski, S. A. Jones, and A. J. Charlton, "Application of cryoprobe H-1 nuclear magnetic resonance spectroscopy and multivariate analysis for the verification of corsican honey," Journal of Agricultural and Food Chemistry, vol. 56, no. 14, pp. 5451–5456, 2008.

31. P. Sandusky and D. Raftery, "Use of selective TOCSY NMR experiments for quantifying minor components in complex mixtures: application to the metabonomics of amino acids in honey," Analytical Chemistry, vol. 77, no. 8, pp. 2455–2463, 2005.

32. DIN 10752, Untersuchung von Honig; Bestimmung des Wassergehaltes, Refraktometrisches Verfahren, Berlin, Germany, 2012.

33. D. W. Lachenmeier, W. Frank, E. Humpfer et al., "Quality control of beer using high-resolution nuclear magnetic resonance spectroscopy and multivariate analysis," European Food Research and Technology, vol. 220, no. 2, pp. 215–221, 2005.

34. R. A. van den Berg, H. C. J. Hoefsloot, J. A. Westerhuis, A. K. Smilde, and M. J. van der Werf, "Centering, scaling, and transformations: improving the biological information content of metabolomics data," BMC Genomics, vol. 7, article 142, 2006.

35. M. Lolli, D. Bertelli, M. Plessi, A. G. Sabatini, and C. Restani, "Classification of Italian honeys by 2D HR-NMR," Journal of Agricultural and Food Chemistry, vol. 56, no. 4, pp. 1298–1304, 2008.

36. P. Truchado, I. Martos, L. Bortolotti, A. G. Sabatini, F. Ferreres, and F. A. Tomas-Barberan, "Use of quinoline alkaloids as markers of the floral origin of chestnut honey," Journal of Agricultural and Food Chemistry, vol. 57, no. 13, pp. 5680–5686, 2009.

37. G. Beretta, E. Caneva, L. Regazzoni, N. G. Bakhtyari, and R. Maffei Facino, "A solid-phase extraction procedure coupled to H-1 NMR, with chemometric

analysis, to seek reliable markers of the botanical origin of honey," Analytica Chimica Acta, vol. 620, no. 1-2, pp. 176–182, 2008.

38. D. Bertelli, M. Lolli, G. Papotti, L. Bortolotti, G. Serra, and M. Plessi, "Detection of honey adulteration by sugar syrups using one-dimensional and two-dimensional high-resolution nuclear magnetic resonance,"Journal of Agricultural and Food Chemistry, vol. 58, no. 15, pp. 8495–8501, 2010.

39. J. A. Donarski, D. P. T. Roberts, and A. J. Charlton, "Quantitative NMR spectroscopy for the rapid measurement of methylglyoxal in manuka honey," Analytical Methods, vol. 2, no. 10, pp. 1479–1483, 2010.

CHAPTER 17

NEAR-INFRARED SPECTROSCOPY AS AN ANALYTICAL PROCESS TECHNOLOGY FOR THE ON-LINE QUANTIFICATION OF WATER PRECIPITATION PROCESSES DURING DANHONG INJECTION

Xuesong Liu, Chunyan Wu, Shu Geng, Ye Jin, Lianjun Luan, Yong Chen, andYongjiang Wu

INTRODUCTION

Danshen and Honghua are two traditional herbal medicines commonly used in China for the treatment of cardiovascular diseases. Danhong injections are prepared from aqueous Danshen (Salvia miltiorrhiza Bunge) and Honghua (Carthamus tinctorius L.) extracts. Danhong injections are applied to activate circulation of the blood and resolve stasis to promote regeneration [1], and the formula is applied extensively to tens of millions of patients for the prevention and treatment of coronary artery disease [2] in clinics in China. The crude extracts of Danshen and Honghua contain both fat-soluble and water-soluble compounds. The fat-soluble compounds, such as tanshinone, are primarily antibacterial and antiphlogosis [3]. Meanwhile, the water-soluble ingredients, such as phenolic acids, are the medicinal components in the Danhong injection and promote blood circulation and resolve stasis to promote regeneration [4]. The Danhong injection quality depends on separating and purifying the crude Danshen and Honghua extracts. The industrial production of the Danhong injection uses alcohol precipitation initially to purify the crude extract. This process removes some macromolecular substances such as proteins, polysaccharides, tannins, and pigments [5]. However, certain fat-soluble compounds, primarily tanshinones, are removed via subsequent water precipitations [6].

As a separation technology for producing botanical medicines, water precipitation has been researched less than ethanol precipitation. This work uses water precipitation to purify the ethanol precipitation solution for a Danhong injection to describe its separation characteristics. During the water precipitation process, several active ingredients more or less precipitated with impurities via encapsulation. To data, few studies have examined the loss of active ingredients, such as danshensu (DSS), protocatechuic aldehyde (PA), rosmarinic acid (RA), and salvianolic acid B (SAB) [7, 8]. Danshensu [3-(3,4-dihydroxyphenyl)-2-hydroxy-propanoic acid; DSS] is the major biologically active and water soluble component in Danshen. In vitro and in vivo experiments have demonstrated that DSS has antioxidant, antiatherosclerotic, and endothelial cell protective effects [9, 10]. PA inhibits migration and proliferation of vascular smooth muscle cells and intravascular thrombosis [11]. RA prevents against memory deficits in ischemic mice [12] and has potent antiviral activity against enterovirus 71 infections [13]. SAB has antipulmonary fibrotic activity [14] and attenuates lung inflammation induced by cigarette smoke in mice [15]. Guaranteeing the final Danhong injection quality requires water precipitation research.

Citation: Xuesong Liu, Chunyan Wu, Shu Geng, et al., "Near-Infrared Spectroscopy as an Analytical Process Technology for the On-Line Quantification of Water Precipitation Processes during Danhong Injection," International Journal of Analytical Chemistry, vol. 2015, Article ID 313471, 10 pages, 2015. doi:10.1155/2015/313471

Detection using on-line, real-time measurements would greatly improve the process efficiency for water precipitation.

Radiation in the near-infrared (NIR) energetic range excites overtone and combination vibrations in the sample material, revealing a high content of chemical as well as physical information at the same time dispersed in the whole spectral area [16]. NIR region covers wavelengths from 780 nm up to 2.5 μm and mainly reflects the overtones and combinations of fundamental vibrations of C-H, N-H, O-H, and S-H bond in organic molecules [17, 18]. Multivariate data evaluation methods were employed to extract the NIR spectral features and to investigate the correlation between the spectral data and the concentration (content) variables measured by the reference assays [19, 20]. Together with multivariate data evaluation methods, NIR spectroscopy has become a popular process control tool with a broad range of applications for precise material characterization [21]. In 2004, the Food and Drug Administration (FDA) issued a guidance document to the pharmaceutical industry regarding the implementation of process analytical technology (PAT) [22]. NIR spectroscopy was applied to develop a fast and reliable quality control system for a pharmaceutical substance to support information obtained through PAT surveillance of its manufacturing process. NIR spectroscopy with multivariate classification as a PAT production control supplement was applied to reliably determine the quality of the end product at minimum measuring effort and to at least partially replace laborious, conventional analysis methods in a long term [23].

As a commonly used PAT tool, NIR spectroscopy is a fast and nondestructive technique; it requires no sample preparation and will not produce any waste products. NIR spectroscopy can be considered as a powerful tool for raw material testing, product quality control, and process monitoring [24–26] in the pharmaceutical industry and, moreover, it has gained wide acceptance by pharmaceutical manufacturers and regulatory agencies. NIR spectroscopy has been investigated and successfully applied to on-line detection during the alcohol precipitation processes for Danhong injections [27]. The variations between the NIR spectra and intermediate quality attributes were investigated. The results indicate the NIR models were suitable for the real-time on-line of alcohol precipitation. A method [28] based on high-performance liquid chromatography combined with chromatographic fingerprint analysis was developed to quantitatively analyze ten Danhong injection components. This outstanding and precise method simultaneously detects multi-indexes, including DSS, PA, RA, and SAB. However, investigation of the variations in the important intermediate quality attributes during water precipitation processes for Danhong injections was not performed.

This paper used NIR spectroscopy for the on-line monitoring of water precipitation processes for Danhong injections. As an analytical process technology, NIR spectroscopy was tested to monitor the quality during water precipitation. DSS, PA, RA, and SAB, the principal bioactive components in the Danhong injection, were selected as the modeling parameters [29].

MATERIALS AND METHODS

Materials

The Danhong alcohol precipitation solution was provided by Heze Buchang Pharmaceutical Co., Ltd. (Heze, Shandong, China). The PA standard was purchased from the National Institute for the Control of Pharmaceutical and Biological Products (Beijing, China). The DSS, RA, and SAB standards were all purchased from Chengdu Must Bio-technology Co., Ltd. (Chengdu, China). The standard purities were above 98%. Other reagents were of analytical grade.

On-Line NIR Monitoring Device

The on-line NIR monitoring device for the water precipitation process is shown in Figure 1. This device consisted of a Bruker Matrix-F Fourier transform NIR spectrometer (Bruker Optic Inc., Germany), peristaltic pump, duplex filters (100-mesh sieve for removing deposit and reducing velocity), two hand valves, and one sampling valve. The on-line measurements used two fiber optic probes designed to transmit NIR radiation through a 2 mm flow cell (Solvias, Germany) connected to the Matrix-F Fourier transform NIR spectrometer. When the water precipitation solution passed through the flow cell, the flow rate, solid impurities, and bubbles significantly influenced the collected NIR spectra [30]. However, avoiding such influences is difficult for in-line methods, for example, inserting the NIR probe into the reactor, as described in [31]. To eliminate these influences during this study, the solution was passed through the duplex filters before entering the circulation loop to ensure an even distribution of the chemical components in the reactor. The peristaltic pump maintained the solution flow rate through the flow cell and reduced the effect of the flow rate on the NIR spectra. Hand valves and sampling valves were designed for sampling during the NIR spectra collection.

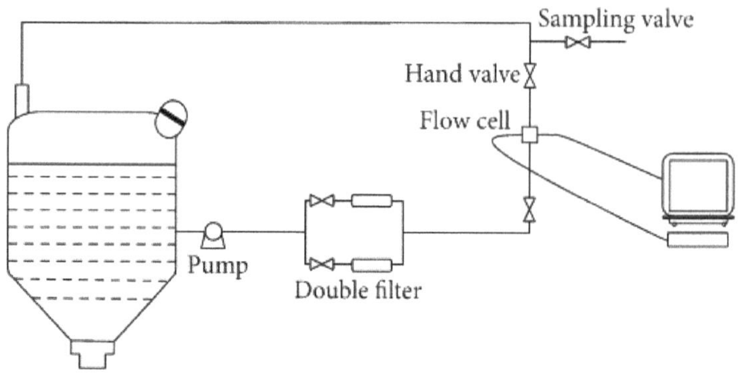

Figure 1: Scheme for the on-line NIR spectroscopy detection device.

On-Line Collection of NIR Spectra and Samples

Spectra were collected from 4500 to 12000 cm^{-1} with a resolution of 8 cm^{-1}. Each spectrum was collected in the absorbance mode using the average of 32 scans. To capture the variation during water precipitation, NIR spectra were collected every 30 seconds in real-time. However, samples were added every 2 minutes via the sampling valve. Meanwhile, the frequency parameter for the peristaltic pump was set to 20.00 Hz to continuously pump water precipitation solution through the flow cell. At the moment, the flow rate in the flow cell was 360 L/h.

The water precipitation experiment was performed five times, and 146 samples were collected. The first batch was used for the preexperiment, the fifth batch was used as a prediction set for validation, and remaining batches were identified as calibration sets for the model. Number of samples in calibration set and in validation set was 89 and 32, respectively. In total, all spectral pretreatments and chemometrics analyses were performed using OPUS software (version 7.0, Bruker, Germany).

High Performance Liquid Chromatography (HPLC) Analysis Method

This paper used HPLC as the reference method to simultaneously quantify the DSS, PA, RA, and SAB.

HPLC Conditions

The chromatographic analysis was performed on an Agilent 1200 HPLC system (Agilent Technologies, USA) equipped with a diode array detector (wavelength range from 190 nm to 949 nm) and an Agilent ChemStation software used for recording chromatograms. All samples were separated on an Agilent Eclipse-C18 analytical column (4.6 mm × 250 mm, 5 μm particle size) at 35°C. The mobile phase was a mixture of (A) methanol and (B) aqueous solutions containing 0.5% (v/v) formic acid. The gradient elution procedure was as follows: initial 9% (A); 0–20 min, linear change from 9% to 39% (A); 20–36 min, linear change from 39% to 47% (A); 36–39 min, linear change from 47% to 90% (A); hold at 90% (A) for 39–45 min. The reequilibration duration between individual runs was 10 min. The mobile phase flow rate was fixed to 1.0 mL/min. The detection wavelength was 280 nm at 0–13 min, was converted to 403 nm at 13–21 min, and reconverted into 280 nm at 21–45 min. The samples during this research were diluted before centrifuging at a rotating speed of 1500 rpm for 10 min, the supernatant was filtered with a 0.45 μm Nylon microfiltration membrane (Beijing Envta Technology CO., Ltd., the location of the producer is room 210 C, Jiaxin Business Building, No. 59, Annei street, Dongcheng district, Beijing, China), and 5 μL of the filtrate was injected into the HPLC system for analysis. One chromatogram including peak assignment is shown in Figure 2.

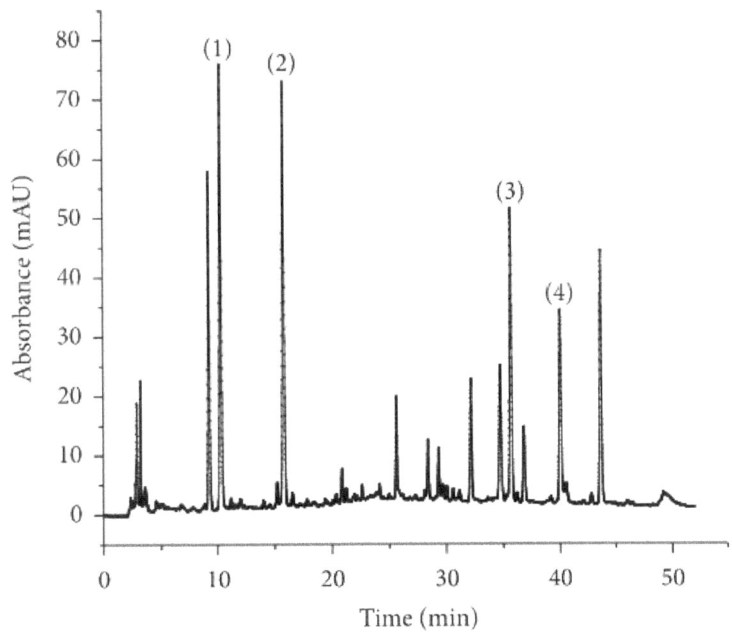

Figure 2: Typical chromatogram of sample. (1) DSS, (2) PA, (3) RA, and (4) SAB.

HPLC Method Validation

The developed HPLC method was validated based on its linearity, precision, stability, and accuracy. Parameters of HPLC method validation 0 10 20 30 40 50 0 10 20 30 40 50 60 70 80 Absorbance (mAU) Time (min) (1) (2) (3) (4) Figure 2: Typical chromatogram of sample. (1) DSS, (2) PA, (3) RA, and (4) SAB. for DSS, PA, RA, and SAB were presented in Table 1. Standard curves were obtained from the linear regression for the peak area versus the respective concentrations for the standard analytes. The regression equations and correlation coefficients (R) were $y = 3.241x - 1.152$ ($R = 0.9999$), $y = 22.697x - 2.542$ ($R = 0.9999$), $y = 8.554x - 1.459$ ($R = 0.9999$), and $y = 5.984x - 64.335$ ($R = 0.9999$) for DSS, PA, RA, and SAB, respectively. Good linear relationships were achieved for the range from 9.84 to 147.6, 1.13 to 22.68, 4.83 to 57.98, and 52.80 to 1056 μg/mL for DSS, PA, RA, and SAB, respectively. In this calibration range, six standards (concentrations) were applied and two consecutive injections were performed

Table 1: HPLC method validation for DSS, PA, RA, and SAB.

Ingredients	Regression equations	R	Range (μg/mL)	RSD of repeatability	RSD of stability	Average recoveries
DSS (mg/mL)	$y = 3.241x - 1.152$	0.9999	9.84~147.6	0.18%	0.88%	101.2%
PA (mg/mL)	$y = 22.697x - 2.542$	0.9999	1.13~22.68	1.08%	0.38%	98.2%
RA (mg/mL)	$y = 8.554x - 1.459$	0.9999	4.83~57.98	0.46%	0.60%	99.2%
SAB (mg/mL)	$y = 5.984x - 64.335$	0.9999	52.80~1056	0.10%	0.17%	100.8%

To determine the method repeatability, a sample (number 12 from the second batch) was randomly selected and analyzed by consecutively injecting six needles under the above HPLC conditions. The relative standard deviation (RSD) in the peak areas for DSS, PA, RA, and SAB was 0.18%, 1.08%, 0.46%, and 0.10%, respectively. These results suggest that the instrument precision was acceptable. Additionally, the stability was tested by analyzing the same sample every 2 h for 12 h at room temperature. The RSD in the peak areas for DSS, PA, RA, and SAB were 0.88%, 0.38%, 0.60%, and 0.17%, respectively, which indicates that samples were stable for 12 h. The accuracy was evaluated using a recovery test via the standard addition method at three concentrations. Accordingly, previously analyzed samples (number 12 from the second batch) were spiked with DSS, PA, RA, and SAB to 0.5, 1.0, and 1.5 times their amount in the sample, respectively. The average recoveries were 101.2%, 98.2%, 99.2%, and 100.8% for DSS, PA, RA, and SAB with RSD below 2.0%. The above validation data indicates the developed HPLC method was acceptable for determining DSS, PA, RA, and SAB.

Chemometrics and Data Analysis

The spectral data were manipulated by identifying the usable spectral regions selecting appropriate preprocessing methods and correlating to the quantitative HPLC data with PLSR to develop the calibration models. During this work, the raw spectra were pretreated using derivatives, straight line subtraction (SLS), vector normalization (VN), standard normal variate (SNV), and multiplicative scatter correction (MSC). Derivatives, including the first derivative (1st Der) and second derivative (2nd Der), were introduced to remove any spectral baseline drift [32]. The SLS corrected the baseline. The VN and MSC are commonly used to eliminate irrelevant information in the spectra from unknown sources such as surface irregularities, distance variation of sample, and detector [33]. Specifically, a spectrum undergoes VN by subtracting the average intensity, and the MSC corrects any multiplicative effects due to scattering via the linear transformation of each spectrum. The SNV was also considered as a scatter correction method [34]. The partial least squares regression (PLSR) helped correlate the pretreated spectral data to the indicator contents to construct the calibration models [35]. To avoid under- or overfitting, the optimum number of latent variables (LVs) in the PLSR models was determined via the Leave-One-Out cross-validation method [36]. The PLSR computations were performed using the OPUS software (version 7.0, Bruker, Germany).

The predictive capabilities of the developed PLSR models were estimated via the coefficient of determination (R^2), root mean square errors of calibration and prediction (RMSEC and RMSEP, resp.), relative standard errors of calibration and prediction (RSEC and RSEP, resp.), root mean squares error of cross-validation (RMSECV), and ratio of prediction to deviation (RPD) for the calibration set. An excellent model generally has low RMSEC, RMSEP, and RMSECV; high R^2 ; and a small difference between the RMSEC and RMSECV. Moreover, the RMSEP value should be close to the RMSEC value.

RESULTS AND DISCUSSION

Results of HPLC Determination

The dynamic course of the DSS, PA, RA, and SAB concentrations during the water precipitation process for Danhong injection is shown in Figure 3. The time evolution curves from the water precipitation process were divided into two phases based on the zero minute. One phase referenced the water-adding stage before 0 min, while the other was the thermostatic stage from 0 to 100 min. For the five batches, the first batch began its water precipitation process two days later than the others. Thus, the first batch was different from the other four batches for DSS and SAB and indicated the water precipitation process should be performed immediately after the alcohol precipitation. The DSS, PA, RA, and SAB concentrations at the end of the water precipitation are summarized in Table 2. The average DSS, PA, RA, and SAB concentrations for the four parallel batches were 1.28, 0.260, 0.564, and 2.07 mg/mL, respectively. The RSD values were 2.19%, 4.54%, 3.13%, and 7.35% for the DSS, PA, RA, and SAB concentrations, respectively.

Table 2: DSS, PA, RA, and SAB concentrations for the final water precipitation (mg/mL).

Ingredients	Batches					Average	RSD (%)
	Fist	Second	Third	Fourth	Fifth		
DSS (mg/mL)	1.45	1.32	1.26	1.27	1.29	1.28	2.19
PA (mg/mL)	0.257	0.272	0.248	0.252	0.267	0.260	4.54
RA (mg/mL)	0.541	0.582	0.541	0.560	0.572	0.564	3.13
SAB (mg/mL)	2.71	2.22	1.93	1.95	2.19	2.07	7.35

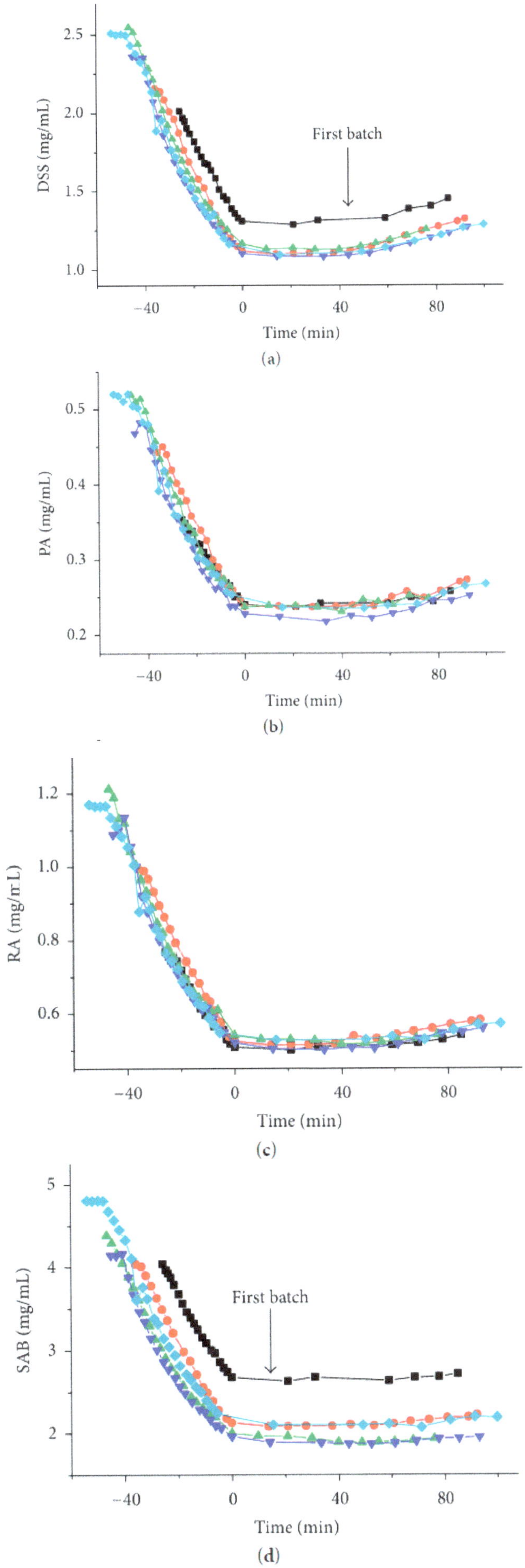

Figure 3: DSS (a), PA (b), RA (c), and SAB (d) concentrations measured by HPLC.

Selection of NIR Spectral Regions

Raw NIR spectra for the water precipitation solutions ranging from 4000 to 12000 cm^{-1} were acquired during the monitoring period as shown in Figure 4. The 4000 to 4600 cm^{-1} region had some noise, which was caused by the optical fiber absorption [37]. Simultaneously, the noise was significantly enhanced from 4600 to 5450 cm^{-1} and 6100 to 7500 cm^{-1} (strong signals from water) [38] due to the flow rate of the water precipitation solution through the flow cell. Additionally, the region from 9400 to 12000 cm^{-1} exhibited low intensities and a low signal-to-noise ratio [39]. Furthermore, spectral regions with absorbances equal to or high than 1.5, such as 4000 to 5450 cm^{-1}, were considered inappropriate for the spectral analysis due to their zero transmissivity, and they were considered saturated [36]. Hence, removing the abovementioned regions may improve the calibration model accuracy. The remaining regions, 5450–6100 cm^{-1} and 7400–9500 cm^{-1}, were utilized for the DSS, PA, RA, and SAB models. To confirm the choice of spectral region was valid, the correlation coefficients for the spectra were investigated as shown in Figure 5. The coefficients for the selected regions were generally above 0.6.

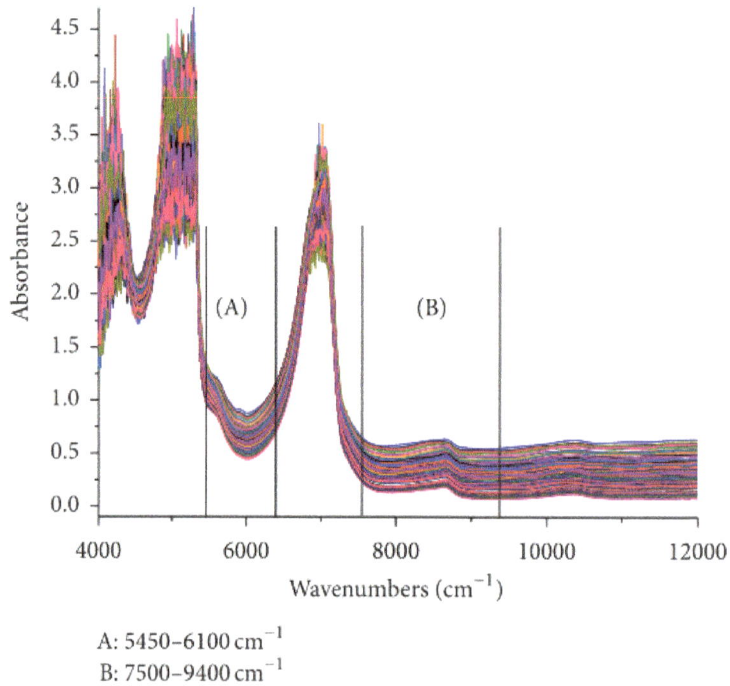

Figure 4: Raw NIR spectra for the Danhong injection water precipitation process.

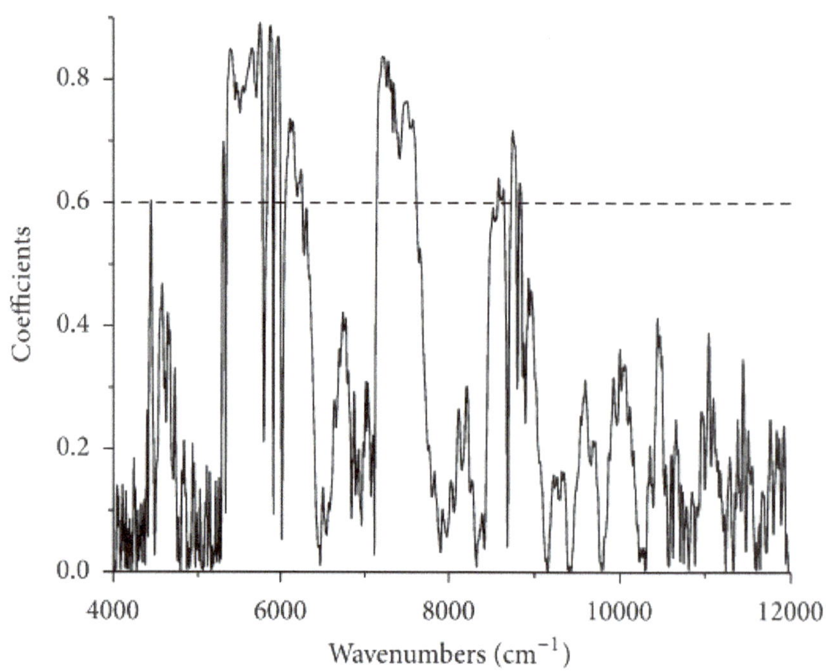

Figure 5: Correlation coefficients for the 1st Der NIR spectroscopy.

Spectral Data Pretreatments

Different spectral pretreatments were investigated to optimize the calibration performance. Several preprocessing methods were tested with the NIR spectra, including VN, MSC, constant offset elimination, first derivative, SLS, and their combinations. The VN and MSC are commonly used to eliminate irrelevant information in the spectra from unknown sources such as surface irregularities, distance variation of sample, and detector. The SNV transformation was applied to correct for light scattering and to reduce the changes in the path length. The 1st Der can reduce peak overlap and eliminate constant and linear baseline drifts. The SLS corrected the baseline. The detailed description of the techniques can be found in [40, 41]. During the water precipitation process, the flow rate influences the spectral baseline, and derivatives not only reduce peak overlap, but also eliminate linear baseline drifts [42]. However, the second derivative operation decreased the signal-to-noise ratio. Therefore, the first derivative was used to eliminate the spectral differences from baseline shifts. To avoid enhancing the noise, all of the derivative spectra were smoothed with a 17-point Savitzky-Golay filter [43]. The first derivative spectra with a 17-point Savitzky-Golay smoothing pretreatment, obviously reflected the quality attributes investigated in the 5450–6100 cm^{-1} and 7700–8700 cm^{-1} regions. The pretreated spectra were depicted in Figure 6. Figure 6(a) showed the first derivative spectra with a 17-point Savitzky-Golay smoothing pretreatment from 4000 to 12000 cm^{-1}. The spectral regions of wavenumbers (1) and wavenumbers (2) in Figure 6(a) were from 5450 to 6100 cm^{-1} and 7700 to 8700 cm^{-1}, respectively, as shown in the Figures6(b) and 6(c).

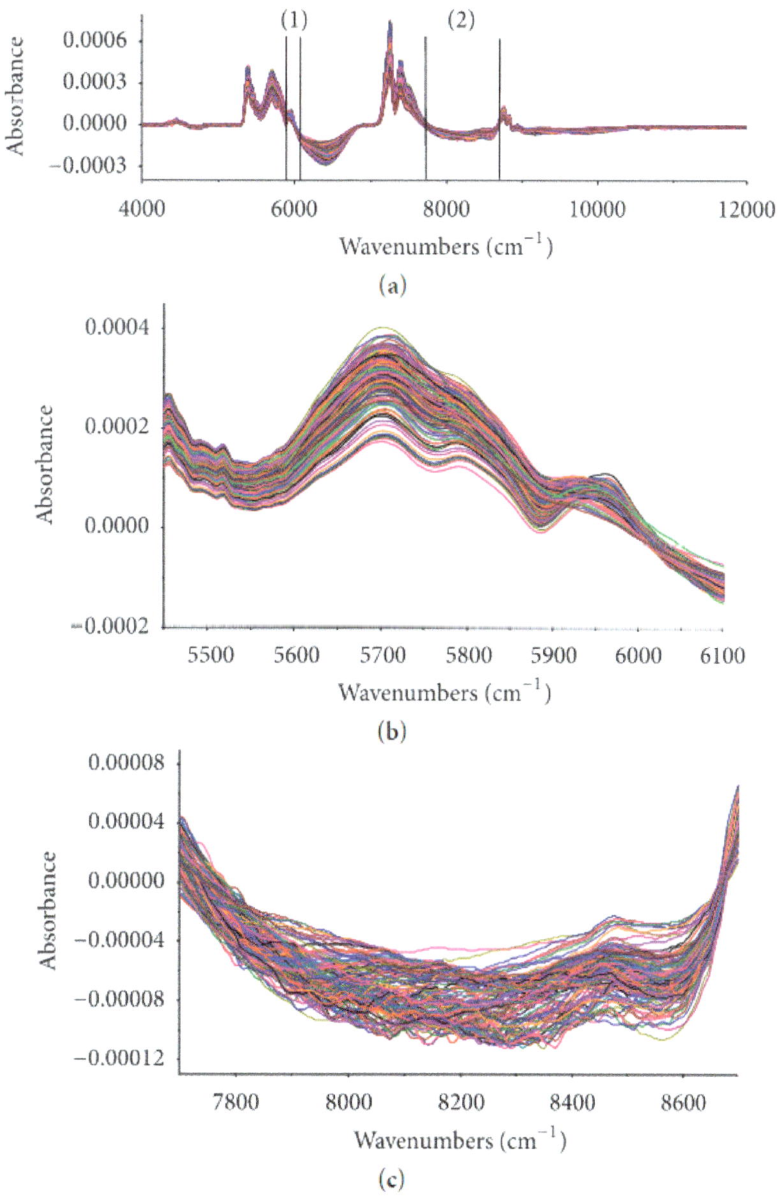

Figure 6: First derivative spectra with a 17-point Savitzky-Golay smoothing pretreatment. (a) The first derivative spectra with a 17-point Savitzky-Golay smoothing pretreatment from 4000 to 12000 cm^{-1}; (b) the wavenumbers (1) from 5450 to 6100 cm^{-1}; (c) the wavenumbers (2) from 7700 to 8700 cm^{-1}.

Comparing several different preprocessing techniques for modeling (results summarized in Table 3) indicates the first derivative preprocessing method yielded the best DSS, PA, RA, and SAB models.

Table 3: Parameters for the PLSR models with different spectral pretreatment methods.

Pretreatments	R^2				RMSEC (mg/mL)				RMSECV (mg/mL)			
	DSS	PA	RA	SAB	DSS	PA	RA	SAB	DSS	PA	RA	SAB
Raw spectra	98.08	97.81	97.76	95.97	0.0642	0.0141	0.0312	0.177	0.0684	0.0150	0.0333	0.1886
SLS	87.26	88.99	89.22	91.15	0.1540	0.0341	0.0716	0.263	0.1619	0.0356	0.0748	0.2765
VN	97.61	97.64	97.57	95.86	0.0716	0.0147	0.0340	0.180	0.0763	0.0157	0.0363	0.1905
1st Der	98.20	98.14	98.06	96.10	0.0619	0.0130	0.0304	0.174	0.0651	0.0137	0.0320	0.1806
2nd Der	81.38	80.63	83.88	85.45	0.2000	0.0420	0.0876	0.337	0.2024	0.0425	0.0887	0.3411
1st Der + SLS	89.20	89.58	90.79	88.84	0.1520	0.0308	0.0662	0.295	0.1577	0.0320	0.0687	0.3042

SLS: straight line subtraction.
VN: vector normalization.
1st Der: first derivative.
2nd Der: second derivative.
R^2: coefficient of determination.
RMSECV: root mean squares error of the cross-validation.
RMSEC: root mean square error of the calibration.

Establishing the Calibration Models

Four calibration models were established using the first derivative pretreatment in the regions from 5450 to 6100 cm^{-1} and 7700 to 8700 cm^{-1}. The regression plots between the measured and predicted values using HPLC and NIR, respectively, for the DSS, PA, RA, and SAB concentrations in calibration and validation sets are depicted in Figure 7. Based on the data in Table 3, the R^2 , RMSECV, and RMSEC values for the four first derivative pretreatment models were superior to the spectra using other preprocessing methods. Parameters of selected NIR models for DSS, PA, RA, and SAB were shown in Table 4. The R^2 values were all above 0.96. The calculated RSEC values were 3.71%, 3.72%, 3.91%, and 5.72% for DSS, PA, RA, and SAB, respectively. The RPD values were 7.49, 7.34, 7.18, and 5.06 for DSS, PA, RA, and SAB, respectively. Generally, RPD values above three are considered good for prediction purposes [44, 45]. Based on the above results, the four models were reliable and could accurately predict the quality attributes, and the water precipitation process can be monitored via on-line NIR using the developed quantitative models.

Table 4: Parameters of selected NIR models for DSS, PA, RA, and SAB.

Ingredients	R^2	RSEC	RPD of calibration	RSEP	RPD of prediction
DSS	0.9820	3.71%	7.49	3.86%	8.83
PA	0.9814	3.72%	7.34	3.26%	8.39
RA	0.9806	3.91%	7.18	3.02%	9.25
SAB	0.9610	5.72%	5.06	4.02%	8.22

Figure 7: Calibration models for DSS (a), PA (b), RA (c), and SAB (d).

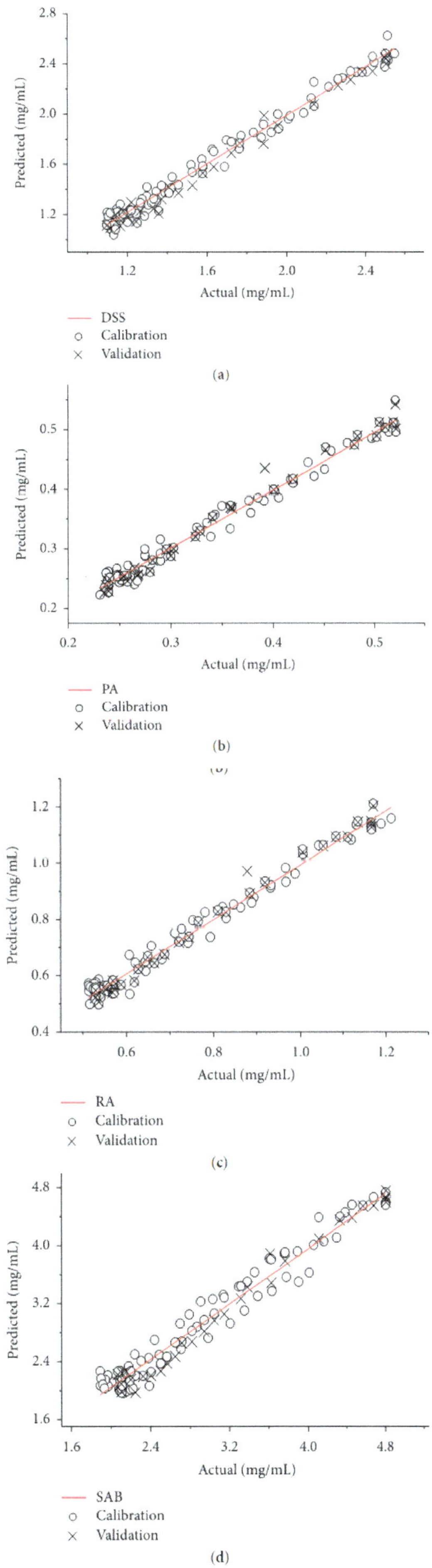

On-Line Quantitative Monitoring

The established NIR models were used to monitor the water precipitation process and predict the on-line DSS, PA, RA, and SAB concentrations in real-time. The RSEP and RPD values were used to assess the accuracy of the predicted results. If the RSEP value was below 20% and the RPD value was above 3.0, the established model accuracy was acceptable [46].

To validate the applicability, the fifth sample batch (the prediction set) was scanned on-line and predicted using the established models. Figure 8 shows the concentration trends predicted by the NIR in real-time and those obtained from the reference assay agreed well. In addition, the RSEP and RPD values (presented in Table 4) were calculated as 3.86% and 8.83 for DSS, 3.26% and 8.39 for PA, 3.02% and 9.25 for RA, and 4.02% and 8.22 for SAB, respectively. Validated results with RSEP values below 10% and RPD values above 3.0 met the practical requirements. All of these data demonstrated the developed models had high predictive accuracy during the on-line monitoring DSS, PA, RA, and SAB for the water precipitation process in Danhong injection. Hence, NIR spectroscopy is a suitable on-line quality control analysis technology for practical applications.

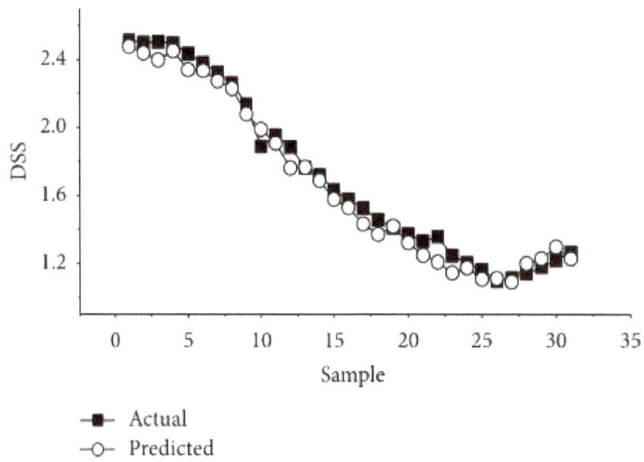

(a)

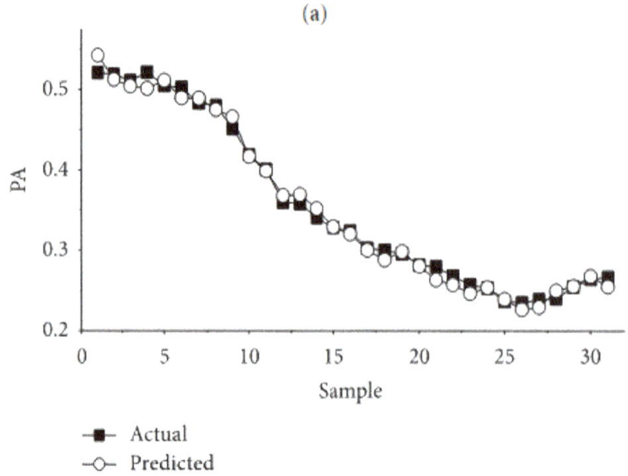

(b)

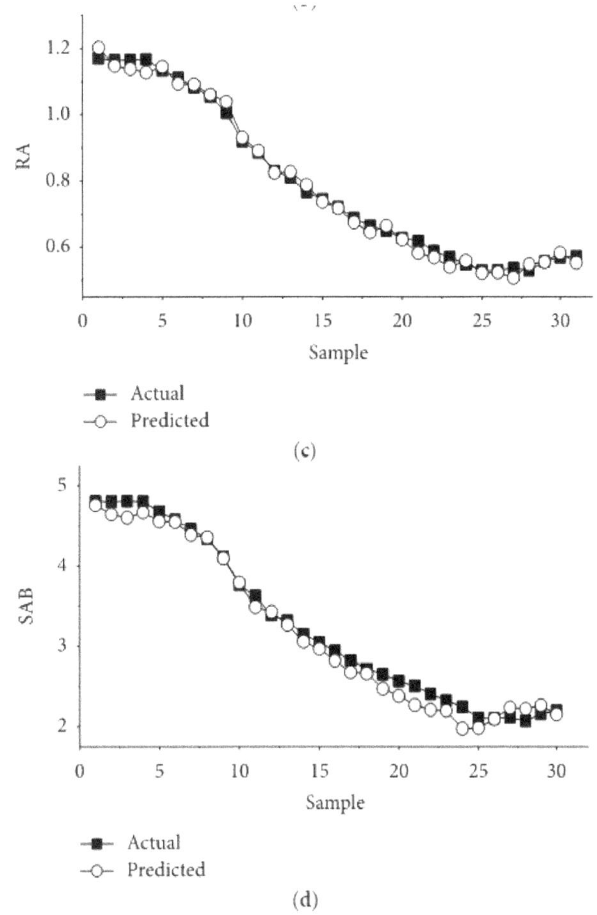

Figure 8: Predicted values for DSS (a), PA (b), RA (c), and SAB (d) obtained using the developed models.

CONCLUSIONS

The present study has demonstrated that NIR spectroscopy provided us with important advantages to determine the concentrations of DSS, PA, RA, and SAB in a quantitative and nondestructive manner and in an extremely short time. Based on the HPLC reference method, the combination of NIR spectroscopy with PLSR was investigated for on-line detection of the compositions of Danhong injection during the water precipitation process. When the established models were used for prediction, these excellent results proved that NIR spectroscopy was valuable for on-line detecting the concentrations of DSS, PA, RA, and SAB. Finally, application of on-line NIR spectroscopy to Danhong injection production enabled us to detect, in real-time, changes in the compositions of Danhong injection during the water precipitation process.

Generally, industrial on-line applications have only rarely been described in scientific publications, although there is a need of further large-scale on-line studies in the pharmaceutical industry to verify reliability and accuracy of NIR spectroscopy under process conditions. However, many of the published studies appear to prove the suitability of NIR spectroscopy as a valuable tool for on-line process control and quality management or at least underline its potential for industrial applications.

CONFLICT OF INTERESTS

The authors have declared no conflict of interests.

ACKNOWLEDGMENTS

This project was supported by the National Science and Technology Major Project of the Ministry of Science and Technology of China (2013ZX09201020). The authors also appreciate the cooperation and support from Heze Pharmaceutical Co. Ltd.

REFERENCES

1. Q. Chen, D. Yi, Y. Xie, W. Yang, Y. Zhuang, and J. Du, "Analysis of clinical use of Danhong injection based on hospital information system," China Journal of Chinese MateriaMedica, vol. 36, no. 20, pp. 2817–2820, 2011.
2. J.-F. Cui, W. Yang, Y.-M. Xie, Y. Sun, Y. Zhuang, and Y.-Y. Wang, "Real-world analysis of concurrent diseases and medicine use among patients with insomnia," China Journal of Chinese Materia Medica, vol. 39, no. 18, pp. 3519–3526, 2014.
3. D.-S. Lee, S.-H. Lee, J.-G. Noh, and S.-D. Hong, "Antibacterial activities of cryptotanshinone and dihydrotanshinone I from a medicinal herb, Salvia miltiorrhiza Bunge," Bioscience, Biotechnology and Biochemistry, vol. 63, no. 12, pp. 2236–2239, 1999.
4. M. Sun, J.-J. Zhang, J.-Z. Shan et al., "Clinical observation of Danhong Injection (herbal TCM product from Radix Salviae miltiorrhizae and Flos Carthami tinctorii) in the treatment of traumatic intracranial

hematoma," Phytomedicine, vol. 16, no. 8, pp. 683–689, 2009.

5. G. Schmourlo, R. R. Mendonça-Filho, C. S. Alviano, and S. S. Costa, "Screening of antifungal agents using ethanol precipitation and bioautography of medicinal and food plants," Journal of Ethnopharmacology, vol. 96, no. 3, pp. 563–568, 2005.
6. H. Q. Huang, Study on production technology of salvia miltiorrhiza injection [M.S. thesis], Shanghai Jiao Tong University, Shanghai, China, 2013.
7. F. Sun, S. L. Su, S. Jiang et al., "Resource components analysis on Salvia miltiorrhiza residues in production process of Danhong injection and its transformation mechanism," Chinese Traditional and Herbal Drugs, vol. 46, no. 16, pp. 2471–2476, 2015.
8. W. X. Zhang, L. Xuan, and J. Ni, "Stability of savianolic acid B in water solution," Journal of Beijing University of Traditional Chinese Medicine, vol. 32, no. 12, pp. 856–858, 2009.
9. Y.-C. Hung, P.-W. Wang, and T.-L. Pan, "Functional proteomics reveal the effect of Salvia miltiorrhizaaqueous extract against vascular atherosclerotic lesions," Biochimica et Biophysica Acta, vol. 1804, no. 6, pp. 1310–1321, 2010.
10. Y. Zeng, J.-X. Song, and X.-C. Shen, "Herbal remedies supply a novel prospect for the treatment of atherosclerosis: a review of current mechanism studies," Phytotherapy Research, vol. 26, no. 2, pp. 159–167, 2012.
11. C. Y. Moon, C. R. Ku, Y. H. Cho, and E. J. Lee, "Protocatechuic aldehyde inhibits migration and proliferation of vascular smooth muscle cells and intravascular thrombosis," Biochemical and Biophysical Research Communications, vol. 423, no. 1, pp. 116–121, 2012.
12. A. A. Fonteles, C. M. de Souza, J. C. de Sousa Neves et al., "Rosmarinic acid prevents against memory deficits in ischemic mice," Behavioural Brain Research, vol. 297, pp. 91–103, 2016.
13. Y. C. Chung, F. J. Hsieh, Y. J. Lin et al., "Magnesium lithospermate B and rosmarinic acid, two compounds present in Salvia miltiorrhiza, have potent antiviral activity against enterovirus 71 infections," European Journal of Pharmacology, vol. 755, pp. 127–133, 2015.
14. M. Liu, M. J. Zheng, H. Y. Xu et al., "Anti-pulmonary fibrotic activity of salvianolic acid B was screened by a novel method based on the cyto-biophysical properties," Biochemical and Biophysical Research Communications, vol. 468, no. 1-2, pp. 214–220, 2015.
15. D. F. Zhang, J. Zhang, and R. Li, "Salvianolic acid B attenuates lung inflammation induced by cigarette smoke in mice," European Journal of Pharmacology, vol. 761, pp. 174–179, 2015.
16. M. Otsuka, "Comparative particle size determination of phenacetin bulk powder by using Kubelka–Munk theory and principal component regression analysis based on near-infrared spectroscopy," Powder Technology, vol. 141, no. 3, pp. 244–250, 2004.
17. M. Blanco and I. Villarroya, "NIR spectroscopy: a rapid-response analytical tool," TrAC—Trends in Analytical Chemistry, vol. 21, no. 4, pp. 240–250, 2002.
18. C. Sandorfy, R. Buchet, and G. Lachenal, "Principles of molecular vibrations for near-infrared spectroscopy," in Near-Infrared Spectroscopy in Food Science and Technology, Y. Ozaki, W. F. McClure, and A. A. Christy, Eds., pp. 1–46, John Wiley & Sons, Hoboken, NJ, USA, 2007.
19. L. Eriksson, E. Johansson, N. Kettaneh-Wold, J. Trygg, C. Wikström, and S. Wold, Multi- and Megavariate Data Analysis, Part I, Basic Principles and Applications, Umetrics Academy, 2006.
20. M. Watari, H. Higashiyama, N. Mitsui, M. Tomo, and Y. Ozaki, "On-line monitoring of the density of linear low-density polyethylene in a real plant by near-infrared spectroscopy and chemometrics," Applied Spectroscopy, vol. 58, no. 2, pp. 248–255, 2004.
21. M. P. A. Ribeiro, T. F. Pádua, O. D. Leite, R. L. C. Giordano, and R. C. Giordano, "Multivariate calibration methods applied to the monitoring of the enzymatic synthesis of amipicilin," Chemometrics and Intelligent Laboratory Systems, vol. 90, no. 2, pp. 169–177, 2008.
22. Food and Drug Administration (FDA), "Guidance for Industry, PAT: A Framework for Innovative Pharmaceutical Development Manufacturing and Quality Assurance," 2004, http://www.fda.gov/downloads/Drugs/GuidanceComplianceRegulatoryInformation/Guidances/UCM070305.pdf.
23. J. Märk, M. Andre, M. Karner, and C. W. Huck, "Prospects for multivariate classification of a pharmaceutical intermediate with near-infrared spectroscopy as a Process Analytical Technology (PAT) production control supplement," European Journal of Pharmaceutics and Biopharmaceutics, vol. 76, no. 2, pp. 320–327, 2010.
24. S. Tosi, M. Rossi, E. Tamburini, G. Vaccari, A. Amaretti, and D. Matteuzzi, "Assessment of in-line near-infrared spectroscopy for continuous monitoring of fermentation processes," Biotechnology Progress, vol. 19, no. 6, pp. 1816–1821, 2003.
25. F. J. S. Nieuwmeyer, M. Damen, A. Gerich, F. Rusmini, K. van der Voort Maarschalk, and H. Vromans, "Granule characterization during fluid bed drying by development of a near infrared method to determine water content and median granule size," Pharmaceutical Research, vol. 24, no. 10, pp. 1854–1861, 2007.
26. G. Mirschel, K. Heymann, T. Scherzer, and M. R. Buchmeiser, "Effect of changes of the coating thickness on the in-line monitoring of the conversion of photopolymerized acrylate coatings by near-infrared reflection spectroscopy," Polymer, vol. 50, no. 8, pp. 1895–1900, 2009.
27. Y. Jin, Z. Z. Wu, X. S. Liu, and Y. J. Wu, "Near infrared spectroscopy in combination with chemometrics as a Process Analytical Technology (PAT) tool for on-line quantitative monitoring of alcohol precipitation," Journal of Pharmaceutical and Biomedical Analysis, vol. 77, pp. 32–39, 2013.

28. X. S. Liu, Z. Z. Wu, K. Yang, H. Y. Ding, and Y. J. Wu, "Quantitative analysis combined with chromatographic fingerprint for comprehensive evaluation of Danhong injection using HPLC-DAD,"Journal of Pharmaceutical and Biomedical Analysis, vol. 76, pp. 70–74, 2013.

29. D. D. Qi, Y. Jia, X. J. Tan, and K. S. Bi, "HPLC determination of 4 components in Danhong powder for injection," Chinese Journal of Pharmaceutical Analysis, vol. 27, no. 8, pp. 1165–1167, 2007.

30. Y. Y. Zhang, J. W. Zhang, and Y. Liu, "Application of on-line quality control for paeoniflorin extraction by NIRS," Journal of Chinese Pharmaceutical Sciences, vol. 41, no. 9, pp. 662–665, 2010.

31. H. X. Huang and H. B. Qu, "In-line monitoring of alcohol precipitation by near-infrared spectroscopy in conjunction with multivariate batch modeling," Analytica Chimica Acta, vol. 707, no. 1-2, pp. 47–56, 2011.

32. R. G. Brereton, Chemometrics: Data Analysis for the Laboratory and Chemical Plant, John Wiley & Sons, Chichester, UK, 2003.

33. J. P. Conzen, Multivariate Calibration. A Practical Guide for Developing Methods in the Quantitative Analytical Chemistry, Bruker Optik, Ettlingen, Germany, 2003.

34. A. Kohler, M. Zimonja, V. Segtnan, and H. Martens, "2.09—standard normal variate, multiplicative signal correction and extended multiplicative signal correction preprocessing in biospectroscopy," inReference Module in Chemistry, Molecular Sciences and Chemical Engineering, from Comprehensive Chemometrics, pp. 139–162, Elsevier, 2009.

35. S. Wold, M. Sjöström, and L. Eriksson, "PLS-regression: a basic tool of chemometrics," Chemometrics and Intelligent Laboratory Systems, vol. 58, no. 2, pp. 109–130, 2001.

36. G. Gong, "Cross-validation, the jackknife and the bootstrap excess error estimation in forward regression logistic regression," Journal of the American Statistical Association, vol. 81, no. 393, pp. 108–113, 1986.

37. W. L. Li, L. H. Xing, L. M. Fang, J. Wang, and H. B. Qu, "Application of near infrared spectroscopy for rapid analysis of intermediates of Tanreqing injection," Journal of Pharmaceutical and Biomedical Analysis, vol. 53, no. 3, pp. 350–358, 2010.

38. M. Streefland, P. F. G. Van Herpen, B. Van de Waterbeemd et al., "A practical approach for exploration and modeling of the design space of a bacterial vaccine cultivation process," Biotechnology and Bioengineering, vol. 104, no. 3, pp. 492–504, 2009.

39. X. Niu, F. Shen, Y. Yu et al., "Analysis of sugars in Chinese rice wine by Fourier transform near-infrared spectroscopy with partial least-squares regression," Journal of Agricultural and Food Chemistry, vol. 56, no. 16, pp. 7271–7278, 2008.

40. J. P. Conzen, Multivariate Calibration. A Practical Guide for Developing Method in the Quantitative Analytical Chemistry, Bruker Optik, Ettlingen, Germany, 2003.

41. H. M. Heise and R. Winzen, "Fundamental chemometric methods," in Near Infrared Spectroscopy: Principles, Instruments, Applications, H. W. Siesler, Y. Ozaki, S. Kawata, and H. M. Heise, Eds., pp. 125–162, Wiley, Weinheim, Germany, 2002.

42. W. J. McCarthy, TQ Analyst User's Guide, Thermo Nicolet, Madison, Wis, USA, 2003.

43. A. Savitzky and M. J. E. Golay, "Smoothing and differentiation of data by simplified least squares procedures," Analytical Chemistry, vol. 36, no. 8, pp. 1627–1639, 1964.

44. P. C. Williams, "Application of near infrared reflectance spectroscopy to analysis of cereal grains and oilseeds," Cereal Chemistry, vol. 52, no. 4, pp. 561–576, 1975.

45. D. Cozzolino, M. J. Kwiatkowski, M. Parker et al., "Prediction of phenolic compounds in red wine fermentations by visible and near infrared spectroscopy," Analytica Chimica Acta, vol. 513, no. 1, pp. 73–80, 2004.

46. Y. J. Wu, Y. J. Jin, Y. R. Li, D. Sun, X. S. Liu, and Y. Chen, "NIR spectroscopy as a process analytical technology (PAT) tool for on-line and real-time monitoring of an extraction process," Vibrational Spectroscopy, vol. 58, pp. 109–118, 2012.

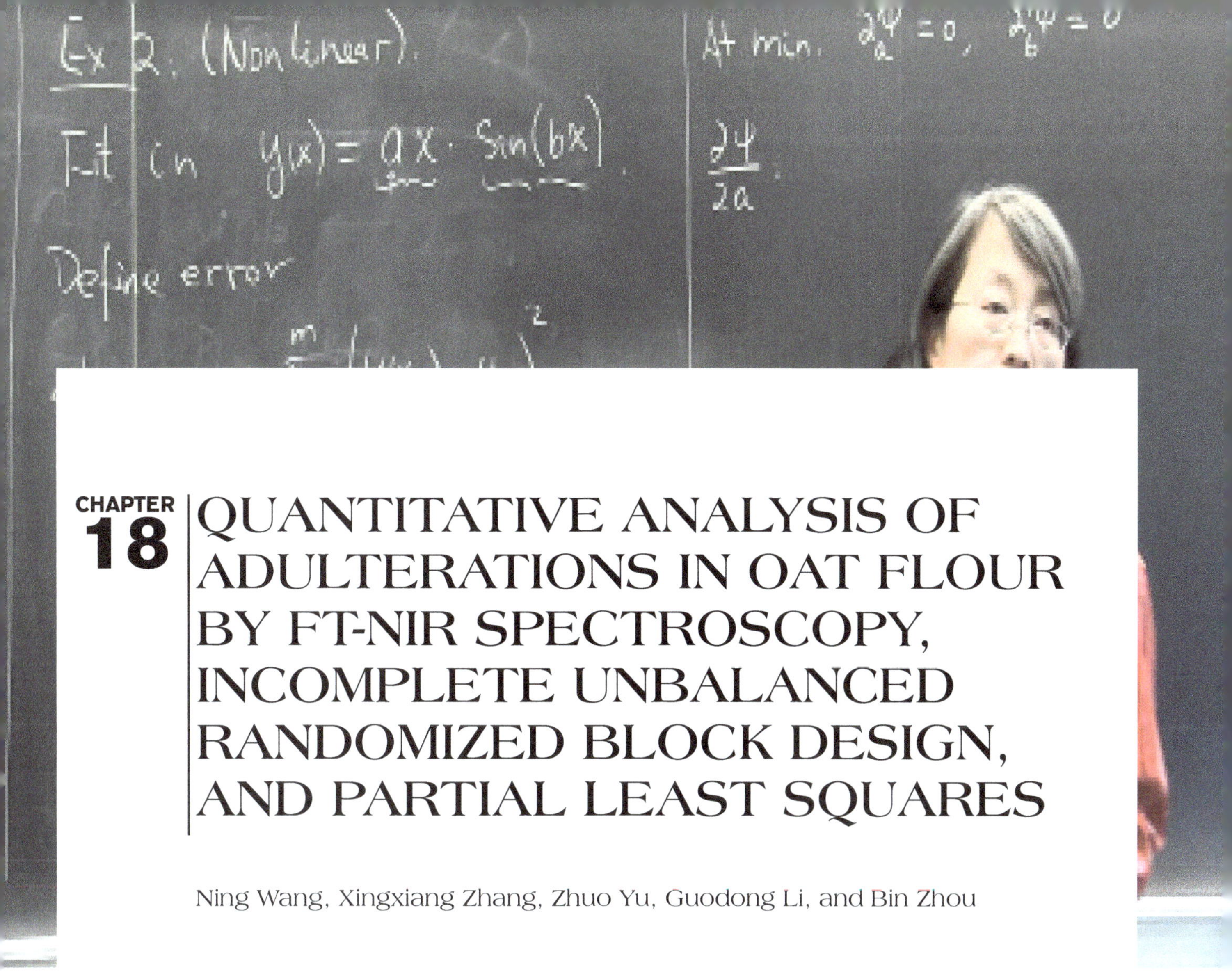

CHAPTER 18

QUANTITATIVE ANALYSIS OF ADULTERATIONS IN OAT FLOUR BY FT-NIR SPECTROSCOPY, INCOMPLETE UNBALANCED RANDOMIZED BLOCK DESIGN, AND PARTIAL LEAST SQUARES

Ning Wang, Xingxiang Zhang, Zhuo Yu, Guodong Li, and Bin Zhou

INTRODUCTION

Food adulteration and fraud have been a common problem in food production since ancient times. Food adulteration is economically motivated and is performed by the addition, substitution, or removal of food ingredients, for example, replacing or diluting high-cost ingredients with cheaper ones [1, 2]. It is an issue that concerns not only consumers, but also food producers, sellers, regulatory agencies, and even the entire food industry chain. For consumers, food adulterations have caused growing concern about health risks, as well as the food quality and nutrition value. A notorious and common phenomenon is the adulterations of raw food materials, which can not only influence the quality of raw materials but also cause potential crisis in further processed foods [3, 4]. Oat is widely utilized for human consumption and food industrial uses. Due to its high nutritional value and characteristic flavor, oat flour plays an important role in the breakfast cereals group and other processed foods as an alternative or supplement to the ordinary wheat flour [5]. As a nonstaple cereal, the yield of oat is much less than wheat, so oat flour is more expensive than wheat flour. For producers and sellers, it is economically profitable to add wheat flour to oat flour. Because the appearances and physical and chemical properties of wheat and oat flours are very similar, rapid and effective methods are required to analyze the adulterations.

For food analysis and quality control, NIR spectroscopy has demonstrated some advantages, including less sample treatment, reduced analysis time and cost, and the feasibility for nondestructive analysis and online analysis. NIR spectroscopy has been widely used for analysis of grains and cereals. Hurburgh et al. [6] combined NIR spectroscopy and principal component analysis to discriminate transgenic grains and nontransgenic grains. Munck et al. [7] applied the NIRS technology to distinguish barley flour with different levels of lysine amino acids. NIR was also successfully used to distinguish corn samples of different genotypes [8, 9]. For quantitative analysis, NIR technology has provided a rapid tool for analysis of different constituents or quality parameters in grain products, including corn dry-milling quality [10], protein content in wheat kernels [11], the ratio of starch amylose content to total grain in corn [12], undried rough rice constituent content [13], kernel rots and mycotoxins in maize

Citation: Ning Wang, Xingxiang Zhang, Zhuo Yu, Guodong Li, and Bin Zhou, "Quantitative Analysis of Adulterations in Oat Flour by FT-NIR Spectroscopy, Incomplete Unbalanced Randomized Block Design, and Partial Least Squares," Journal of Analytical Methods in Chemistry, vol. 2014, Article ID 393596, 5 pages, 2014. doi:10.1155/2014/393596

[14], and protein, moisture, dry mass, hardness, and other residues of wheat [15] and so on [16].

Considering the large number of samples in market shelf and small private retailers, NIR is a convenient and economic technique for analysis of adulterations in oat flour. This paper was aimed at developing a rapid method for analysis of potential wheat flour added to oat flour using NIR spectrometry and chemometrics. Considering the composition variations of oats from different producing areas, an incomplete unbalanced randomized block design [17] was performed to ensure the generalization performance of multivariate calibration models.

MATERIALS AND METHODS

Collection and Preparation of Samples

Pure oat flour and wheat flour objects were made from intact grains. A set of oat and wheat kernels harvested in 2013 were collected from domestic markets. Oat kernels were produced in Hebei (15), Henan (17), Gansu (17), Shanxi (18), and Qinghai (13). Wheat kernels were produced in Henan (11), Hebei (10), Jiangsu (8), Anhui (10), Shandong (11), Shanxi (7), and Heilongjiang (8). The kernels were dried in the sun and milled by a crusher. All the particles were filtered through a 200-mesh sieve.

Adulterated oat flour samples were made by mixing the oat flour with different levels of the wheat flour. The doping levels were 5%, 10%, 15%, 20%, 25%, 30%, 35%, 40%, 45%, and 50% (w/w). In order to obtain a representative nut not too large sample set, an incomplete and unbalanced randomized design [17] was performed. Because this paper was focused on quantitative analysis of wheat flour in oat flour, the producing areas of oat and wheat were considered to be two blocking factors. In this way, 120 adulterated objects were prepared for developing calibration models with 10 doping levels each having 12 objects. Another 100 adulterated objects (10 doping levels each having 10 objects) were prepared for model validation by mixing oat and wheat flour objects that were different from those used for preparation of training samples.

The NIR diffuse reflectance spectra of impacted powders were measured on a Bruker-TENSOR37 FTIR spectrometer (Bruker Optics, Ettlingen, Germany). The working range of spectrometer was 4000–12000 cm^{-1}. The spectra were measured using a PbS detector with an internal gold background as the reference. The instrument resolution was 4 cm^{-1} and the scanning interval was 1.929 cm^{-1}, so each spectrum contained 4148 wavelengths. Each spectrum was the average of 64 scans. For each object, three spectra were measured by stirring the powder and the average spectrum was taken.

Data Preprocessing and Multivariate Calibration

All the data preprocessing and chemometrics models were performed using MATLAB 7.0.1 (Mathworks, Sherborn, MA). Smoothing was used to remove random noise in the data and improve the signal-to-noise ratio (SNR). In this work the S-G polynomial fitting algorithm [18] was used for smoothing for its simplicity and effectiveness. Taking second-order derivative (D2) of spectra was performed to enhance spectral resolution and remove linear baseline shifts. The D2 spectra were also computed by S-G polynomial fitting algorithm because this method can avoid degradation of SNR compared with direct differencing. Standard normal variate (SNV) [19] transformation was performed to reduce the spectral variations caused by scattering and uneven sizes of particle.

Partial least squares (PLS) models were developed using the raw and preprocessed spectra. An important problem when performing PLS is the overfitting of models. In this paper, Monte Carlo cross-validation (MCCV) [20] was used to select the number of PLS components. MCCV can avoid the risk of overfitting by multiple random splitting of the training objects and having a higher percent of leave-out objects for prediction.

RESULTS AND DISCUSSION

The raw NIR spectra of 120 adulterated oat flour objects are shown in Figure 1. Some of the absorbance peaks can be assigned as follows [21]: (1) the peak at 4318 cm^{-1} caused by the combination absorbance of $-CH_2$deformation and various C–H stretching; (2) the wide peak at 4748 cm^{-1}, overlapping of combination of C=O stretching and peptide group deformation and combination of N–H stretching and peptide group deformation; (3) the peak at 5167 cm^{-1}, combination of O–H stretching and O–H deformation; (4) 5629 cm^{-1}, the first overtone of symmetrical and asymmetrical C–H stretching in $-CH_2$; (5) the wide peak at 6802 cm^{-1}, overlapping of the first overtone of O–H stretching (~6900 cm^{-1}) and the first overtone of N–H stretching (~6500 cm^{-1}); (6) 8329 cm^{-1}, the second overtones of C–H stretching in various groups; and (7) 9970 cm^{-1}, the second overtones of N–H stretching or the third overtones of C–H stretching. The spectral interval 9000–12,000 cm^{-1} has no significant peaks, so this interval was not used for developing calibration models.

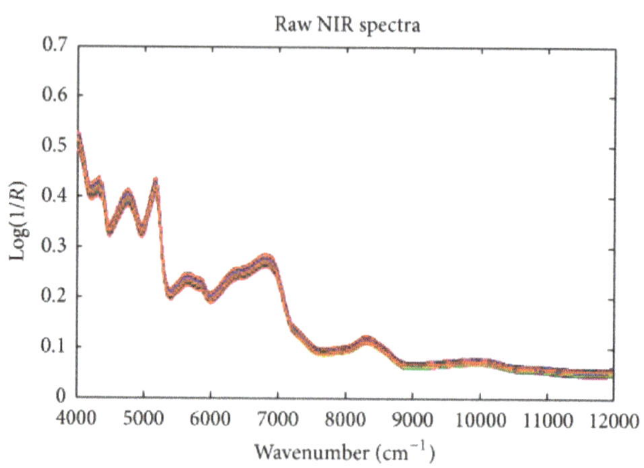

Figure 1: The raw NIR spectra of 120 adulterated oat flour objects with doping levels ranging from 5% to 50% (w/w).

Smoothed, D2, and SNV spectra were shown in Figure 2. Seen from Figure 2, the D2 spectra can remove most of the backgrounds and the peak resolution was largely improved by taking D2 spectra. D2 spectra also obtained much detailed and high-frequency information. SNV transformation can remove most unwanted variations. Multivariate calibration models were developed with PLS to predict the levels of wheat flour. The number of PLS components was estimated using F-test of MCCV. In this work, random splitting of the training set was performed for 100 times and each time 70%

of the training objects were used for developing a PLS model and 30% for prediction. The pooled predicted residual sum of squares (PRESS) was computed using different numbers of PLS components. Finally, F-test was performed to select the fewest PLS components with a PRESS value not significantly higher than the minimum PRESS value. As recommended by the original literature, the significance level of the F-test was set to be 0.25 [22, 23].

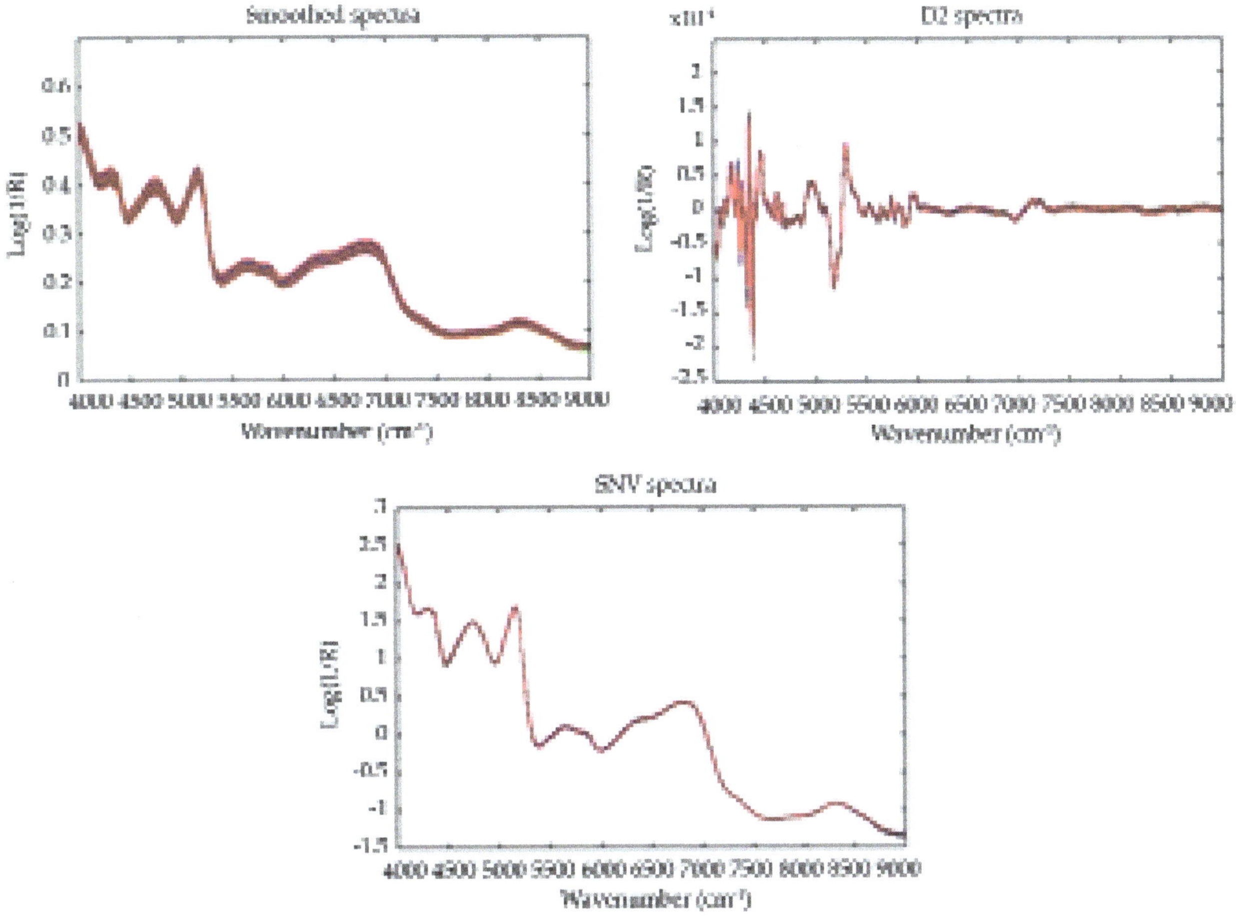

Figure 2: The smoothed, second-order derivative (D2) and standard normal variate (SNV) NIR spectra of 120 adulterated oat flour objects with doping levels ranging from 5% to 50% (w/w).

Based on differently preprocessed spectra (9000–4000 cm^{-1}) the calibration and prediction results of PLS were demonstrated in Table 1. Seen from Table 1, the model complexity of D2-PLS and SNV-PLS was reduced by one component compared with the PLS model using raw data. Moreover, seen from the model complexity of PLS models (3 or 4 components), the mixtures of wheat flour and oat flour investigated in this work were not a simple two-component system and spectral variations caused by different geographical origins had made it necessary to perform multivariate calibration. The root mean squared error of MCCV (RMSEMCCV) value was slightly reduced by smoothing (2.274), taking D2 (1.921), and SNV transformation (1.981) compared with PLS with raw data (2.388). For prediction, the lowest root mean squared error of prediction (RMSEP) of 1.975 was obtained by D2-PLS model. The training and prediction results by D2-PLS are demonstrated in Figure 3. For all of the PLS models, the differences between RMSEP and RMSEMCCV values were insignificant, indicating that both the training and test sets obtained by IURB design were representative to include composition variations caused by different origins of oat and wheat flour.

Table 1: Quantitative analysis of wheat flour in oat flour by PLS.

Preprocessing	N^a	RMSEMCCV (%)	RMSEC (%)	RMSEP (%)
Raw data	4	2.388	2.115	2.213
Smoothing	4	2.274	2.102	2.244
D2	3	1.921	1.781	1.975
SNV	3	1.981	1.842	2.054

[a]The number of significant PLS components.

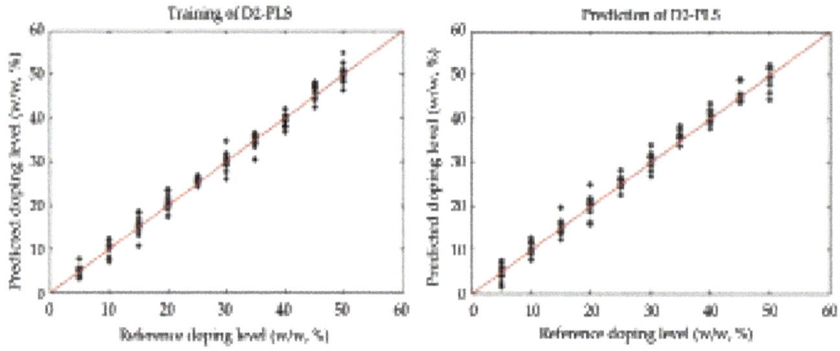

Figure 3: The training and prediction of contents of wheat flour in oat flour by three-component D2-PLS.

CONCLUSIONS

Multivariate calibration models were developed by PLS for analysis of wheat flour in oat flour from different geographical origins. The results demonstrated that a three- or four-component PLS model can accurately predict the levels of wheat flour in oat flour. Moreover, IURB design was shown to be useful to obtain representative training and test sets to include the composition variations caused by different producing areas. The developed PLS models will have a good generalization performance and are useful for quantitative analysis of oat flour in domestic market.

CONFLICT OF INTERESTS

Ning Wang, Xingxiang Zhang, and Bin Zhou declare no conflict of interests regarding the publication of this paper.

ACKNOWLEDGMENT

The authors thank Science Funds of Aviation (201329Q2001) for the financial support.

REFERENCES

1. J. H. Kalivas, C. A. Georgiou, M. Moira, I. Tsafaras, E. A. Petrakis, and G. A. Mousdis, "Food adulteration analysis without laboratory prepared or determined reference food adulterant values," Food Chemistry, vol. 148, pp. 289–293, 2014.
2. M. Woolfe and S. Primrose, "Food forensics: using DNA technology to combat misdescription and fraud," Trends in Biotechnology, vol. 22, no. 5, pp. 222–226, 2004.
3. N. Rai, D. Banerjee, and R. Bhattacharyya, "Urinary melamine: proposed parameter of melamine adulteration of food," Nutrition, vol. 30, pp. 380–385, 2014.
4. J. L. Dorne, D. R. Doerge, M. Vandenbroeck et al., "Recent advances in the risk assessment of melamine and cyanuric acid in animal feed," Toxicology and Applied Pharmacology, vol. 270, no. 3, pp. 218–229, 2013.
5. S. Serranti, D. Cesare, F. Marini, and G. Bonifazi, "Classification of oat and groat kernels using NIR hyperspectral imaging," Talanta, vol. 103, pp. 276–284, 2013.
6. C. R. Hurburgh, G. R. Rippke, C. Heithoff, S. A. Roussel, and C. L. Hardy, "Detection of genetically modified grains by near-infrared spectroscopy," in Proceedings of the PITTCON 2000—Science for the 21st Century, p. 1431, New Orleans, La, USA, March 2000.
7. L. Munck, J. Pram Nielsen, B. Møller et al., "Exploring the phenotypic expression of a regulatory proteome-altering gene by spectroscopy and chemometrics," Analytica Chimica Acta, vol. 446, no. 1-2, pp. 171–186, 2001.
8. M. R. Campbell, J. Sykes, and D. V. Glover, "Classification of single- and double-mutant corn endosperm genotypes by near-infrared transmittance spectroscopy," Cereal Chemistry, vol. 77, no. 6, pp. 774–778, 2000.
9. Y. Rui, Y. Luo, K. Haung, W. Wang, and L. Zhang, "Discrimination of transgenic corns using NIR diffuse reflectance spectroscopy and back propagation (BP)," Spectroscopy and Spectral Analysis, vol. 25, pp. 1581–1592, 2005.
10. R. L. Wehling, D. S. Jackson, and B. R. Hamaker, "Prediction of corn dry-milling quality by near-infrared spectroscopy," Cereal Chemistry, vol. 73, no. 5, pp. 543–546, 1996.
11. S. R. Delwiche, "Protein content of single kernels of wheat by near-infrared reflectance spectroscopy,"Journal of Cereal Science, vol. 27, no. 3, pp. 241–254, 1998.
12. M. R. Campbell, S. R. Mannis, H. A. Port, A. M. Zimmerman, and D. V. Glover, "Prediction of starch amylose content versus total grain amylose content in corn by near-infrared transmittance spectroscopy,"Cereal Chemistry, vol. 76, no. 4, pp. 552–557, 1999.
13. S. Kawamura, M. Natsuga, and K. Itoh, "Determination of undried rough rice constituent content using near-infrared transmission spectroscopy," Transactions of the American Society of Agricultural Engineers, vol. 42, no. 3, pp. 813–818, 1999.
14. N. Berardo, V. Pisacane, P. Battilani, A. Scandolara, A. Pietri, and A. Marocco, "Rapid detection of kernel rots and mycotoxins in maize by near-infrared reflectance spectroscopy," Journal of Agricultural and Food Chemistry, vol. 53, no. 21, pp. 8128–8134, 2005.
15. D. Özdemir, "Genetic multivariate calibration for near infrared spectroscopic determination of protein, moisture, dry mass, hardness and other residues of wheat," International Journal of Food Science and Technology, vol. 41, no. 2, pp. 12–21, 2006.
16. H. Huang, H. Yu, H. Xu, and Y. Ying, "Near infrared spectroscopy for on/in-line monitoring of quality in foods

and beverages: a review," Journal of Food Engineering, vol. 87, no. 3, pp. 303–313, 2008.

17. V. K. Gupta, A. Das, and A. Dey, "Universal optimality of block designs with unequal block sizes,"Statistics and Probability Letters, vol. 11, no. 2, pp. 177–180, 1991.

18. A. Savitzky and M. J. E. Golay, "Smoothing and differentiation of data by simplified least squares procedures," Analytical Chemistry, vol. 36, no. 8, pp. 1627–1639, 1964

19. R. J. Barnes, M. S. Dhanoa, and S. J. Lister, "Standard normal variate transformation and de-trending of near-infrared diffuse reflectance spectra," Applied Spectroscopy, vol. 43, no. 5, pp. 772–777, 1989.

20. Q. S. Xu and Y. Z. Liang, "Monte Carlo cross validation," Chemometrics and Intelligent Laboratory Systems, vol. 56, no. 1, pp. 1–11, 2001.

21. L. Xu, P. Shi, Z. Ye, S. Yan, and X. Yu, "Rapid analysis of adulterations in Chinese lotus root powder (LRP) by near-infrared (NIR) spectroscopy coupled with chemometric class modeling techniques," Food Chemistry, vol. 141, no. 3, pp. 2434–2439, 2013.

22. D. M. Haaland and E. V. Thomas, "Partial least-squares methods for spectral analyses. 1. Relation to other quantitative calibration methods and the extraction of qualitative information," Analytical Chemistry, vol. 60, no. 11, pp. 1193–1202, 1988.

23. D. M. Haaland and E. V. Thomas, "Partial least-squares methods for spectral analyses. 2. Application to simulated and glass spectral data," Analytical Chemistry, vol. 60, no. 11, pp. 1202–1208, 1988.

INDEX

A

Acid detergent lignin (ADL) 85
Assessment risk 60
Atmospheric mercury depletion events\" (AMDEs) 107

B

Bihor (BH) 178
Bio-ethanol 125
Biomass Bubbling Fluidized Bed (BBFB) 136
Biomass Combined Cycle (BCC) 136
Bistrița Năsăud (BN) 178

C

Chemical Looping Combustion (CLC) 147
Chemometrics 17, 18
Compartmental 42, 43, 44, 45, 47
Correlation optimized warping (COW) 121

D

Dynamic time wariping (DTW) 121

E

Enzyme hydrolysis 126

F

Fermentation 125, 126, 127, 129, 130, 131, 132, 133
Fluorescence lifetime imaging microscopy (FLIM) 18
Food and Drug Administration (FDA) 196, 206
Fresh Fruit Bunch (FFB) 141

G

Gas chromatography (GC) 185
Gaseous mercury compounds (RGM) 107
Gas phase mercury (RGM) 110
General Algebraic Modeling System (GAMS) 135
Generalized additive models (GAMs) 1
Generalized linear models (GLMs) 1
General Public License (GPL) 33
Genetic algorithm (GA) 83, 84
Genetic programming (GP) 75
Geographical information system (GIS) 60, 61

H

Harmful wastes 59
Heteronuclear multiple bond correlation (HMBC) 192

J

Journal of Statistical Software (JSS) 18

K

Kernel Fisher discriminant (KFD) 77

L

Large solar Photovoltaic (LPV) 136
Least squares (LS) 74
Linear discriminant analysis (LDA) 185
Linear models (LMs) 1
Liquid chromatography systems (LC/MS) 18

M

Mixed Integer Linear Programming (MILP) 135
Monte Carlo cross-validation (MCCV) 210
Multiple linear regression (MLR) 21, 74, 86

N

Natural Gas Combined Cycle power plant (NGCC) 136
Near-Infrared (NIR) 177
Near-infrared spectra (NIR) 23
Nuclear magnetic resonance (NMR) 186

O

Oxygen carrier (OC) 147, 150, 153

P

Palm Oil Mill Effluent (POME) 141, 142, 144
Partial least-square (PLS) 83

Partial least squares-genetic programming (PLS-GP) 192
Partial least squares (PLS) 121
Partial least squares regression (PLSR) 21
Particle-bound mercury (PHg) 107
Potato Dextrose Agar (PDA) 142
Prediction to deviation (RPD) 198
Principal component analysis (PCA) 74, 83, 85, 181
Principal component regression (PCR) 21, 74, 121, 185
Principal components (PCs) 181
Protocatechuic aldehyde (PA) 195
Pulverized Coal (PC) 136

R

Relative standard deviation (RSD 198
Renewable Energy (RE) 135
Response Surface Methodology (RSM 126
Ridge regression (RR) 74
Root mean squares error of cross-validation (RMSECV) 198

S

Saccharomyces cerevisiae 126
Salvianolic acid B (SAB) 195
Separable nonlinear least squares 53
Signal-to-noise ratio (SNR) 210
Site-specific natural isotopic fraction NMR (SNIF-NMR) 192
Spectral correlative chromatography (SCC) 168, 172, 173, 174
Spectral correlative chromatography (SCC) was 168
Spectroscopy 17, 18, 33, 34, 35, 36, 38, 41, 42, 53, 55
Standard normal variate (SNV) , 120
Subwindow factor analysis (SFA 168
Subwindow factor analysis (SFA) 168, 174
Superposition model 34, 38, 40
Support vector machine (SVM) 85

T

Total ion chromatogram (TIC) 168

U

United State Department of Agriculture (USDA) 70

V

Visible wavelengths (VIS) 70

W

Waste potato mash 126, 127, 128, 129, 130, 132
Weighted sum method (WSM) 63